Agricultural research for development

The Namulonge Cotton Research Station from the air.

Frontispiece

COTTON RESEARCH CORPORATION

Agricultural research for development

THE NAMULONGE CONTRIBUTION

EDITED BY

M. H. ARNOLD

Formerly Director of the Cotton Research Station, Namulonge
and member of the National Research Council of Uganda

CAMBRIDGE UNIVERSITY PRESS

CAMBRIDGE

LONDON · NEW YORK · MELBOURNE

Published by the Syndics of the Cambridge University Press
The Pitt Building, Trumpington Street, Cambridge CB2 1RP
Bentley House, 200 Euston Road, London NW1 2DB
32 East 57th Street, New York, NY 10022, USA
296 Beaconsfield Parade, Middle Park, Melbourne 3206, Australia

First published 1976

Printed in Great Britain at the
University Printing House, Cambridge
(Euan Phillips, University Printer)

Library of Congress cataloguing in publication data
Main entry under title:
Agricultural research for development.
At head of title: Cotton Research Corporation.
Includes bibliographical references and indexes.
1. Underdeveloped areas – Agricultural research.
2. Agricultural research – Administration.   3. Cotton
Research Station, Namulonge, Uganda.   4. Cotton research.
I. Arnold, M. H.   II. Cotton Research Corporation.
S540.A2A3   630'.7'206761   75–31400
ISBN 0 521 21051 8

CONTENTS

The colour plates fall between pages 116 and 117

CONTRIBUTORS

Dr M. H. Arnold,
Plant Breeding Institute,
Maris Lane,
Trumpington,
Cambridge, CB2 2LQ

S. J. Brown,
Plant Breeding Institute,
Maris Lane,
Trumpington,
Cambridge, CB2 2LQ

A. B. Hearn,
CSIRO Division of Plant Industry,
Cotton Research Unit,
PO Box 59,
Narrabri,
New South Wales 2390,
Australia

Dr N. L. Innes,
National Vegetable Research
Station,
Wellesbourne,
Warwickshire V35 9EF

E. Jones,
Rice Research Station,
Rokupr,
Sierra Leone

R. G. Passmore,
Institute for Agricultural Research,
Samaru,
PMB 1044 Zaria,
Kaduna State,
Nigeria

Dr W. Reed,
11 Wilberforce Road,
Wisbech PE 13 2EU,
Cambridgeshire

Dr D. A. Rijks,
c/o PNUD,
BP 256,
Niamey,
République du Niger

F. E. Tollervey,
Ministerio de Asuntos Campesinos
e y Agricultura,
Asesores Britanicos e n
Agricultura Tropical,
Casilla 359,
Santa Cruz,
Bolivia

FOREWORD

SIR JOSEPH HUTCHINSON

The Cotton Research Station at Namulonge in Uganda was planned as part of a radical deployment of the research resources of the Cotton Research Corporation (then the Empire Cotton Growing Corporation). By 1939, work on cotton at Barberton in South Africa had elucidated the major factors governing successful cotton growing in British Colonial territories, and in Trinidad had revealed the genetic constitution and physiological behaviour of the range of species of *Gossypium* that make up the world's cottons. The next stage was to be an integration of the practical programme of the Barberton station with the long-range research of Trinidad, and for this a location in an important cotton-growing country was sought.

After the war the Uganda Government invited the Corporation to set up a station in that country, and a Uganda landlord offered an admirable site. Development began, but it soon became evident that the integration of Barberton and Trinidad was in no way an adequate concept on which to build. First, the old dominance of Barberton as the parent station of the Corporation's work in Africa was fading, with the growing strength and experience of the teams in other African territories, and second, the kind of academic studies carried on in Trinidad were neither possible nor desirable in the new circumstances.

A new field of interest emerged, and was to dominate the station's programme. This was first recognized by Professor Sir Frank Engledow. Surveying the site for the new station on his first visit, he said 'I would like to see this place *farmed* for five years'. Devising a highly productive farming system on the ancient Uganda soils took, not five years, but fifteen, but that enterprise turned out to be central to the whole research project.

Thus, though Namulonge grew out of Barberton and Trinidad, it developed uniquely, and made a major contribution to research in tropical agriculture. The concept of a central station gave way to the emergence of a group of territorial stations working on equal terms. Academic research on cotton genetics was replaced by studies of the application of genetics and statistics to breeding techniques, and studies of crop physiology were integrated with advanced climatology and investigations of water use.

Pest and disease work were more closely related to what had gone before, since a multitude of pests and a major disease had been limiting factors in cotton production that the Corporation's staff had done much to overcome.

Soil science had only a small place in the original plans. This was a cotton station, and soils are not peculiar to one crop. Hence they tend to be overlooked in commodity-based research. Nevertheless, the Cotton Research Corporation had a tradition of studying cotton in the context of the cropping systems of which it is a part, and when it became clear that soil fertility was at the heart of the problem of raising productivity, the support of the Nuffield Foundation (and, later, the Ministry of Overseas Development) was enlisted to increase the soil fertility research programme.

The work on a farm scale of devising a high productivity farming system may be described as exploratory rather than experimental, but it was an essential part of the programme. Engledow wanted the place farmed. Parnell, the first Director, held that experimental work should be conducted in the context of a substantial farming enterprise. Namulonge was planned on this basis, and such essential – but not experimental – activities as soil conservation, first by tied ridges and later by contour bunding, and the use of cattle in exploiting land 'rested' under leys, formed a framework within which the experimental work went on. The evidence from which insecticidal control of cotton pests was formulated was primarily farming experience, and not data from precise experimentation. Indeed, all of these non-experimental activities interacted with the research programme and contributed much to agricultural wisdom in Uganda.

The various strands were woven into a fabric of research and demonstration that was peculiar to Namulonge, and was of real distinction. Seeing it, some years after I had left the Station, I urged the Director and his team to write up their own account of what thay had done. This book is their response. It sets out new approaches to old disciplines, and it shows the way to a break-through in the productivity of the ancient soils of equatorial Africa. But its most important contribution is in the demonstration of the strength and effectiveness of a multi-racial team when it is closely knit and well led.

PREFACE

Writing about the results of scientific research is governed by many factors. For example, pressure to maintain the flow of publications, in order to justify the costs of a research programme or to enhance the promotion prospects of an individual, tends to limit papers to single aspects of essentially complex problems; pressure to write reports at intervals unrelated to the progress of the work gives rise to disjointed accounts of the thinking involved; pressure to keep publications short robs authors of the chance to philosophize.

The writing in this book has been subject to no such constraints. Although it reviews the work of the Namulonge Cotton Research Station, we have seen it as a book neither on cotton nor on Uganda. Rather we have used it as an opportunity to reflect upon the disciplines in which we worked, calling upon our experience at Namulonge and at other stations in the tropics where we served. Naturally, many of the details concern cotton but, wherever possible, we have related our findings to the broader principles involved.

Our aim has been to produce, against a background of African development, a synthesis of research results that might otherwise have remained unreported, or obscured in diverse publications. The book is intended for use at universities and research institutions, as well as by those concerned with problems of the developing countries, and with the administration of technical aid. The contents of the chapters, although interrelated, are delimited by the discipline or subject concerned. In order to help the reader to distinguish those parts of the writing that are of more general interest from those that are more specific to cotton, each chapter has been divided into sections with carefully chosen subtitles. Some of the more detailed material has been included as appendices. The reader concerned with integrating the work more closely with other aspects of the local agriculture should refer to *Agriculture in Uganda* edited by J. D. Jameson (2nd edn, Oxford University Press, 1970).

Distilling the essence of some twenty years' agricultural research has been no easy task, and there is little doubt that we have all become wiser in the attempt. We have drawn attention to the more positive findings of our work, exposed some of its limitations and opened the way for further research. In relating their own findings to the Namulonge contribution, perhaps others, with similar experience elsewhere, might be tempted to do likewise.

Acknowledgements. I am grateful to Sir Joseph Hutchinson for his encouragement in producing this volume and for commenting on the entire typescript. The ideas expressed by all contributors have benefited from discussions with colleagues both past and present, but to mention only some by name would be to do others an injustice. More specifically, we are indebted to: Miss Anne Tallantire for identifying the less common weed species mentioned in Chapter 9; Dr J. T. Abrams for information on nitrite poisoning in cattle; Dr P. C. J. Payne for a discussion on the design of chisel ploughs; H. M. Arundale for comments on the co-operative movement in Uganda; T. K. Thorp, K. R. M. Anthony and B. J. S. Lee for providing cotton production data presented in Chapter 10; my wife for helping with the bibliography; Mrs Joan Green for tracing some of the references; and Mrs Dorothy Moore, Mrs Barbara Jones and Mrs Marjory Innes, who shared the typing. S. J. Brown took most of the photographs; I am indebted to him and B. C. Allen for preparing them, and to the Director of the Plant Breeding Institute, Cambridge for permission to use dark-room facilities. Much of the work reported in this book formed part of the Uganda agricultural research programme and was financed jointly by the Government of Uganda and the Cotton Research Corporation. Part of the work on soil productivity was financed by grants from the Nuffield Foundation and the Ministry of Overseas Development.

June 1975   M.H.A.

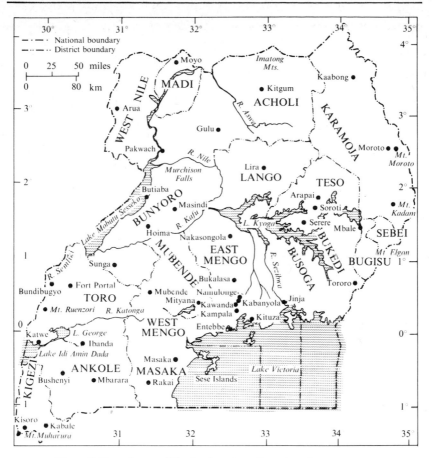

Map of Uganda – political, showing all major place-names.

1

Namulonge

M. H. ARNOLD

Origin of the Cotton Research Station

The Namulonge Cotton Research Station was conceived in the early
1940s as a project in central or regional research. It was designed to
provide information on the more universal problems of cotton-growing
in Africa, for staff working at centres where there was little immediate
prospect of developing extensive research programmes. It also provided the opportunity of combining, at a single station, the more
fundamental lines of research developed by the Cotton Research
Corporation in Trinidad with the more applied approach that it had
encouraged at Barberton, in South Africa. The organization of the
Cotton Research Corporation and the principles governing the work of
its staff in the developing countries have been described in detail by
Hutchinson & Ruston (1965). Here we are concerned only with the
development of the central research station.

In 1942, following negotiations with all concerned, the Corporation
announced its intention of closing down the stations at Trinidad and
Barberton and of building a new research station in Uganda. Centrally
placed in relation to the other cotton-producing countries needing
research results, Uganda was the largest single producer of cotton in
the Commonwealth, excluding India. Under the aegis of the Uganda
Government, the owner of a large estate had offered to lease to the
Corporation some 900 ha of land at Namulonge, which had the advantages of being near both a commercial centre, Kampala, and a university
college, Makerere. It was also located in what was at that time one of
the most important cotton-growing areas in Uganda.

Progress in developing the site was delayed because of the second
world war, but a start was made in 1945 and, although numerous
difficulties were encountered, many of the buildings had been completed
by the time the station was officially opened on 9 November 1950.
Twenty-one years later, after much had been accomplished in the face
of rapidly changing circumstances, the station and all its assets were
handed over to the Government of Uganda.

This book reviews, through the eyes of a small number of those

Figure 1.1. The official opening of the Namulonge Cotton Research Station by the then Governor of Uganda, Sir John Hathorn Hall, in November, 1950.

concerned, the research projects upon which staff at Namulonge focussed their attention. It attempts to do so in the broader context of the scientific principles on which the work was based. To understand the background, however, it is necessary first of all to describe the development of the station, and some of the factors that influenced the course of the research programme.

Initial finance and staffing

Funds for the research station were provided from a variety of sources. Capital expenditure, totalling £221 500 was met by grants from organizations in the United Kingdom. The largest contribution of £103 250 came from the British Government's Colonial Development and Welfare Fund; the Cotton Research Corporation provided £78 250; £25 000 came from the Cotton Industry War Memorial Trust; £15 000 was contributed by the Raw Cotton Commission. As far as running costs were concerned, agreement was reached that for a ten-year period starting from 1947, the Corporation's contribution would be supplemented by subventions from the Colonial Development and Welfare Fund and from the governments of the countries that the station was designed primarily to serve, namely Kenya, Malawi, Nigeria, Sudan, Tanzania and Uganda. At a later stage, the Uganda Government made a special contribution of £90 000 to offset rising capital and recurrent costs.

Figure 1.2. The formal handing-over ceremony on 5 January 1972. Professor Sir Joseph Hutchinson, CMG, FRS, Vice President of the Cotton Research Corporation, hands the deeds of the station to H.E. The President of Uganda, General Idi Amin Dada, while the author looks on.

As facilities were developed and houses became ready for occupation, experienced research staff were transferred to Namulonge from stations in other countries. These officers formed the core of a team of some twelve professional staff, together with locally-recruited support staff and farm workers totalling some 250 employees, which had been built up by about the time the station was officially opened.

The regional research programme

The research programme was imaginative and far reaching. There were the overriding problems of African agriculture; of how the soils could be improved while preserving the benefits of the traditional shifting cultivation; of how cotton could be fitted into the system without disrupting the essential food supply; of the climate; of attempting to understand the erratic occurrence of tropical rainfall and how its distribution could be characterized in terms that would be meaningful for agricultural planning. There were problems of the genetic adaptation of cotton varieties to the range of new environments; of breeding for increased yields, for resistance to pests and diseases; and of ensuring that the quality of the lint was such that it would find a ready place in world markets.

A lot was tackled and some striking results were quickly produced. Those that were more generally applicable were readily adopted by workers in other countries. There was, for example, the development of more suitable techniques for calculating the reliability of rainfall distribution, for measuring soil moisture and for measuring genetic resistance to bacterial blight. There was the release, in comparatively early days, of the important breeding stock, Albar 51, which contributed so widely to the genetic improvement of cotton varieties in Africa. But, even as the benefits of a central research station were beginning to accrue, so the seeds of its demise were unwittingly being sown. How, in succeeding years, for example, could the Corporation convince the administrations of newly emerging nations that they should continue to subscribe to a central station, when the first call on their funds would have to be for urgent local development? Governments wanted Corporation staff to work in their own countries and, in meeting such wishes, the Corporation came to recognize that each country must develop its own integrated research programme, in which aspects of the more fundamental work on cotton would be complementary, rather than secondary, to the central programme in Uganda. Moreover, at Namulonge, costs were rising. The Corporation's income was fully committed and either more money had to be found or changes made.

A change of concept

From discussions with the governments of the countries concerned it became clear that the station could not continue to function indefinitely as a regional research centre, but would have to become more closely associated with Uganda. The Uganda Government was in accord with this view and, from 1957 onwards, became progressively more involved with the Corporation's activities at Namulonge.

Briefly, the changes that occurred between 1957 and 1966 had four main aspects. First, Uganda agreed to make larger contributions to the cost of the Corporation's work; second, there was a progressive re-orientation of the work programme to meet, more directly, the immediate needs of Uganda; third, co-operative schemes were developed to make use of the surplus land at Namulonge; and, fourth, there was a gradual disengagement from Namulonge's responsibilities for regional research. None of the changes to the research programme was sudden or dramatic. All were phased over relatively long periods so that projects already embarked upon could be rounded off and the general continuity of the research approach would not seriously be disrupted.

These changes paved the way for the most important step of all. In 1967, the Corporation formally ratified the intention it had expressed in former years of presenting the station, its library and all other assets, as a free gift to the Uganda Government on 1 January 1972. A joint Working Party was set up to work out the details involved in the

Figure 1.3. Clearing trees and scrub using a crawler tractor during the late 1940s.

hand-over, to make recommendations for safeguarding the future of cotton research and to plan further development of the station. The report of the Working Party was submitted to the Uganda Government in November 1969, and approved without amendment. It provided a mandate for the continued employment of all locally-recruited staff at Namulonge, for the continuation, after the hand-over, of technical assistance from the Corporation and for the progressive Ugandaniza-tion of the research posts. It established the principle of organizing cotton research on the basis of a country-wide unit, with officers stationed at the major research centres, and its headquarters remaining at Namulonge. It laid down that the work at Namulonge should be widened in scope so that the station could continue to develop as one of the main agricultural research stations in Uganda. It gave recognition to the developments that had taken place over the previous ten years, which had affected both the Namulonge estate and the cotton research programme.

The Namulonge estate

When the lease for the land at Namulonge was being negotiated, the owner pointed out that the Corporation should not attempt to acquire only the good land, but should be prepared to take the 'bones with the fat'. Furthermore, it was thought that the area available for arable crops would be seriously restricted by the steepness of the slopes of many of the hillsides. When the farm was developed, however, it soon became

Figure 1.4. A view of the research station in 1971, looking south-eastwards towards the Nakyesasa Livestock Husbandry Experiment Unit, which can be seen in the background.

apparent that the total area, amounting to 908 ha, gave far more usable land than was necessary for the cotton experiments and an excellent opportunity arose of investigating some of the problems of large-scale farming. The arable land was organized into a number of farming units designed to test different farming systems, in which emphasis was placed on the maintenance of soil fertility by resting the land under grass for three years, and on methods of soil conservation in relation to mechanical cultivation.

By 1957, some 360 ha had been cleared and laid out as strips for arable cultivation, or designated as permanent pastures. There remained uncleared a block of 280 ha at the eastern end of the station, a block of about 80 ha at the western end and some 190 ha of swamps and their undeveloped margins.

One of the first major problems encountered concerned the resting phase of the rotation. Where the grass was managed by grazing, it was soon found that the local *Nganda* cattle were relatively inefficient in converting grass to meat or milk and that if production was to be intensified, a programme of cross-breeding with exotic stock would be essential. A successful technique of artificial insemination was developed and a breeding programme started, but it was soon recognized that a major programme of livestock improvement would be outside the terms of reference of the Corporation and, accordingly, after consultations with the Ministry of Agriculture an approach was made to the Uganda Veterinary Department to assist in this work. The Corporation offered to loan, free of charge, the 280 ha of uncleared land at the eastern end of the station, known as 'Nakyesasa' for the purpose of carrying out joint experiments in animal breeding and husbandry. A small research unit was set up under the administrative direction of the Veterinary Department. Policy for the unit was formulated, in its early years, by a small standing committee chaired by the Director of the Cotton Research Station and having as its members the Chief Research Officers of both the Departments of Agriculture and Veterinary Services. The Corporation provided 166 cross-bred animals, at nominal cost, to start the breeding programme, and built an additional senior staff house so that the officer-in-charge of the unit could be accommodated on the station. So, in September 1958, the Livestock Husbandry Experiment Unit at Nakyesasa officially came into being. It grew rapidly and soon became known throughout Uganda for the important contribution it has made to the development of the livestock industry.

Although the development of the Nakyesasa end of the station meant that a further 280 ha of land were efficiently used, the scale of operations on the remainder of the farm was still greater than necessary for the revised research programme. It was obvious that additional projects could easily be fitted into the available land and, after consultations with the Department of Agriculture, plans were made for a unit of its seed multiplication scheme to be developed at the western end of the

Figure 1.5. The Namulonge estate at the time of the hand-over in 1972, comprising the research station, the seed multiplication unit and the livestock husbandry experiment unit.

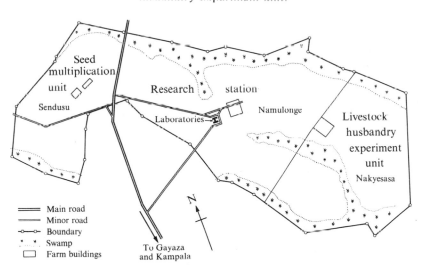

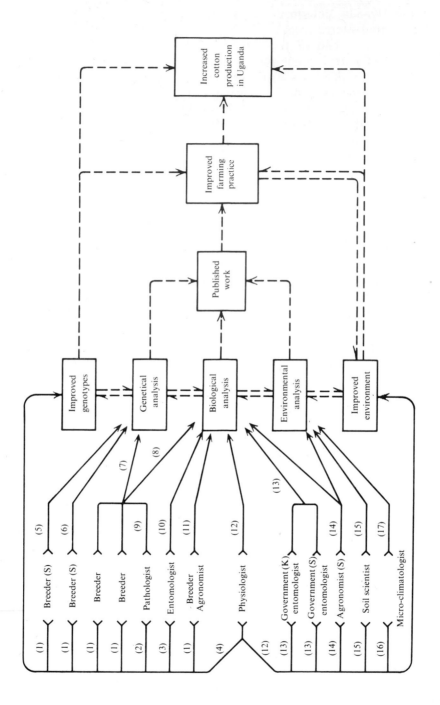

Figure 1.6. A diagrammatic representation of the cotton research programme in Uganda, as defined in 1966. (K) and (S) represented staff at Kawanda and Serere respectively; the remainder were based at Namulonge. Research projects are shown numerically and are described below. Co-operative projects are indicated by the joined lines, with the more routine aspects of the work to the left of the diagram and the more fundamental aspects, together with the aims, shown to the right.

Brief descriptions of research projects shown in Figure 1.6.

(1) The main cotton breeding programme. This covers a comprehensive programme of hybridization, selection and testing, leading to annual issues of pedigree seed. Recent developments have been the closer integration of the Serere and Namulonge programmes assisted by regular meetings of the staff concerned. District variety trials have been put on a country-wide basis with common control varieties. Two new varieties, SATU and BPA, are being established and these are expected to make a substantial contribution to increased production.

(2) Screening for resistance to bacterial blight at Namulonge with assistance to staff at Serere and Ukiriugru.

(3) A new programme of screening for susceptibility to lygus attack under field conditions. In the first season's work considerable differences between genotypes were recorded.

(4) Fibre testing for all countries in which Corporation staff work. Approximately 3000 samples are now tested annually.

(5) A detailed investigation of the effects of reduced seed coat fuzz on the components of yield and quality. Oil content of seeds will also be studied.

(6) A study of the components of yield in Albar Acala hybrids.

(7) Continuation of work on the genetics of resistance to bacterial blight.

(8) A programme of investigation into the causes of variation in the host-parasite relationship with bacterial blight of cotton.

(9) More detailed studies on problems arising from (7) and (8).

Clones of cotton have been established in order to minimize genetic variation in host plants.

(10) A programme of investigation into the bases of host plant attraction to insect pests of cotton. Work on *Earias* sp. and lygus is in progress, the latter being used to follow up the field screening.

(11) Work on the chemical control of weed populations in arable crops.

(12) Crop physiological studies on a background of population and sowing date trials.

(13) An extensive programme by Government entomologists on the chemical control of cotton pests.

(14) Work at Serere on populations, sowing dates and herbicides.

(15) The Nuffield programme, starting as an investigation of nutrient cycles in arable and resting crops, has now been merged with the fertilizer work. Recent observations have indicated possible deficiencies in trace elements under certain circumstances and these are included in current experiments.

(16) Collaboration with the Meteorological Department and the Water Development Department in the collection and collation of meteorological data on a country-wide basis, advice on irrigation projects, and continuation of work on the calculation of cotton crop forecasts using sowing date, acreage and rainfall data.

(17) Work on micro-climatology in Uganda, South Arabia and the Sudan on factors affecting evaporation from cotton crops. A large volume of data has been collected and this will be written up before embarking on new work.

station, covering an area of land that became known as 'Sendusu'. (The name 'Sendusu' was given originally only to a small area of land of some 20 ha, in the middle of the area concerned.) Development of the Unit was held up for some time through lack of staff and funds, but cropping was started in the first rains of 1970 and by the end of the year new farm buildings and houses for the farm manager and his assistant were under construction. The unit was to be concerned mainly with food crops but would be able to multiply seed of other crops, including cotton, when necessary.

The central area of the estate, about 400 ha, was further developed from 1966 onwards as the main area for the Corporation's research programme, and it was more specifically to this central area that the name Namulonge came, in later years, to apply. The layout of the station, showing its three main areas, is illustrated in Figure 1.5.

Revised research programme

When the Cotton Research Station was built, about two-thirds of the cotton crop in Uganda were produced in Buganda and Busoga. In the ensuing years, however, cotton was rapidly supplanted by coffee as the major cash crop in Buganda and parts of Busoga, and maintenance of cotton production at former levels became dependent upon expansion of the crop in the north and east of the country. Accordingly, by arrangement with the Department of Agriculture, greater emphasis was placed on the Corporation's programme of work in these areas. One Corporation officer was posted to Serere in 1956 and a further two in later years. Thus, during the period when staff at Namulonge were gradually being reduced, an effective team was being created at Serere to work closely with Departmental staff in tackling problems in the north and east of the country.

The revised research programme was described in the *Progress Report* for 1965–66 (Arnold 1967) from which the diagram shown in Figure 1.6 is reproduced. The programme was designed to integrate the work in different disciplines on a countrywide basis. At the same time it attempted to strike a balance between the immediate requirements for applied research and the desirability of continuing some of the more fundamental investigations that had formed part of the regional research programme. This framework formed the basis of the programme for the ensuing six years, but allowed ample flexibility for the details to be modified as new ideas emerged and circumstances changed. The policy was to define the aims of the research and to decide where the main effort should be put. Provided that the main programme always took priority in time and resources, staff were given freedom to choose additional problems that appealed to them, in order to penetrate more deeply into specialized aspects. In this way the division of research into categories such as 'applied' and 'fundamental' was avoided and all research staff shared in the routine work of the station as well as deriving stimulus from the more challenging lines of research.

The concept of an integrated team approach was reinforced in practice by various group activities. Occasionally these would take the form of organized discussion groups, sometimes with participation from visiting scientists. Additional discussions were held as regular monthly meetings, which began with a tour of part of the research station farm or a group of field experiments. A Land Rover, and later a farm trailer, were modified for this purpose so that up to twenty-five participants could stand on the back and air their views about a particular experiment from a position of vantage.

Such organized discussions were continually supplemented, however, by informal debates that arose frequently and naturally in a community that lived on the station and generated its own recreational activities and social life. Many of the freest exchanges of views took place over a glass of beer in the club-house after a game of tennis, squash or golf.

In these ways scientists in one discipline gained a good knowledge of work in the others. The cross-fertilization of ideas and the criticisms of outlook that occurred undoubtedly helped to avoid some of the pitfalls in agricultural research that is too narrowly channelled. In such circumstances, typical of research stations in the developing countries, communication comes easily and naturally. There is a total involvement in the life and work of the research station which generates an atmosphere quite different from that on research institutes in the developed countries, where home-life is sharply demarcated from work and where conscious, sustained effort is essential to maintain effective communication among research workers in different disciplines.

Interchange of ideas within the group is not enough, however. Research must be kept relevant to national needs. The extent to which the results of agricultural research in developing countries can be transformed into increased production is largely dependent on the extent to which the grower can modify his methods. This, in turn, is related to a complex of economic and sociological factors with which planners and extension workers constantly grapple. No quick or easy solutions to these problems in Africa have yet been found. Their existence makes it essential for the research worker to keep in touch with both the grower and the field officer; and an important aspect of the organization of research is to provide the means of bringing this about.

Journeys through the cotton areas to talk to growers and extension workers on the spot formed an important part of the activities of

Figure 1.7. Visits to the main cotton producing areas to talk to growers and extension workers on the spot formed an important part of the activities of all research workers.

each research worker. Equally, visitors to the station were actively encouraged through open days and visits organized by government officials and educational establishments. Although adding to the workload of the research worker, time spent on all such activities must be regarded as an important investment in the mechanism for the eventual implementation of results.

The lessons of Namulonge

The Namulonge story provides lessons that have wide implications for the organization of agricultural research, and for the administration of technical aid in the developing countries. It shows the advantages that accrue from focussing the attention of a group of highly trained specialists on the problems of a single crop so that each, working against a background of the others' findings, is able to fit his own investigations into a broader pattern. The greater understanding that emerges not only makes for deeper penetration into the principles involved, but also creates a spearhead of knowledge with which to attack problems that lie outside the immediate realm of the particular crop. Thus, as well as contributing to development, Namulonge made its contribution to agricultural science.

Essential to any success that was achieved, however, was the continuity of thought that resulted from long-term funding (latterly,

Figure 1.8. A party of visitors being shown round the station.

quinquennial), a large measure of autonomy in administration and a reasonable degree of continuity in staffing. The size of the research team was also important, perhaps to a greater extent than was recognized in determining its structure. A group of from twelve to twenty specialists is large enough to give breadth of outlook but small enough to develop unity of purpose.

But Namulonge was not wholly or continuously successful, and we should consider its limitations as well as the contributions it made. The original aim of providing a central research service failed, as we have seen, because the concept did not take fully into account the national aspirations of the co-operating countries, the diversity of their needs or the inevitable re-ordering of priorities that accompanied their self determination. As far as research for Uganda was concerned, the concentration of the cotton research team into one location could not be sustained when the balance of cotton production moved to different ecological zones in the north and east of the country. Both of these types of considerations, the regional and the national, are fundamental to projects in the developing countries that aim to make their main contributions through the work of a large, central research station. The attractiveness of such a station as a clearly defined object for raising funds, and as a favourable environment for intensive research effort, has to be weighed very carefully against its possible limitations. How then could the benefits of an approach, such as that developed at Namulonge, be realized without investing in a large, central station?

The answer may lie partly in an extension of the concept which gave rise to the creation of a *research unit* to carry on the cotton research in Uganda. The concept envisages a basic administrative structure that provides a framework within which research units can operate. The framework is formed by the main, national research stations (and other research institutions) together with their supporting substations and district variety-trial centres. Within this framework, separately-accounting research units can be created, augmented, reduced or disbanded without seriously impairing the overall framework, or disrupting its administration. The research institutions become the pegs on which the research projects are hung. The overall research programme can rapidly and easily be adjusted to fit national priorities.

The research unit would comprise a group of specialists, each working on a different aspect of the same crop at the station giving the best facilities, in terms of resources and ecological conditions, to tackle the particular problem. The unifying effect of a common interest in a single crop, reinforced by belonging to the same separately funded unit under one leader, would help to preserve the spirit of a team approach.

In addition to its flexibility, other advantages of the system would be its suitability for project financing and hence its potential for attracting financial assistance both from aid-giving organizations and direct from the local, agricultural industry. Furthermore, the research unit structure makes it possible to keep research teams to a reasonable size for effective internal co-ordination, and provides opportunities for research leadership that are not too exacting for the individual officer.

Attempts to co-ordinate and direct research through committees, however structured, are no substitute for effective, on-the-spot leadership. In research projects that are funded from external aid and staffed initially by expatriates, a great deal of effort is usually devoted to recruiting local 'counterparts' or 'understudies'. Neither of these terms is particularly useful; a creative research worker has no exact 'counterpart' and to call a man an 'understudy' is to imply inferiority. It is better to build up teams in which local staff and expatriates work side-by-side, and to defer decisions on who should replace whom until the natural abilities of the local staff have had time to develop, and those most suitable for positions of leadership have been identified.

In these various ways, attempts to solve the problems that developed at Namulonge, produced answers that might well find application to other crops and in other situations.

2

Agrometeorology

D. A. RIJKS

Introduction

Agrometeorology is an applied science that is concerned with the relationships between weather and agricultural production. It describes and analyses the effects of the weather on vegetation, soil, open water and animals as well as the reciprocal effects of these 'surfaces' on their atmospheric environment. It seeks to formulate the relationships that exist, and then to apply this knowledge both to increase the benefits that can be derived from natural resources, and to minimize the disadvantages that are caused by adverse meteorological conditions.

Agrometeorology is thus a science of the boundary layers between the airmass on the one hand and crops, animals, soil and water on the other. It makes use of the existing knowledge of meteorology, plant physiology and soil science. Adequately defined, quantitative assessments of the processes involved, and of the phenomena described, are essential aspects of such an applied science.

The Namulonge agrometeorology programme evolved from the early work on factors limiting cotton production. This work made extensive use of existing knowledge and techniques in the fields of plant physiology and soil physics, but at the time there was no equivalent body of information available in the field of tropical meteorology. Therefore, the early agrometeorology programme contained not only work on the relationship between crop and weather, but also considerable effort devoted to the description of the atmospheric environment near the ground. As knowledge of meteorology in the tropics increased, the differences between temperate and tropical zones in meteorology and crop environment became increasingly evident, and pointed to the need for special studies to describe and characterize the atmospheric environment of the tropics.

Consequently, this chapter on agrometeorology deals with the description of the atmospheric environment as well as with some of the work on the relationships among weather, crops and soils, which might equally well have been dealt with in the chapters on soil chemistry or crop physiology. Their inclusion here reflects primarily the way in which the work at Namulonge developed but also serves to emphasize the interdependence of these disciplines.

General description of the climate

Two belts of high pressure girdle the earth roughly at latitudes of 23° N and 23° S. At the surface of the earth, these zones often appear broken up into more or less clearly developed cells of high pressure. Their geographical location is quasi-constant apart from a seasonal movement from north to south and back again, following 'the movement of the sun'. At the surface of the earth, air flows away from these cells of high pressure. Owing to the rotation of the earth and conforming with the laws of inertia, the airstream away from these high pressure cells is deflected, on the northern hemisphere to the right, and on the southern hemisphere to the left. The region where the airmasses originating from the areas of high pressure in the northern and southern hemispheres converge, is called the inter-tropical convergence zone (ITCZ). In East Africa, the ITCZ is located approximately on the equator in the months of April and October, and follows the sun on its annual north–south 'movement' with a time lag of about four to six weeks. Apart from these seasonal movements, the ITCZ is subject to irregular movements depending on the degree of development of the high pressure cells to the north and to the south, as well as on diurnal movements. The latter are of relatively minor importance in determining the general weather pattern of the area.

At the ITCZ, air moves upwards under the combined forces of convergence and intense heating by the sun, giving rise to cooling by expansion. At a certain height, depending on the water-vapour content of the air, the water vapour in the air will condense and, eventually, precipitation may occur.

It follows that in Uganda the rainy seasons at latitudes near the equator are centred on April and October. More northward, where the ITCZ 'arrives' later and 'departs' earlier, the two rainy seasons are centred on May and August–September, while in the very northern part of the country (as in the southern part of the Sudan) the two seasons merge into one longer rainy season, leaving the period from December to March as the dry season. In the extreme south of the country, the converse occurs, and the main dry season falls in the months of June to August.

It is evident that the rainfall distribution and reliability, as well as the sunshine, wind, temperature and evaporation regimes, are all related to the dynamic situation of the ITCZ relative to the place of observation.

Given that the basic weather pattern is determined by the dynamics of the ITCZ, there are two major modifying influences on the weather in Uganda. The first is the effect of the topography, notably that caused by the mountains and high plateau areas in the West and East. The second is the effect of the large body of water formed by Lake Victoria. This latter effect is noticeable almost every day as far as 100 km inland from the shore-line depending on the general direction of the main flow of air over the area; but its overall influence extends over most of the southern part of the country.

Mean annual rainfall in Uganda varies from 500 mm in the north-east to 2100 mm in the Lake Victoria area. Mean duration of sunshine varies from five hours per day in the south-west to almost nine hours per day in the north-east. Run-of-wind is on average about 160 km per day, most of it occurring during the daytime. Altitude has an important influence on the temperature regime; generally speaking, the north-eastern half of the country experiences a moderate to warm, and the south-western half a moderate, regime of temperatures, with local exceptions associated with high altitude. A comprehensive description of the climate of Uganda has been presented by Griffiths (1968, 1972) and by Jameson & McCallum (1970).

At Namulonge, the main feature of the climate is the presence of two rainfall seasons, the first and most reliable from March to May, and the second from September to November. Mean annual rainfall (1947–68) is about 1230 mm. Mean monthly maximum temperatures range from 25 to 30 °C and minimum temperatures from 13 to 17 °C. Afternoon dewpoint values range from 12 to 20 °C; dew is observed frequently, persisting sometimes until 9 or 10 a.m. Mean monthly duration of sunshine varies from four to eight hours per day. Annual total evaporation from open water is slightly in excess of annual total rainfall. The seasonal distribution of evaporation is discussed later.

Description of the atmospheric environment

Methods of data collection

There are essentially two levels of data collection. The first is the collection of the data necessary to describe particular relationships and to express these relationships in a manner that permits their wider application. It involves the accumulation of many, often detailed, data at a single location. This is data collection on a meso- or micro-meteorological scale. Such data serve to elucidate and quantify the relationship between crop water use, for example, and a number of meteorological parameters that are more generally available. At a later stage these commonly available data are used to make inferences (such as on water infiltration rates in relation to rainfall intensity) that can be applied elsewhere in the same broad ecological zone.

The second level of data collection consists of the country-wide collection of standard (macro) meteorological data. The early network of stations for the collection of macro-meteorological data consisted of a large number of rainfall observation posts and about seven synoptic stations, run by the East African Meteorological Department (EAMD), as well as a few climatological observation posts at the major agricultural research stations. During the 1960s, the establishment of several hydrometeorological development schemes was proposed. It therefore became necessary to complement the rainfall observation network with a network for the measurement of potential evaporation. For its implementation, close co-operation among the meteorological, hydrological and agricultural services was established, and this finally

Figure 2.1. Stations recording meteorological data for the calculation of Penman estimates of evaporation (after Rijks, Owen & Hanna 1970).

Figure 2.2. The Namulonge meteorological station; altitude 1150 m, latitude 0° 32′ N, longitude 32° 35′ E.

resulted in a fairly uniform pattern of instrumentation and station layout. Procedures for observation were also standardized as well as methods of analysing, interpreting and publishing data, not only within Uganda, but also in Kenya and Tanzania.

The Uganda agrometeorological network

As outlined above, the Uganda agrometeorological network expanded from seven to about twenty-four stations over the period from 1960 to 1970 (Figure 2.1). Its primary purpose was to furnish estimates of potential evaporation, that could be combined with the rainfall data already available, to furnish detailed estimates of the water balance in all regions of Uganda. The instruments and the programme of observations also permit, however, the accumulation of data on the temperature, radiation and wind regimes in each area. The distribution of the stations over the country, initially somewhat uneven, improved as more and more organizations became aware of the use that could be derived from the data, and co-operated in what finally became an integrated national effort. The coverage at the end of the decade was in many respects very satisfactory (Rijks, Owen & Hanna 1970).

The basic layout of each station was a 30×30 m enclosure under natural ground cover, usually short grass, in which were installed the following: a standard 5 in. raingauge; a meteorological screen with maximum, minimum, wet bulb and dry bulb thermometers; a Campbell–

Figure 2.3. A Gunn–Bellani radiation integrator. These instruments, which were calibrated against a Kipp solarimeter at Namulonge, were used for estimating solar radiation at other meteorological stations in Uganda.

Table 2.1. *Criteria in checking observations on form 602*

1. Dry 9.00 a.m.	⩾ wet 9.00 a.m.	±0.5 °C
2. Dry 3.00 p.m.	⩾ wet 3.00 p.m.	±0.5 °C
3. Max	⩾ dry 9.00 a.m.	±0.5 °C
4. Max	⩾ dry 3.00 p.m.	±0.5 °C
5. Min	⩽ dry 9.00 a.m.	±0.5 °C
6. Min	⩽ dry 3.00 p.m.	±0.5 °C
7. Max reset	= dry 9.00 a.m.	±0.5 °C
8. Min reset	= dry 3.00 p.m.	±0.5 °C
9. Max	⩾ min	
10. 5 in. gauge	= Dines	±2 mm when ⩽ 25 mm
		±4 mm when ⩾ 25 mm

11. All data present.
12. All entries to correct number of decimal places, i.e. all temperatures to one decimal place; rainfall and evaporation to one decimal place if measured in millimetres, to two decimal places if measured in inches; run of wind in units; cups added to or taken out of evaporation pan in units; sunshine hours and radiation to one decimal place.

Stokes sunshine recorder; a Gunn–Bellani radiation integrator; an anemometer. Where funds were available, a recording raingauge, a thermo-hygrograph and a United States Weather Bureau class A evaporation pan were also installed. Details of the arguments that led to the choice of the station layout and to the instrumentation have been described elsewhere (Rijks 1968*a*).

In the earlier years, observers were trained at Namulonge. In the latter part of the decade, training courses were offered by the EAMD, so that homogeneity in observing procedures was achieved not only on a national, but even on an East African scale. Similarly, the instructions for observers and the record sheets were established uniformly for the three East African countries in a joint effort by the national and East African Community services involved. The layout of these record sheets was adapted to the ultimate transfer of the data onto cards for computer analysis. The record sheets were completed in triplicate by the observers. One copy was kept at the meteorological site, two copies were forwarded to departmental HQ, checked for instrument and observer error according to a standard set of criteria (Table 2.1) and finally one of these was forwarded to EAMD HQ in Nairobi, where the data were transferred to computer cards, stored and kept available for reference. Ten-day and monthly totals and means were published annually by the EAMD.

There was no centrally organized service for the regular recalibration and maintenance of the instruments, mainly because this would involve the pooling of recurrent expenditure from various sources and departments. Neither did the EAMD have a budget for the maintenance of instruments installed by other organizations. A central maintenance and calibration service would nevertheless have been an advantage.

Analysis and evaluation of basic and special observations

Agricultural research workers always, in the first instance, relied on the basic standard observations, such as rainfall and screen temperature, to understand the various processes observed in agriculture, but sometimes such observations were inadequate to explain phenomena, such as the variation in water stress experienced by crops, in sufficient detail. In these cases more detailed measurements were made, and subsequently related back to standard observations. Some of such studies undertaken by staff at Namulonge are reviewed below.

Rainfall: totals and probability

The simplest and usually the first approach to describing the rainfall regime of an area is the use of annual means. In Uganda, such data for about 200 rainfall stations are presented in a mean annual rainfall map, published in the *Atlas of Uganda* (1967).

In agriculture, the reliability of rainfall is often more important than the mean, and therefore maps of the minimum amounts of rainfall to be exceeded in nine years out of ten and four years out of five have also been prepared and published (EAMD 1961). However, even the use of these probability data for the planning of agricultural practices and development is recognized to be of rather limited value. An obvious further refinement is the use of mean monthly data, maps of which have also been published in the *Atlas of Uganda*. Nevertheless, even monthly means proved to be rather erratic for the scheduling of agricultural practices, especially in those parts of Uganda where two crops are grown each year. In these equatorial regions, where there are two seasons of rainfall separated by two dry seasons, knowledge of the duration and distribution of rain within the rainfall season, and of its date of onset, is of prime importance. The effective length of each of these two growing seasons, already short when compared with the single growing season in many temperate climates, is determined to a large extent by the balance between rainfall and crop water requirement at the beginning and the end of each of the two seasons. There are thus four critical periods each year, when success or failure may depend on relatively small changes in the amount of water available, or on the time at which it is available. Manning (1950) therefore proceeded to determine the reliability of monthly rainfall totals and introduced the concept of confidence limits. He selected 9:1 limits (a deficit once in twenty years) as appropriate for subsistence farming. For more advanced agricultural systems, where a greater measure of security has been reached, he suggested 4:1 or 1:1 limits. He thought that these limits might also be appropriate for farming systems where the introduction of cash crops was contemplated.

Manning (1956*a*) later modified this technique to cover three-week moving totals, a length of period better adapted to agricultural practices in Uganda. The choice of the length of the period to suit local conditions is obviously related to the general pattern of rainfall distribution, potential evaporation, soil water storage and crop water requirement.

Describing a drought month as one in which the rainfall expectation was less than 50 mm more often than once in four years, Manning concluded that north of 2° N and south of 2° S there were single bimodal rainy seasons, whereas the stations between 2° N and 2° S had two distinct rainy seasons (Figure 2.4).

Figure 2.4. 4 : 1 confidence limits of peak date and seasonal quantity of rainfall for latitudes 8° N to 5° S. Minimum expectation for nine years in ten shown by blocking. Rainy seasons shown by joining months ⩾ 2 in. for 3 years in 4 (from Manning 1956*a*; reproduced with permission from the Royal Society of London).

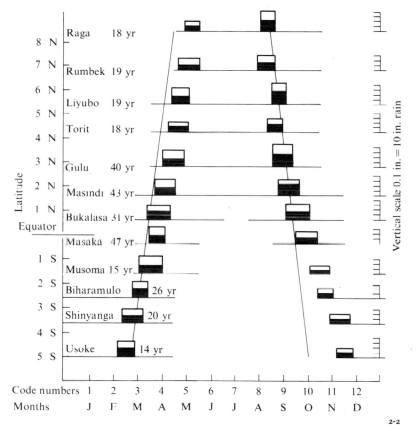

He also noticed that within this narrow equatorial belt there was a further division of regions. Rainfall expectation at stations west of 32° E was greater in the second (July to December) rains than in the first whereas to the east of 32° E the converse applied. The boundary between these two areas is characterized by a rather unreliable rainfall distribution (see Manning 1956*a*, map 1). Since 1971, satellite pictures have become available that provide synoptical support for the existence of a 'veer' of southerly Atlantic air over the Congo basin in the latter part of the year, which would help to explain such a phenomenon. In Figure 2.4 can be observed a relatively slowed-down 'outward' movement of the rains, both on the northern hemisphere in April–May, and on the southern hemisphere in October–November.

Manning proceeded to calculate confidence limits of three-week moving totals of rainfall for more than 100 stations in Uganda, an example of which is given in Figure 2.5. This information was then used, amongst other purposes, to select the optimum sowing period for cotton (and other crops), by comparing minimum rainfall expectation with water requirements of the crop, at the time of sowing, and during leaf-area development and flowering.

Manning's basic concept has continued in use in Uganda and many other countries, although the units have been changed to millimetres, the periods to ten-day or five-day moving totals, and the procedures adapted for use on computers (Walker & Rijks 1967; Thorp 1973). In addition, proposals for a different statistical treatment of the basic data have been suggested (Glover & Robinson 1953; Brunet-Moret 1969).

From an analysis of successive twenty-year runs of the data for 120 consecutive years from Padua (Italy), Manning concluded that fifteen

Figure 2.5. Confidence limits of rainfall distribution at Namulonge. Unbroken line: twenty-day totals at ten-day intervals, 1970. Broken lines: confidence limits (1 : 1) calculated from data from 1950 to 1969.

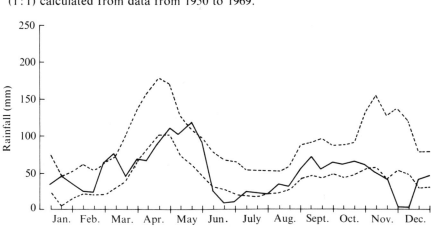

to twenty years would constitute an adequate sample of the data to allow valid extrapolation to be made, with the statistical techniques he used. However, rainfall regimes in Africa have often been thought to exhibit cycles, or sequences of wetter and drier years, which are explicable in geophysical terms. If such cycles were shown to exist, knowledge about them could contribute to the formulation of specific practices in agricultural management. This would be especially useful if ways could be found to predict the onset of the sequences of wet and dry seasons.

Nevertheless, the existence of such cycles has not yet been proved. Statistical work has been hampered by the fact that there are very few stations for which more than sixty years' data are available in the tropics. Most stations have reliable records of, at best, about thirty years. A very thorough examination of these data has been made by H. Mörth (unpublished) in East Africa as well as by Giraud (1973) in West Africa, and no statistical evidence for the existence of such cycles has as yet been demonstrated. Analysis of the data will undoubtedly continue, also because the fluctuations in lake and river levels, both in East Africa (Lamb 1966) and in West Africa (Juton 1971) show variations that might possibly be associated with certain, pre-established rhythms in rainfall regimes.

Rainfall intensity

Apart from totals and reliability, the intensity of rainfall was recognized to be of great relevance in the development of mechanized farming systems in the tropics. Farbrother (1951) developed a sensitive rate-of-rainfall recorder, when he noticed that the existing instruments, designed for use in temperate zones, failed to provide measurements with sufficient details of the rainfall intensities that occurred in tropical storms. He measured peak rainfall intensities exceeding 330 mm per hr for short periods. Of 490 storms in the period from 1950 to 1955, more than 300 had peak rates greater than 25 mm per hr and 74 had peak rates greater than 100 mm per hr, but few storms had rates of more than 12 mm per hr that were sustained for more than 30 min. Analysis of a six-year run of his data showed that a logarithmic relationship existed between instantaneous peak rates and the total amount of rainfall in a storm (Hutchinson, Manning & Farbrother 1958a) and a negative relationship between the logarithm of duration and peak rate. It was also observed that rain tended to fall in shorter and more intense storms in the second rainfall season (September–October) than in the first. But while intensity of storms is higher in the second rains, the frequency of storms greater than 12 or 25 mm is greater in the first rainfall season (Rijks 1968c).

Having measured intensity of rainfall, Farbrother next set out to determine the relative rates of runoff and percolation under various intensities of rainfall on bare and mulched soils and on a *Cynodon* turf. He observed percolation rates of up to 90 mm per hr on grass mulch and turf, as compared with 18 or 20 mm per hr on bare soil, and his measurements showed that up to 60 per cent of the total rainfall in storms of more than 6 mm could be lost as runoff from bare soil. On grass mulched plots, or on turf, however, loss from runoff would be as little as 8 per cent. This work was of prime importance in estimating the availability of water for crops and in determining the ultimate layout of the farm fields, cultivation methods and other farm practices.

Influence of rainfall on local topography

Six raingauges at different locations on the 1000 ha farm at Namulonge were read for a series of years. In individual storms, important differences were observed which might have decisive effects on the timing of particular farming operations and possibly on crop yields. However, differences observed among gauges in monthly totals were small and did not follow a systematic pattern. There is thus no evidence of an effect of the local ridge and valley topography on measured rainfall distribution.

Wind

Run of wind is fairly steady throughout the year and from year to year. Strong winds that cause problems in crop production are infrequent. Special studies of the wind regime have therefore not been made. The average run of wind is about 125 km per day and most of this occurs during the daylight hours. Nights are very often calm. During storms, gusts of wind occur, but widespread damage caused by such gusts is seldom observed although, locally, damage to maize crops (and, indeed, to buildings) occurs in some seasons in the path of convection storms.

Namulonge is situated in an area where the influence of Lake Victoria on the wind regime is occasionally felt, usually because lake breezes may drive showers that originate on the Lake in a northerly direction. This effect does not extend very far beyond Namulonge. Its extent is demonstrated by the map that shows the distribution of the number of raindays around the Lake (*Atlas of Uganda* 1967).

Hail and lightning

Damage to crops caused by hail storms occurs relatively frequently in the cotton season, but rarely during the first rains. Instantaneous damage to cotton crops may look severe. Lakhani (1973) reported measurements of the extent of the damage. He followed the growth of the crop throughout the season and concluded that the effect on yield was relatively minor.

Lightning damage to cotton, maize and bean crops is observed from time to time. When lightning strikes, the crop may be killed in a more or less circular area of six to twenty metres' diameter around the centre of impact of the lightning.

Sunshine and solar radiation

Solar radiation is the most important source of energy for the processes of photosynthesis and evaporation, and for heating the air and soil. Measurements or estimates of the receipt of solar radiation make it possible to calculate rates of potential photosynthesis and potential evaporation. Knowledge of these two parameters is of great relevance in planning new agricultural systems and in understanding the limitations on production of existing ones.

Measurement of solar radiation has for many years been confined to those centres where suitable measuring and recording equipment could be installed and adequately maintained. At Namulonge, a Callendar recorder was in operation from 1949 to 1959. In 1959 this instrument was replaced by a Kipp solarimeter. A second Kipp solarimeter was installed at Kabanyolo (the Makerere University farm) in 1960. From 1960 onwards, Gunn–Bellani radiation integrators, simple, robust and reasonably accurate instruments (Pereira 1959), became more generally available, and a network was built up covering most regions of Uganda. In co-operation with agricultural research centres in Kenya and the then Federation of Rhodesia and Nyasaland, a series of tests on the best method of exposure of Gunn–Bellani radiation integrators was carried out. The results (Rijks 1966) showed that the well-type mounting, as traditionally used in East Africa, was as sensitive and satisfactory as any of the other methods proposed. Continuous collection and observations from one of the Gunn–Bellani instruments over a five-year period showed that a gradual decline in sensitivity, of about 2 per cent a year, occurred and that regular calibration was essential. Based partly on these country-wide measurements of radiation, a map of mean monthly potential evaporation was eventually compiled.

In the absence of measurements, solar radiation can be estimated from records of duration of sunshine or of cloudiness. The latter relationship has been derived by Farbrother for a number of stations in Uganda (see Woodhead 1967). The relationship shows a clear curvilinearity, and the accuracy of the derived estimates of radiation is somewhat varied, owing to the fact that all types of cloud are lumped together in the cloudiness observations, irrespective of their height and relative transparency for radiation.

The relationship between duration of sunshine and solar radiation is linear, except for the narrow range from zero to periods of up to 0.5 hr sunshine per day, and estimates of solar radiation for ten-day and monthly means can be derived from this relationship with reasonable precision. Calculation of this relationship for many stations in Uganda (Rijks & Huxley 1964; Rijks 1968d) showed that no single set of coefficients in the Ångström-type relation was generally applicable, because local climate has a marked influence on the attenuation of solar radiation. Use of an inappropriate set of coefficients can easily cause errors in the calculated rates of potential evaporation of up to 20 per cent on the basis of monthly totals (Farbrother 1960c).

Duration of sunshine is also an important agroclimatic parameter in its own right, which, in addition to the planning of crop production, is used in many other fields, such as crop processing and animal husbandry. Maps showing the mean monthly duration of sunshine in Uganda have been published elsewhere (Rijks et al. 1970). Diurnal patterns of solar radiation and sunshine were studied and described by Farbrother (1960c) and by Dale (1960) in connection with work on relative turgidity and stomatal aperture.

Temperature and humidity

The temperature regime at Namulonge has not been a major limitation for the production of the crops traditionally grown, and has therefore not been the subject of a great deal of study. Temperatures are very steady throughout the year and from year to year. Mean monthly maximum temperatures are only very rarely outside the range 25–30 °C, mean monthly minimum temperatures seldom outside the range 14–17 °C. The diurnal range in temperature is about 10 ± 3 °C in the wet season and about 15 ± 4 °C in the dry season. It seldom exceeds 20 °C.

Katabatic flow of air into the valleys may occur in the Buganda area and the lowest temperatures observed at the valley floors in extreme conditions were about 7 °C. Presumably, this is one of the reasons why valley floors are rarely cultivated and never inhabited.

Mean dewpoint values of 9.0 a.m. are in many months about 2 °C higher than the mean minimum temperatures. Dew is regularly observed on pastures and field crops, and often persists until midmorning. Important differences between a.m. and p.m. dewpoint values are usually only observed in the months of January, February and March. The consistent presence of water vapour in the atmosphere, through its buffering effect on the loss of outgoing long-wave radiation during the night, contributes, no doubt, to the relatively high minimum temperatures that are observed.

One of the consequences of the small diurnal range of temperature, coupled with the relatively low radiation values, is that the rate of potential photosynthesis is not as high in Uganda as it can be in temperate zones during the summer.

In the context of the calculations for the Penman evaporation estimates, the values of mean air temperature (T_a) and mean vapour pressure (e_d) have been calculated in different ways. Mean air temperature can be calculated from $(T_{max} + T_{min})/2$ or from $(T_{a.m.} + T_{p.m.})/2$. At Entebbe, these two estimates were compared with the value of the mean obtained from twenty-four hourly observations. Mean air temperature from $(T_{a.m.} + T_{p.m.})/2$ appeared to yield the closer approximation. At Gulu a similar conclusion was reached when a comparison was made with five three-hourly observations per day. Similarly, mean dewpoint-temperature was reasonably accurately estimated from $(T_{d\ a.m.} + T_{d\ p.m.})/2$ (Channon 1968).

As a consequence, estimates of actual and saturation vapour pressure are best obtained from a.m. and p.m. observations of dry-bulb and dewpoint temperatures. Other methods of calculation, especially those involving mean relative humidity, were shown to yield considerably different values (up to 20 per cent) of calculated vapour pressure deficit in the air.

Standardization of the methods of calculating humidity and of the type of tables to be used would be desirable. Differences in dewpoint of up to 2 °C can commonly be observed, simply as a result of the use of different methods of measurement and calculation (Rijks 1968*d*).

Evaporation

Interest in evaporation from open water (or potential evaporation) at Namulonge, and at many other agricultural research centres around the world, has been aroused for two main reasons. First, because there is a general interest in the water balance of land and water surfaces in certain regions. (This is one of the reasons why the network of stations where evaporation could be measured was established in Uganda, to complement the network of stations with raingauges.) Second, because potential evaporation seems to be a useful reference parameter in studies of actual and potential evapotranspiration from crops. Measurements of the actual volumes of water used by crops are often limited in value, owing to their dependence on the atmospheric conditions under which the measurements were made. When such measurements are related to a more or less independent and reproducible estimate of the energy and of the sink available for evaporation, such as pan evaporation or preferably a Penman estimate, the exploitation of the meaning of such results at other places, to a certain extent, becomes possible.

Potential evaporation has been estimated in various ways. One group of methods uses observations from evaporation pans, Piche evaporimeters or from instruments recording radiation or sunshine. These observations are easily made but suffer from a number of disadvantages. Evaporation pans are seldom exactly similar in design, and almost never exposed in a uniform way. They may suffer from an oasis effect in a dry environment which may exist perhaps for only part of the year. Results obtained from different areas are therefore not always comparable, and there is no certainty that they are good estimates of potential evaporation. Piche evaporimeters only take changes in net radiation into account in a very indirect way. This may be no disadvantage in an area with an almost constant radiation pattern, but such areas are rare in equatorial Africa. Radiation records, on the other hand, do not sufficiently take the effects of daily changes in run of wind and in vapour pressure deficit into account.

The second group of methods employs formulae in which commonly observed meteorological parameters are used. Of the general formulae, the one proposed by Penman (1948, 1956) has been found to be the most reliable by agricultural research workers under a wide range of conditions, especially in East Africa (Anon. 1965), because it takes into account the effects of net radiation as well as those of wind speed and vapour pressure deficit. An account of the principles underlying the Penman formula and associated physical measurements is given by Monteith (1973).

At Namulonge, Farbrother (1954) and Kibukamusoke (1960) compared the relative merits of various instruments. Farbrother compared 24-hr totals of evaporation from a class A pan with those from a recording (sunken) pan and observed good agreement between the data for rainless days. On raindays, however, important differences (a mean of 2 mm per day for the 141 raindays of the year 1953–54) were observed. Detailed analysis of the records showed that different amounts of water splashed into and out of the pans during rainstorms, so that evaporation could not simply be calculated by allowing for the amount of rainfall as measured in a raingauge. Later, a second sunken pan was surrounded by a water jacket, and the estimate of evaporation of raindays was greatly improved.

As it became evident from experience elsewhere in East Africa (Sansom 1954; Anon. 1962) that observations from evaporation pans were often incomparable owing to different exposures, or subject (to a varying degree in different sites and seasons) to the effects of advective energy, estimates of potential evaporation using the Penman formula became more and more commonly used. There were two attitudes about the calculation of potential evaporation from Penman's formula. One was that it should always be calculated from the equation with Penman's original coefficients, so that estimates from all parts of the world would be derived in a similar way. Supporters of this attitude are prepared to sacrifice some accuracy for comparability, but it is debatable whether the parameters thus compared are in fact estimates of potential evaporation. The other attitude is that potential evaporation should be estimated by using measured values of radiation and locally verified coefficients where available. Thus, approximate physical quantities take the place of uniformly derived statistics.

A programme of verification of the empirical coefficients was started at Namulonge. Some of the results concerning the radiation balance have been described above. Results concerning the coefficients in the Dalton term of the equation (Rijks & Walker 1968) and concerning net long-wave loss have been collected and are being analysed and prepared for publication. This work gave rise to similar work in the Sudan Gezira, where the coefficients in the back radiation term were found to be dominantly influenced by the overlying airmass. Very important numerical differences in the coefficients were observed between the dry season conditions and the rainy season conditions (Rijks 1968b). It was also in the course of the calculations for the Penman estimates that the different methods of calculating mean air temperature and mean dewpoint, as outlined above, were evaluated.

Agrometeorological surveys

The major part of the agrometeorological knowledge that has been collected has thus been concerned with the water relations of crops and the water balance of smaller or greater units of land. The assessment of the water balance of a region requires information on three main topics: rainfall, evaporation, soil and vegetation. Summaries of the rainfall and evaporation data for Uganda have been published in the *Atlas of Uganda*, or in specialized publications of the EAMD (1959, 1961), or in agrometeorological surveys (Rijks & Owen 1965; Rijks *et al.*, 1970). On an East African scale, these surveys were complemented by those of Kenya and Tanzania (Woodhead 1968a, b; Dagg, Woodhead & Rijks 1970) and of the highlands of Eastern Africa (Brown & Cocheme 1969). The water balance at a few centres has been calculated and published in the surveys. For many other centres the water balance can be easily calculated, using the information on rainfall and evaporation that has been described above, when information on soil and local vegetation becomes available. For very detailed studies, reference can also be made to the numerical information stored at EAMD HQ or to Manning's graphs of confidence limits of rainfall kept at Namulonge.

Relationships among atmosphere, soil and crop

The central theme in the study of the relationships among atmosphere, soil and crop at Namulonge has been the water balance of crops, first of cotton only, but later of all the crops that took their place in the rotation. The evolution of the water balance throughout the course of the rotation not only affects the crop during its growing cycle, but also concerns the state of the soil at and after harvest and up to the start of the next crop, and thus it has a profound influence on cultural operations and the fertilizer regime. Indeed, it determines to an important extent, the potentialities and impossibilities of certain choices in the agricultural system. Furthermore its influence reaches beyond the direct realms of agriculture into those of hydrology and the development of water resources.

In tropical climates, where temperature rarely exerts a limiting influence on crop production, crops are often grown from the beginning of a rainfall season. This means that, depending on the length of both the crop growing cycle and the season of rainfall, crops are often sown on a virtually empty soil water profile, whereas the peak of the rainfall distribution coincides with the period of peak water requirement. This is particularly true in areas with a single, rather short, rainfall season and in those equatorial areas with a bimodal rainfall pattern, such as at Namulonge where, given the altitude and the temperature regime, each of the two rainfall seasons in a year is only just long enough for the crop cycle to be completed. A little more latitude exists in the northern part of Uganda where the single rainfall season is rather long.

This phenomenon is fundamentally different from that in certain temperate climates, where the soil profile is filled up by the winter rains; establishment of the crop is, as a rule, completed on a full soil water profile; and soil water shortage may occur only towards the end of the growing-cycle of a crop. A further fundamental difference from agriculture in temperate zones is that a two-crop-a-year system lacks the rest period (the dead season) that is characteristic of the winter in temperate zones, which can buffer the effects of variations in the cropping season, caused by such things as late sowing. In a one-crop-a-year system there is an independent chance each year of starting the crop at the right time. In a two-crop-a-year system a delay in any one season may put a heavy strain on management resources as well as on stored soil water, if one wishes to start the next season's crop at the proper date.

The seasonal pattern of crop water use

Year-to-year variation in the yield of the cotton crop has been a point of both major concern and serious investigation from the very first years that Corporation staff worked in Uganda. Manning set up a series of date-of-sowing experiments at many locations in Buganda and, in his 1946–47 report, tried to explain the observed variation in yield as being a function of sunshine and rainfall. In 1951–52 Farbrother and Manning concluded from their work over several seasons that soil water must be one of the main factors in accounting for this variation. They then set out to obtain the seasonal pattern of water requirements of cotton in order to match this to the expected pattern of rainfall as described by Manning's confidence limits, so that a rational choice of optimum sowing dates for all areas in Uganda might be made.

The problems in assessing the use of water by the cotton crop were large. Gravimetric sampling of soil water, the recognized most reliable method and obvious first choice, proved unwieldy as a research tool, when repeated samples at frequent intervals, on many plots with different treatments, were required. Plots were just not big enough to allow a sufficient number of soil samples to be taken, without unduly interfering with the uniform watering of the plots and crop growth. Farbrother therefore developed the electrical resistance technique with new nylon and stainless steel resistance units which were improved over the years and finally perfected to provide reproducible observations of stored soil water, free from influences of salinity and fertilizer applications (Farbrother 1957). Using these measured changes in stored soil water, he calculated actual crop-water use (ET) and then proceeded to formulate a model of the seasonal water requirement of cotton, which was closely related to the development of leaf area, for conditions when soil water was not limiting (Hutchinson *et al.* 1958*b*). He verified from soil water data, and by a convincing use of measurements of the relative turgidity of cotton leaves, that observed departures from this model were caused by a lack of available water, in periods when rainfall was insufficient. He further demonstrated that such departures did not occur if supplementary irrigation was given to the crop during periods of drought.

While the magnitude of the quantitative differences between evaporation from open water (the Penman E_o) and the seasonal pattern of crop water use thus derived were criticized (and this criticism was later shown to be largely valid), the main conclusions of the work on the dominant influence of crop development on the seasonal pattern of water use was ultimately accepted, and has become a recognized feature of almost all later work on crop water use on seasonal crops throughout the world.

In describing the seasonal pattern of water use, Hutchinson *et al.* (1958*b*) followed Penman's thesis that potential crop water use (E_t) is, in its physical aspects, determined by the same meteorological factors that determine evaporation from open water (E_o). They therefore expressed the seasonal pattern of crop water use in terms of the ratio ET/E_o. They observed an initial increase in the leaf area index (LAI, acres of leaf area per acre of soil) which was approximately logarithmic up to a maximum leaf area of 1.8, and a subsequent departure from this exponential curve, after about 14 weeks of plant growth. This pattern of leaf-area development was approximately concurrent with the increase in ET/E_o. It appeared later from further experiments in which plants were grown at different spacings that percentage ground cover would have been a better guide than LAI. They then set up the hypothesis that water use could be estimated from the product of LAI and E_o over a considerable period of the life of the plant, up to the point where the ratio ET/E_o reached the value 1.8, if energy was not limiting; but they recognized that, in fact, energy often became limiting when the value of the ratio reached 1.4, and that actual evaporation in excess of this value was rarely observed. The numerical value of their figures was subsequently questioned and, in a re-examination, Farbrother (1960c) calculated that his early estimates of E_o could easily have been up to 15 per cent too low because they were derived from calculated rather than from measured radiation data. Application of this correction to his original data yields a figure of about 1.2 for the ratio of ET to E_o, a value that has been observed by many others (e.g. Rijks 1965) working on similar crops. Such a 20 per cent increase in evaporation from a fully grown, tall crop over that from open water can be explained by the increase in turbulence at the evaporating surfaces. Farbrother (1962) measured a similar maximum value of 1.2 over actively-growing elephant grass during a period of an increasing water balance.

The number and nature of the limitations to 'successful translation in space or time' that Penman (1948) expressed about his formula, have often been overlooked when his formula has been applied elsewhere, and Penman explicitly reformulated these limitations in 1956 when potential evaporation was defined as the 'evaporation from a short green crop, actively growing, completely covering the ground and adequately supplied with water'.

Thus the work described by Hutchinson *et al.* (1958b) emphasized the role that availability of water and complete ground cover (or at least a uniform interception of radiation and subsequently a uniform repartitioning of the heat budget) play in determining the rate of evaporation. This aspect was subsequently described in detail by Rijks (1967b, 1968b, 1971) who measured the changes in radiation interception and the repartitioning of the heat budget throughout the development of an irrigated crop. These measurements confirmed the dominating influence of degree of crop cover on evaporation.

Penman's (1956) definition of potential evaporation also states that potential evaporation can only be achieved if a crop is actively growing. An illustration of the importance of this point was provided by measurement of the evaporation from a papyrus swamp (Rijks 1969a). The papyrus was fully grown and completely covering the ground and definitely not short of water and there was, furthermore, a complete cover of undergrowth. However, the papyrus stand was six years old and contained a large number of dried out heads. Its upper surface, where the bulk of the incoming radiation was intercepted, was definitely not 'actively growing'. The measured ratio of ET to E_0 was 0.60.

In the course of the years, the techniques used for the measurements required in this work were elaborated and refined. In the later studies, measurements of the radiation balance and the micro-meteorological techniques were used in addition to the nylon-unit measurements and the original observations obtained by gravimetric sampling.

The effects of water shortage on crop performance

A large volume of data on crop development and growth, associated with different natural or imposed soil water regimes, was accumulated over the years by Farbrother but, in spite of its significance, little of it has been published. Farbrother (1955) confirmed the close correlation between soil water and relative turgidity, established by Weatherley, and concluded (1956) that water stress caused an almost immediate stop in dry-matter production, followed by a check in the production of fruiting points, the severity of which he found later (1958) to depend on the amount of crop already set, and finally followed by the shedding of 'matchhead' buds. Subsequent replenishment of soil water did not result in additional crop. Farbrother (1961*b*) developed the 'composite plant diagram' to describe plant growth as a function of environmental conditions and successfully applied the use of such diagrams to the analysis of the causes of crop success and failure. Rijks (1965) observed a similar sequence of effects of water stress on growth of cotton in South Yemen but with intervals between the different stages reduced to about two weeks, presumably because climate and conditions of soil water shortage were more severe. He also observed that maturing bolls were shed when water stress became extreme. Dale (1960) measured rapid stomatal closure, from which there was no recovery, soon after the occurrence of water stress, as measured by relative turgidity of the leaves, and he suggested that this was the cause of the decrease in production of dry matter. Nyahoza (1966*a*) concluded from his greenhouse work that atmospheric humidity was relatively unimportant and that only soil water conditions influenced plant response. Hearn (1967, 1969) reviewed the state of knowledge, and a detailed account of the effects of water shortage on crop performance can be found in Chapter 4.

The soil water balance

The role of the soil water balance in determining the pattern of water use has been outlined above. At Namulonge, a calculation of available water in the soil was originally started to determine the frequency of water shortage that a permanent pasture would have experienced during the eighteen years from 1951 to 1968. Information of this kind was necessary to estimate the need for fodder storage that would exist in different parts of the country upon the introduction of highly productive, exotic cattle.

The calculations used daily totals of rainfall and Penman estimates of open-water evaporation as inputs. For calculation purposes, total stored soil water available to the grass was inferred never to lie outside the limits of 0 and 200 mm water. It was assumed that any water in excess of 200 mm would be lost as deep drainage. On days with rain, crop water use was estimated to be $0.85\ E_o$. On days without rain, crop water use was estimated to be $0.85\ E_o$, as long as stored soil water was equal to or greater than 100 mm and to be $0.45\ E_o$ when stored soil water was less than 100 mm. A diagram of the results is shown in Figure 2.6. The results of this study served as a background for the discussion on rates of potential grass production, and for the calculation of the frequency of occurrence of poor and plentiful grass supply. It also

Figure 2.6. Calculated soil water status under grass at Namulonge for the years 1951 to 1968. Water status is shown as mm in the top 2.0 m. The method of calculation and the assumptions made are described above. (Data prepared by E. Jones using a computer programme written with assistance from D. Mann, Department of Mathematics, Makerere University.)

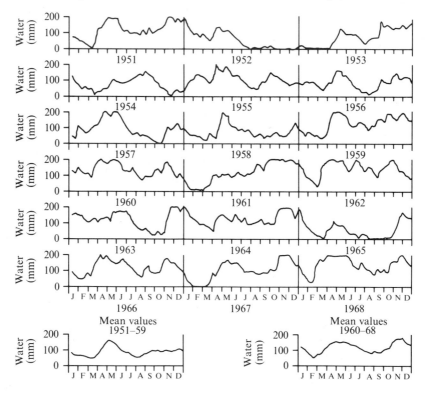

indicated the variation in rainfall in terms of water available for plant growth, rather than as millimetres of rain.

This soil water balance approach was later adapted by Thorp (1973) to calculate the critical soil water deficit for cotton. To convert the calculated value of Penman's *ET* to estimated water use by cotton, he used a factor that increased from 0.4 to 1.2 during the first eighty days of crop growth, remained at a value of 1.2 for the next eighty days, and finally decreased to 0.4.

Knowledge of the timing of shortages and excesses of water stored in the soil has also been used in the study on nutrient availability and to understand soil physical and chemical processes, which are described in detail in Chapter 3.

Applications of agrometeorological knowledge

The main application of agrometeorological knowledge is in planning farming practices and schemes for agricultural development. One dominant aspect in the development of raingrown agriculture is the selection of sowing periods for annual crops.

Optimum sowing date is most usefully defined as the date on which a crop must be sown in order to be likely to experience the most favourable combination of factors determining yield. Manning (1949) recognized that, under natural conditions, the water balance and insect attacks were the two foremost factors influencing yield in cotton. It was this conclusion that led to the effort invested in the analysis of rainfall regimes and crop water use, and to the extensive work on crop spraying, developed by the Uganda Department of Agriculture. When widespread insect control became possible, and quantitative knowledge of rainfall probabilities, and rates of potential evaporation, became available for the whole country, the choice of sowing dates (found by trial and error in some areas) could be confirmed or changed as necessary. Moreover, the sowing date in new areas and for new crops (sometimes using a pattern of crop water need already determined in other countries) could be reasoned instead of guessed. Under some rainfall regimes that allow a fair latitude in the choice of sowing date, when only the water-balance criteria are employed, further help for the selection of an optimum date can be provided by knowledge of the temperature and sunshine regimes both after sowing and at harvest. The rainfall regime recorded at Balindi (Rijks 1967a) provides an example of this type. Furthermore, spacing is associated with the choice of a particular sowing date (and vice versa) because it affects the rate of increase in LAI and hence the water balance.

Agrometeorological knowledge also affects the planning of day-to-day farming operations. The application of fertilizer as a function of the water balance of the crop (see Chapter 3) and the choice of cultural operations (Hutchinson *et al.* 1958*a*) and their timing (Rijks 1968*c*) are examples of their application in a system of raingrown agriculture (Chapter 9). In irrigated agriculture calculation of a water balance may be kept up from day to day, so that the timing and amount of irrigation can be decided for individual fields and crops according to their needs. This procedure has been tested at the Mubuku irrigation scheme and has resulted in yields as heavy as those in any other regime of irrigation, at a smaller expenditure of water and a lower cost of labour for irrigation (Rijks & Harrop 1969). The crop factors with which potential evaporation was multiplied to obtain the water requirement of the crop were derived from the ratio of ET/E_o under potential evaporation conditions. For application of such results elsewhere, knowledge of the temperature regime is required because this determines the rate of development of the crop and the length of its growing-cycle and thus the pattern of crop-water use, and the crop factors.

The method of applying irrigation water, according to the requirements as calculated from meteorological data and the pattern of crop-water use, has subsequently been applied to crops grown at the Atera and Labori irrigation schemes in Uganda.

Crop forecasts

Manning (1949) laid the foundation for a method of predicting the Uganda cotton crop every year before buying starts. The method is based on the correlations that have been found to exist between yield on the one hand and acreage, mean sowing date and rainfall during the season, on the other. J. M. Munro (unpublished) grouped the rainfall data into three periods: sowing and early growth, full crop cover and flowering, ripening and harvest, and he calculated regression equations separately for each district in Uganda, over twenty-five years of data. Estimates of the crop to within a few per cent were obtained with this method as long as there were no significant changes in the agricultural system, but upon the introduction of new, improved varieties, and the increased use of insecticides, the estimate started to fall short of reality (Rijks 1968*d*; Arnold 1969*a*).

Measurements of the radiation balance

The impetus to make measurements of the components of the radiation balance came from the need to get an unequivocal assessment and understanding of the amounts of water used by the crop, net radiation being the only source of energy for evaporation in non-advective conditions. The components of the radiation balance are measured as incoming short-wave radiation, reflected short-wave radiation and total net radiation. From these measured components, net long-wave loss can be obtained by calculation. The measurement of solar radiation has been dealt with above. Rijks (1965, 1967b, 1968b) measured the other components and the factors affecting their magnitude.

The seasonal pattern of the reflection coefficient (i.e. the ratio between total solar radiation reflected and received by a surface over a tall annual crop, such as cotton) was very much associated with the development of the crop (i.e. by the degree of soil cover) and by the degree of wetness of the soil surface in the period before full soil cover was attained. Row orientation played a role during the period when plants met within the rows but not between the rows. Solar elevation was not found to be a very important variable, understandably so as its zenith variation is not very great in the tropics, and crop water status was important in so far as it permitted the measuring instruments to 'see' more bare soil when plants were wilted than when they were fully turgid. Over a short-grass crop the influence of crop water status was much more marked (Rijks 1969b) both because the grass leaves 'rolled' thus showing the dead, brownish, vegetative matter underneath, and because they changed colour to a greyish green.

The diurnal pattern showed a more pronounced increase in reflection coefficient over a tall crop in the early morning and late afternoon than over grass, an apparent effect of different optical crop roughness at low solar elevations. However, as solar radiation intensities are low during these periods of the day, this difference has little influence on the radiation balance and on crop water use. Afternoon values were sometimes lower than morning values because of the occurrence of afternoon wilting, with an effect of leaf orientation as described above.

Net long-wave losses (which are mainly determined by the properties of the overlying airmass) were, like the reflection coefficients, most significantly affected by LAI and by the soil water status. Other crop and soil factors played relatively minor roles. In addition to the factors enumerated above, soil heat flux was affected by the orientation of the rows, particularly during the period when the crop had not met between the rows.

Totals of net radiation, which numerically express the energy available for potential evaporation, are thus determined by all of the above factors. It is evident that both crop development affecting soil cover (and the water status of the crop and soil) and purely meteorological factors, play important roles in determining the rates of potential and actual evaporation through their influence on the radiation balance. Row orientation could, in some cases, explain the presence of a midday depression in actual water loss from a row crop.

Crop temperature studies

In the early years, little attention was paid by staff at Namulonge to the temperature regime *per se*, as far as the cotton crop was concerned. Attention has been focussed, from the beginning, on a study of the water balance which was shown, first by the statistical relation between rainfall and crop yield, and later by physical and crop physiological analyses, to have a preponderant effect on crop performance. Temperature was paid secondary attention because of its importance in determining the rate of development of the crop and hence water use. Its importance has been recognized by Hearn (1969) in his analysis of growth and development of cotton in different environments, and by Arnold & Brown (1968) in their studies of genotype–environment interactions of bacterial blight in cotton.

Experiments to study the response of cotton to different temperature regimes were initiated by Rijks (1962, 1964) and Thorp (1973) but facilities for this work proved to be inadequate so that no general conclusions were drawn.

Conclusions

Agrometeorological studies at Namulonge were directed to the analysis of certain aspects of an agricultural system that makes optimum use of existing meteorological conditions for crop production and that minimizes, as far as possible, the influence of adverse meteorological conditions.

A major problem, affecting crop production, diagnosed and defined in the very first years at Namulonge, was the need to obtain quantitative knowledge of the water balance that a raingrown crop could be expected to experience throughout its life cycle. This water balance depended principally on the supply of water by rainfall and on the loss of water by evaporation from crop and soil. Distribution and reliability of rainfall were quantitatively defined by the use of confidence limits. These were calculated for the whole country and the method has found widespread application elsewhere in the tropics. The loss of water by evaporation from annual crops was defined in terms of potential evaporation from open water, and a crop factor, that expresses the effect of the development of the crop on total water loss. Penman's formula was shown to be a sound basis for the calculation of potential evaporation, in the tropics as well as in England, and the necessary precautions for its extrapolation in time and space were identified. Potential evaporation was calculated for Uganda, and eventually for the whole of East Africa, and its calculation in many other parts of Africa is under way. The effect of crop development and degree of soil cover on actual water use was clearly demonstrated and expressed quantitatively, and this has been a basis for many more elaborate studies and refinements, for many crops, all over the world. The method has been extended for use in perennial crops.

The possibility of quantitative assessment of the water balance has found practical application in the selection of sowing dates and in the choice of crops and varieties to be introduced for the development of new agricultural areas all over Africa. The calculation of the water requirements of crops has furthermore been an essential step in the planning of the development of irrigated agriculture and in the conduct of day-to-day irrigation practices.

Quantitative knowledge of the water balance was originally recognized to have an application mainly in the explanation of its effect on plant water status. Later, there followed the use of the concept in soil physical and soil chemical studies. Its impact on the sequence of various operations and possibilities of production in a multi-year agricultural system is becoming more and more felt. In other countries, its effect on the development of plant diseases and on entomological processes has been shown to be a dominant factor. Through its multiple uses and its applicability as a quantitative concept, the water balance has become an essential and reliable tool in planning agricultural production in the tropics.

3
Soil productivity

E. JONES

Introduction

From the late 1940s, when land clearing was started at Namulonge, until about the mid-1950s, levels of crop yields appeared to be satisfactory. During the following decade, however, they dropped markedly. Inadequate solar radiation, poor soil structure and a touch of 'muck and magic' were some of the explanations offered by interested observers to account for this loss of productivity. Although fertility was declining, field experiments failed to show that fertilizers were effective either in raising or in maintaining crop yields. Nevertheless, by 1972, yields were double and treble their best levels of the early 1950s and fertilizer use was an essential component of the farming system. This intensification of agriculture, which has become essential in the tropics as pressure on the land increases, forms the farming background to the account that follows.

Soil productivity depends upon the availability of water and nutrients: the key to its understanding lies in the balance between their supply and demand. The soils at Namulonge are capable of supplying the demands of the traditional farming system (witness the centuries of farming in Buganda) but after ten to fifteen years they failed to meet the demands of a more intensive farming system (described in Chapter 9). The consequent decline in yields served a useful purpose, however, in that it emphasized the need for studies on soil productivity to encompass the whole farming system and its place in the ecosystem, rather than to be limited to the more usual aims of identifying the nutrients that limit the yields of individual crops, and of finding ways of supplying them.

Allan (1965) has described the traditional agriculture of the area of Uganda, referred to as the 'fertile crescent' by Jones (1972), as an intensive form of shifting cultivation which is made possible by the natural fertility of the soil, and the well-distributed rainfall. Bananas and, more recently, coffee are the major crops of the system. The arable crops are represented by a range of tropical food crops and cotton, grown in an unorganized rotation. Cropping can continue on the same area of land for up to ten years, or until production has fallen very low, when the land is abandoned to naturally regenerated vegetation for five or more years. This haphazard rotation has been investigated by staff of the Department of Agriculture and standardized to the officially recommended 3:3 rotation: three years in arable crops followed by three years in rest, preferably grass for grazing (although the original recommendation did not include grazing). Most of the studies of soil fertility and agronomy in Uganda have been carried out in the context of this rotation. Within the area of the fertile crescent, its effectiveness in maintaining crop yields has not been established experimentally, but experience on Government farms has led to the view that without fertilizer inputs, this rotation maintains fertility at a medium to low level. At Namulonge, over a number of cycles of the rotation, crop yields tended to fall.

The 3:3 rotation exhibits a rhythm of fertility similar to that associated with the traditional shifting cultivation system in which crop yields tend to fall during the arable period and recover after the period in grass. Direct experimental evidence for the rhythm is difficult to obtain, because of the very large season-to-season variation in productivity, but several sources of information are summarized in Table 3.1.

The soil fertility problem

Prior to 1945 most of the work on soil fertility in the fertile crescent was aimed at improved management of the crops with special emphasis on developing a rotation. Economic factors precluded any serious consideration of using inorganic fertilizers. After 1940, the need to raise the geneal level of productivity was accepted, and the use of inorganic fertilizers on an experimental basis was introduced. This resulted in a large number of experiments at Namulonge and throughout the area (Manning & ap Griffith 1949; Le Mare 1953; Mills 1953*a*; Stephens 1969*a*). The conclusions drawn from these experiments were that for the area of the fertile crescent no consistent increase in arable-crop yields could be expected from the application of fertilizers except from nitrogen applied to cereal crops in unusually wet seasons (Manning & ap Griffith 1949) and fertilizers applied to old arable land (Stephens 1969*a*). Despite these conclusions the general yield levels of all crops were low in relation to the potential yield that might be expected from the climatic regime. It was generally considered that the only certain means of maintaining fertility was to rest the land in some form of long-term vegetation at regular intervals or to use farmyard manure (Stephens 1970).

Such was the position in the late 1950s when Le Mare initiated a research programme into the rhythm of fertility that occurs within the

Table 3.1. *Summary of published information on the effect of the number of years in arable crops on the yields in kg per ha of subsequent crops in Uganda*

Seed cotton

Source △	(2)	(1)	(2)	(4)	(6)	
	Farm fields 1957–58	Rotation expt 1956–57	Rotation expt 1957–58	Three years' cropping, 1965, 1965, 1967	No. of years from rest	Mean for 5 centres
New land	976	—	—	—	1	475
Second-year land	781	—	—	—	3	446
First year after ley	695	1580	1292	1193	5	186
Second year after ley	—	1159	1172	1265		
Third year after ley	399	1128	1120	879		

Other crops

Source △	(2)	(5)	(3)	(4)	(6)			
	Farm fields 1957–58. Maize	Expt 1966/67. Hybrid maize	Rotation expt 1958. Beans	Expt 1965/66/67. Beans	No. of years from rest	Mean for 4 centres. Maize	Mean for 6 centres. Beans	Mean for 3 centres. Groundnuts
New land	2410	—	—	—	1 or 2	2060	620	1504
Second-year land	2320	—	—	—	3 or 4	2155	433	1048
First year after ley	1934	2960	1008	1237	5 or 6	1480	364	850
Second year after ley	—	2146	727	1032				
Third year after ley	1130		683	903				

'New land' and 'second-year land' refers to land cleared from bush.

(1–5) *Progress Reports from Experiment Stations, Uganda*: (1) 1956–57, p. 14; (2) 1957–58, pp. 15, 16, 20; (3) 1958–59, p. 17; (4) 1965–66, p. 37, 1966–67, p. 41, 1967–68, p. 52; (5) 1966–67, p. 43, 1967–68, p. 53.
(6) Stephens (1969a).

3:3 rotation. The aim of the study was to elucidate the causes of the rhythm. The hypothesis selected as the basis of the programme was the re-cycling of nutrients within the profile during the course of the 3:3 rotation. This is shown diagrammatically in Figure 3.1. The results obtained at Namulonge confirmed the existence of the cycle for all the major nutrients (Jones 1968, 1972). The quantities of nutrients involved in the cycle provided a basis for formulating a mixture of fertilizers supplying five major nutrients and indicated the amounts that should be applied to each crop (see Tables 3.9 and 3.10). The fertilizer mixture was tested over a period of six seasons (three years) and its use consistently increased the yields of maize, beans and cotton (Jones 1972). Furthermore, the use of inorganic fertilizers in the arable phase was shown to benefit the following grass phase (Jones 1971). Thus on Namulonge farm, which is reasonably representative of the surrounding areas, inorganic fertilizers tested in formal field trials in the 1950s gave inconsistent results, whereas a mixture of the same fertilizers used in the late 1960s consistently increased yields of the same crops over a period of three or more years. This inference is further illustrated in the report on farm yields in Chapter 9.

Figure. 3.1. Nutrient re-cycling within a soil profile through a ley–arable rotation or under shifting cultivation.

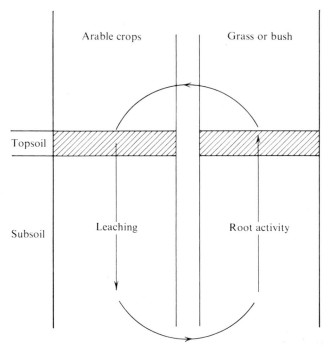

Table 3.2. *Soil physical data*

Variation in mechanical composition found in Namulonge soils

Depth (cm)	Gravel content (%)[a]			Mechanical analysis of fine earth[b]			Apparent[c] density (g per cm³)	Available water[c] (mm per 30 cm)
	1	2	3	Coarse sand (%)	Silt (%)	Clay (%)		
0–30	23	2	9	29±8	13±2	33±8	1.37	50.8
30–60	46	1	30	26±6	12±4	44±15	1.47	40.2
60–90	16	1	43	24±6	12±3	49±5	1.24	44.6
90–120	9	1	18	21±6	14±4	47±8	1.20	50.5
120–150	5	2	11	20±6	16±5	44±12	—	—
150–180	4	31	9	23±8	17±4	38±10	—	—
180–210	4	62	6	23±10	18±3	36±9	—	—
210–240	8	55	8	23±11	20±4	36±8	—	—
240–270	6	31	6	25±11	18±5	31±8	—	—
270–300	6	15	6	24±12	20±7	34±9	—	—

[a] Gravel content of three separate profiles.
[b] Sand, silt and clay mean values and standard errors for five profiles.
[c] Apparent density and available water are after de Jong (1973) and Hearn (1970) respectively.

The remainder of this chapter will attempt to explain the anomaly of consistent crop responses to fertilizers on Namulonge farm in the late 1960s, compared with little response in the 1950s. In order to do this it will be necessary to describe in some detail the chemical and physical conditions of the soil, the climate for the period 1950 to 1970 and the possible effects of climate on soil productivity.

Soils

The soils at Namulonge represent a transition zone between red and yellow ferrallitic soils derived from a basement complex (Radwanski 1960; Harrop 1962). Two soil catenas *Buganda* and *Mirambi* together with their intermediates occur on the farm. Up to the end of 1971, the arable-crop agriculture was restricted to the upper and middle members of the catenas, because the lower members occur either on steep slopes or in permanent swamps. As would be expected from such a mixture of soils, some variation in physical and chemical properties occurs over very short distances. Some indication of the range of mechanical compositions encountered in the soils is given in Table 3.2.

The gravel content of the soil surface provides a visual indicator of soil variability. It can be seen from Table 3.2 that gravel zones can occur at any depth down to 300 cm. Shallow gravel characteristically occurs in the *Mirambi* catena, and deep gravel in the *Buganda* catena. Table 3.2 also shows the sand, silt and clay content of the fine-earth fraction, and it can be seen that an accumulation of clay occurs in the inter-

Table 3.3. Soil chemical data

Mean values and standard errors for nine profiles

Depth (cm)	Carbon (%)	Nitrogen (%)	Phosphorus (total ppm)	Sulphur (total ppm)	pH (CaCl$_2$)	ppm exchangeable		
						Potassium	Magnesium	Calcium
0–30	1.17±0.48	0.095±0.014	396±35	144±46	6.22±0.3	220±54	384±96	1356±328
30–60	0.56±0.19	0.061±0.019	367±45	99±47	5.93±0.5	132±68	229±90	590±248
60–90	0.36±0.10	0.044±0.007	345±41	101±41	5.46±0.4	93±53	198±62	421±113
90–120	0.28±0.06	0.039±0.009	333±44	83±43	5.51±0.4	83±54	172±51	417±123
120–150	0.21±0.06	0.031±0.011	323±49	88±52	5.60±0.4	87±92	155±56	441±147
150–180	0.16±0.08	0.028±0.012	305±47	65±51	5.59±0.3	110±116	121±32	425±217
180–210	0.12±0.05	0.023±0.013	306±57	67±49	5.74±0.5	76±81	123±59	418±191
210–240	0.11±0.06	0.020±0.011	301±70	58±36	5.73±0.5	65±51	105±46	350±209
240–270	0.11±0.04	0.010±0.007	240±74	79±52	5.75±0.5	71±51	103±40	365±283
270–300	0.10±0.03	0.007±0.005	240±61	84±61	5.58±0.6	70±44	99±48	327±258

Table 3.4. *Topsoil compared with subsoil*

Yields of a range of crops in kg per ha showing the
effect of removing 20 cm of topsoil

Crop	Normal soil	Surface soil removed to a depth of 20 cm
Beans	1435	88
Maize	4890	105
Cotton	1840	49
Sorghum	1840	147

The treatments were not replicated; plot size 0.024 ha.

mediate subsoil. This higher clay content is attributed to the eluviation
of clay material from the topsoil (J. Pidgeon, personal communication).
The nature of the clay in these soils has not been thoroughly investigated, but the mode of origin of the soils, their cation exchange
capacity and a few X-ray analyses indicate that the bulk of crystalline
clay material consists of the 1:1 layer minerals (Kintukwonka 1972).
The apparent densities of the soils, given in Table 3.2, as well as their
capacities for holding available water, indicate the presence of a
compacted layer below 30 cm. This was a consistent feature of the soils
by the late 1960s, possibly caused by tractor cultivation.

Table 3.3 shows the distribution of carbon, nitrogen, sulphur and
phosphorus down the profile. All four elements show the highest
content in the top 30 cm of soil. Carbon and nitrogen content decrease
rapidly with increasing depth, whereas the decline of the sulphur and
phosphorus content is more gradual. The sulphur content of the soil
at depths greater than 120 cm is very variable and in some profiles the
content approaches zero, whereas in others the content is similar to
that found in the surface soils. In comparison, the phosphorus content
throughout the profile is less variable. Comparable data for pH,
potassium, magnesium and calcium are also given in Table 3.3. The
pattern of change with increasing depth for all three bases is similar
to that of nitrogen, but variation among profiles is large.

The importance of the enriched topsoil for agricultural production
is illustrated in Table 3.4, which shows the yields of four crops grown
on normal soil, and soil with the top 20 cm of topsoil removed. In each
case the crops growing in the subsoil virtually failed.

In simplified terms, the Namulonge soils may be considered as deep
mineral soils, where the mineral matter is largely an inert residue of
earlier weathering, overlain by a thin topsoil 15–45 cm deep, enriched
by the biological accumulation of nutrients. Thus the fertility of the soils
in terms of nutrient supply is restricted to the surface layer, which is
exposed to the natural vagaries of the climate. An essential step in

understanding the problems of productivity, therefore, is to consider the possible and known effects of climate on the physical and chemical properties of the topsoil. The climate *per se* has been discussed in Chapter 2; this chapter will attempt to present some of the many effects of climate on soil productivity.

Factors influencing soil productivity

The only true measure of soil productivity is the yield of crops grown on the soil. But the yield of any crop is an integrated product of soil factors including water, combined with other environmental factors such as pests and diseases. Biological factors of the environment will not be discussed in this chapter although their possible effects must always be borne in mind in any consideration of a practical farming problem. For the present purpose, discussion of the factors affecting soil productivity will be confined to the existing physical and chemical conditions of the soil, and the rainfall as a source of water for crop growth. The most important of these is the pattern of water supply for crop growth.

The bimodal rainfall distribution at Namulonge allows the production of two crops in each year. Many combinations of two crops per year are possible, but the combination that makes the fullest use of the twelve months is that of maize in the first rains and cotton in the second. Although this sequence of crops has practical limitations (Chapter 9), it was followed for many years at Namulonge. It represents a most intensive use of the environment by arable crops and will be used as the basis for discussing factors influencing soil productivity.

Water supply

The simplest approach to describing the general pattern of water supply for crops, is to compare the total rainfall received in the crop season, with an estimate of the water lost by soil and crop during the season. Such a comparison is given in Table 3.5 for a period of twenty years. From the data presented, it is clear that in the first cropping season, when maize is normally grown, adequate amounts of water were received in every year except 1953. In contrast, in the second cropping season, when cotton is grown, there were eight years out of the twenty when the total rainfall was inadequate to meet the calculated requirements of the crop. Six out of the eight water-deficient seasons occurred during the 1950s. This, as will be seen later, has important implications in understanding work on soil productivity at Namulonge.

The data in Table 3.5 do not take into account different distributions of rainfall from one season to the next. In Table 3.6 an attempt has been made to represent the general pattern of water status in the 0–30

Table 3.5. *Rainfall (mm) from 1951 to 1970 in relation to crop water requirements*

| Year | Maize season | | Cotton season | | Total rainfall (mm) |
	Rain (mm)	Deviation from crop requirement[a]	Rain (mm)	Deviation from crop requirement	
1951	665	+269	845	+229	1510
1952	485	+89	313	−303	798
1953	354	−42	686	+70	1040
1954	411	+15	596	−20	1007
1955	599	+203	552	−64	1151
1956	502	+106	616	0	1118
1957	845	+449	508	−108	1353
1958	504	+108	582	−34	1086
1959	617	+221	681	+65	1298
1960	770	+374	585	−31	1355
1961	450	+54	1138	+522	1588
1962	752	+356	830	+214	1582
1963	601	+205	817	+201	1418
1964	490	+94	854	+238	1344
1965	400	+4	692	+76	1092
1966	739	+343	629	+13	1368
1967	534	+138	882	+266	1416
1968	784	+388	769	+153	1553
1969	667	+271	540	−76	1207
1970	413	+17	405	−211	828
Mean	579	+183	676	+60	1255

[a] Crop water requirements were calculated using the average E_T values for the years 1965–70 and standardized crop factors (see Table 3.6).
Water requirement for the maize season = 396 mm; for the cotton season = 616 mm.

cm depth of soil at the end of each standard (meteorological) ten-day period under a maize–cotton sequence for twenty years. The assumptions and methods used in deriving the data, together with the reasons for partitioning the seasons are given in the footnote to the table. The ten-day periods when the topsoil was theoretically at wilting point are labelled '0', and the periods when leaching was calculated to have occurred are designated 'L'. In the unlabelled periods, the topsoil was calculated to contain some available water and it is assumed that crops could grow. The calculations have been restricted to the 0–30 cm depth of soil because it is in this region that the bulk of roots of both crops occur. With cotton, there is evidence that when all the available moisture in this depth is used up, the growth of the plant slows down (Farbrother 1956; Hearn 1972*b*; Thorp 1973). Furthermore this depth represents the most concentrated source of crop nutrients (see Table 3.3). For the purpose of the discussion the two cropping seasons will be dealt with separately, although in practice the water supply of one

Table 3.6. *Topsoil water status for twenty years*

The occurrence of dry topsoil (0) and leaching conditions (L) under a maize–cotton cropping sequence during a twenty-year period. Months are divided into standard ten-day intervals

	Jan.			Feb.			Mar.			Apr.			May			Jun.			Jul.			Aug.			Sep.			Oct.			Nov.			Dec.		
	1	2	3	1	2	3	1	2	3	1	2	3	1	2	3	1	2	3	1	2	3	1	2	3	1	2	3	1	2	3	1	2	3	1	2	3
1951	\	\	\	\	\	\	\	\	L	\	\	L	L	L	\	\	\	\	\	\	\	\	\	\	\	\	\	L	\	\	\	\	\	\	\	\
1952	\	\	O	O	O	O	\	O	O	\	\	L	L	\	\	\	\	\	\	\	\	O	O	O	O	O	O	O	O	O	O	O	O	O	O	O
1953	\	\	O	O	O	O	O	O	O	O	O	\	L	\	\	\	\	\	\	\	\	\	O	O	\	O	O	\	O	O	L	O	L	L	O	\
1954	L	L	\	\	L	L	\	\	\	L	\	\	\	\	\	\	\	\	L	\	\	O	\	\	\	O	O	O	O	\	\	O	\	\	\	L
1955	\	L	\	\	\	\	\	\	\	\	L	L	L	L	L	\	\	\	\	\	\	\	L	\	O	O	\	O	\	O	O	O	\	O	O	\
1956	\	L	\	\	\	\	\	L	\	L	O	\	L	L	L	\	\	\	\	\	\	\	O	\	O	O	\	O	L	O	\	O	\	\	L	\
1957	\	\	\	\	\	\	L	L	\	L	L	L	L	L	\	\	\	\	\	\	\	\	O	O	O	\	O	\	\	L	O	\	O	\	\	\
1958	\	L	\	\	\	\	\	L	L	L	L	L	L	L	L	L	\	\	L	\	\	\	O	O	\	O	O	\	L	O	\	\	L	\	L	\
1959	\	\	\	\	\	\	L	L	L	L	L	L	L	L	L	L	\	\	L	\	\	O	\	\	O	O	L	\	O	L	L	O	L	O	\	O
1960	L	\	\	\	\	\	\	\	\	L	L	L	L	\	L	\	L	L	\	\	\	\	\	L	\	\	L	L	L	\	L	O	L	\	\	\
1961	O	\	\	\	\	\	\	\	\	\	\	\	\	\	\	\	\	L	\	\	\	\	\	\	O	\	L	L	L	L	L	O	L	\	\	\
1962	\	\	\	\	\	\	L	L	L	L	L	L	L	L	L	L	\	\	L	L	\	\	O	\	O	O	L	\	O	L	\	O	O	\	\	\
1963	L	L	\	\	\	\	L	\	L	\	L	L	L	L	\	L	\	\	L	\	\	\	\	\	L	\	L	L	\	L	L	\	L	L	O	\
1964	\	\	\	O	\	L	\	O	\	\	L	\	L	\	\	\	\	\	\	\	\	\	\	\	O	\	\	\	L	\	\	\	L	\	\	\
1965	\	\	\	O	\	L	\	O	L	\	L	L	L	\	\	L	\	\	\	\	\	O	\	O	O	\	L	L	L	L	L	\	\	\	\	\
1966	O	\	O	O	\	O	\	\	O	L	L	L	L	\	\	L	\	\	\	L	\	O	\	O	\	O	\	O	O	L	L	\	L	L	\	\
1967	\	O	O	O	\	L	O	\	L	L	L	L	L	L	\	\	L	\	L	\	\	\	O	L	\	O	O	O	O	\	\	O	\	\	L	\
1968	O	\	\	O	\	\	\	O	L	L	\	\	L	\	L	\	\	\	\	\	\	\	\	\	\	O	\	\	\	O	\	\	\	\	\	\
1969	O	\	O	L	\	O	L	\	\	O	O	O	\	\	\	\	\	\	\	\	\	\	\	\	O	\	\	O	\	\	O	\	\	L	\	L
1970	\	\	\	\	\	O	\	\	O	O	\	O	\	\	L	\	\	\	\	\	\	\	\	\	O	O	O	\	O	O	\	O	\	\	\	\

Notes on Table 3.6

In calculating the water status of the topsoil, the actual rainfall received in each ten-day period was used, and it was assumed that:
   (i) At field capacity the topsoil contained 50 mm of available water.
   (ii) The average values for Penman's E_T for each ten-day period in the years 1965 to 1970 were applicable to the whole period.
   (iii) Crops were sown on time.
   (iv) The crop factors shown below applied to each year.

The topsoil was considered to be at wilting point when the balance was 0. Leaching conditions were considered to have occurred when the increment of extra water (after satisfying any soil deficit as well as the crop requirement) exceeded 50 mm in any one ten-day period, or two consecutive ones.

The period of peak water and nutrient requirement corresponds with thirty days after tasselling in maize and fifty days after peak flowering in cotton.

Crop factors used in the calculations:

Jan.	1	0.4	Cotton harvest	Jul.	1	0.4	
	2	0.4			2	0.5	
	3	0.4			3	0.5	
Feb.	1	0.4	Maize sown	Aug.	1	0.7	
	2	0.4			2	0.8	
	3	0.5			3	0.9	
Mar.	1	0.6		Sep.	1	1.0	
	2	0.8			2	1.1	
	3	1.0			3	1.2	Peak water
Apr.	1	1.2	Peak water and	Oct.	1	1.2	and nutrient
	2	1.2	nutrient requirement		2	1.2	requirement
	3	1.0	(maize)		3	1.2	(cotton)
May	1	0.8		Nov.	1	1.1	
	2	0.6			2	0.9	
	3	0.6	Maize harvest		3	0.8	
Jun.	1	0.4		Dec.	1	0.6	
	2	0.4	Cotton sown		2	0.6	
	3	0.4			3	0.4	

Figure 3.2. Estimating soil moisture. Resistance units, appearing in the photograph as white rectangular blocks, being lowered into a hole in the ground, previously prepared.

season can sometimes affect the water supply to the crop in the following season.

In the maize–cotton sequence, the growing period for maize extends from about 20 January to 10 June and cotton is grown during the second rains, although in practice it is sown on the tail end of the first rains. In general, during the first cropping season a moist or wet topsoil was maintained throughout the growth period of the maize with a few obvious exceptions: 1953, 1967 and 1970. In two of these relatively dry years, the establishment of the maize crop would have failed, or been seriously delayed, with the result that in these seasons the progression of crop water requirement would have differed from the calculated values shown. In 1970, the maize crop had a dry topsoil for twenty days during the period of peak water and nutrient requirements. Conditions of dry topsoil also occurred for shorter periods in 1954 and 1958. Leaching occurrences during the growth period of a crop are of interest because of the potential loss of nutrients that can occur. It can be seen that leaching conditions did not occur at all in 1954, 1961 and 1970 and occurred to a varying extent at different times in the season in other years. Losses of nutrients by leaching are likely to be most important in the period of maximum nutrient requirements, a situation that occurred in fifteen years out of twenty. In eight years out of twenty leaching conditions occurred near the harvest period, implying wet weather at the time of harvesting. It may be concluded that for the growth of maize, the first cropping season can occasionally be too dry, sometimes ideal and often too wet.

In contrast with the maize season, the cotton growing season presents a very different pattern of calculated water available in the topsoil. At sowing, the topsoil was never short of water and in six years out of twenty it was subject to some leaching. In practice, establishment of the crop is often difficult, because of dry conditions in the upper 5 cm of soil. Nevertheless calculated values indicate that the growth of cotton was never restricted by the supply of water during the period after establishment, a result confirmed by field experience. For the first sixty days after sowing there was water in the topsoil except in the 1952 season. Leaching conditions were also rare during this part of the season.

For the interval starting twenty days before and extending into the period of maximum crop water requirement, however, the topsoil was theoretically at wilting point for ten days or more in eighteen years out of the twenty. This does not mean that there was no rainfall; rather that any rainfall received during each ten-day period was used by the crop before the end of the period. This implies a series of wetting and drying cycles, but in practice the conditions may have been different because a proportion of the crop water requirement would come from below the 30 cm depth. Furthermore, when the topsoil dries to wilting point there will be some delay before the crop can use any subsequent rainfall at the full rate and this would tend to conserve some of the water. Nevertheless, in most years, there seems to be little doubt that the cotton crop cannot depend on obtaining water, and therefore its nutrient requirement, from the 0–30 cm region of the soil profile during this very important phase of growth. It follows that the water content and nutrient status of the soil below 30 cm are very important. In most years, little or no leaching occurs in the early part of the cotton season. Hence the subsoil water required by cotton during September and October is often derived from excess rainfall in the April–May period, i.e. during the latter part of the maize season. Towards the end of the cotton season, extending into the harvesting period, conditions vary from dry topsoil (as in 1953) to severe leaching (as in the early 1960s). These latter conditions are generally unsuitable for harvesting the crop. Conditions during the cotton season may be summarized, therefore, as giving rise to a generally adequate water supply with little leaching risk, in the early part of the season, followed by generally dry topsoil conditions at the period of maximum growth and extremely variable conditions for the harvest period.

In addition to variations within each cropping season, the data in Table 3.6 show an apparent change in the water supply pattern over the years. For the first cropping season, maize during the early 1950s received adequate rain; the late 1950s were wet, the early sixties were average and the late sixties were again wet. In contrast, the pattern for the second cropping season (cotton) shows that all the 1950s were dry in comparison with most of the 1960s. These apparent changes in the weather pattern can have far reaching effects on the supply of nutrients from the soil.

Nutrient supply

Some of the data on major nutrients given in Table 3.3 are presented as kg per ha in Table 3.7. Similar figures for micro-nutrients are not available. A comparison of the nutrient content of the soil with the requirement of crops and grasses shows that the soil contains an abundant supply of all major nutrients. Nevertheless, only small proportions of these nutrients are available to crops and soil productivity is determined more by factors controlling the availability of nutrients than by the total soil content of the nutrients themselves. In this respect, the most important factor is the water status of the soil. Excessive water can result in temporary water-logging, followed by losses of nutrients by leaching; too little water can prevent root activity and uptake of nutrients. Therefore the nutrient aspect of soil productivity is closely related to the water supply pattern discussed in the previous section. In Table 3.6, where the topsoil is shown as being

Table 3.7. *Soil nutrient content*

The quantities of major nutrients, expressed in kg per ha, in the 0–30 cm and 30–90 cm
depths of soil, compared with the requirements of arable and grass crops

	Nitrogen	Phosphorus	Sulphur	Potassium	Calcium	Magnesium
Quantities						
0–30 cm	3800	1580	580	880	5420	1540
30–90 cm	4200	2850	800	900	4040	1710
Total	8000	4430	1380	1780	9460	3250
Requirements						
Arable crop[a]	150	30	20	150	30	15
Grass[b]	200	30	25	200	70	50

[a] Based on the average requirements of good maize and cotton crops.
[b] Requirements of grass for a six-month period.

at wilting point, it is assumed that the crop is not taking up nutrients.
Where leaching conditions are shown, it is assumed that the topsoil
loses a proportion of the nitrate and sulphate and their chemical
equivalent of bases.

Nitrogen and sulphur. Nitrogen and sulphur are stored in the soil in
organic combination and they become available to crops by mineralization of the organic matter, a process that only occurs when the water
content of the soil is above wilting point. Direct measurements of the
rate of mineralization of organic matter are not available, but it has been
shown at Namulonge that over a period of three years (or six cropping
seasons) crop uptake and leaching deplete the top 30 cm of soil of
550 kg of nitrogen and 65 kg of sulphur per ha. These amounts are
equivalent to 91 kg of nitrogen and 11 kg of sulphur per ha per season
and imply an average rate of mineralization per season of the organic-
bound nutrients of 2.4 per cent for nitrogen and 1.9 per cent for
sulphur. These measurements of nitrogen and sulphur losses were made
between 1966 and 1969 which, in terms of rainfall amounts were
reasonably typical. The quantities of nitrogen and sulphur released are
adequate for producing average crops grown without fertilizers, but
are inadequate for the maximum yields possible from the climatic environment. The cropping history of a piece of land, together with
the prevailing weather conditions, can modify this natural supply of
nitrogen and sulphur. The total nitrogen and sulphur content of soil
following a period in grass or bush is ten to twenty per cent higher than
that after three years of arable crops. Therefore, if a fixed proportion
of the nutrients is released by mineralization each season, the newly
cleared land will have a greater supply than old arable land and,
consequently, a higher productivity. A temporary reduction in the

nutrient supply, mainly nitrogen, can occur when bulky crop trash or grass residue is incorporated into the soil. In Uganda, it is recommended that at least six weeks should elapse between burying trash and sowing a crop, but the effectiveness of this period is dependent on the soil being sufficiently moist to allow soil organisms to break down the trash. Experience at Namulonge indicates that six weeks is inadequate especially when elephant grass has been cleared.

Mineralization of nitrogen and sulphur from organic matter can only occur when the soil water content is above wilting point. It can be seen in Table 3.6 that during most seasons, the soil experiences a series of wet and dry periods of varying length. This will result in varying periods of time for mineralization, but the rate of mineralization after a dry period increases in proportion to the length of the dry period (Birch 1960) and this will tend to ensure a similar supply of nutrients in each season. When the nitrogen and sulphur have been mineralized into nitrate and sulphate, their fate will depend in part on the rainfall conditions. If the season is virtually free from leaching, as occurred during the first rains of 1954 and 1961, the nutrients mineralized early in the season, probably in excess of the requirements of the young maize crop, will remain in soil solution until the period of peak nutrient requirement. Losses of nitrogen by denitrification are unlikely to be serious under Namulonge conditions, but little is known about this aspect. In contrast, for example, in 1962 and 1966 leaching occurred early in the season, leaving the crop with only the nutrients released by subsequent mineralization to satisfy its peak demand. Experiments to test the response of maize to nitrogen and sulphur under these two extreme conditions would almost certainly produce divergent results.

Rainfall conditions in any one cropping season can also affect the following season. For example, the latter part of the cotton seasons of 1961, 63 and 64 were excessively wet; leaching occurred and it is conceivable that the following maize crops might have been short of nitrogen and sulphur in their early stages of growth. This situation is in complete contrast with what occurred in most of the seasons between 1952 and 1960 when there was little leaching late in the cotton season.

Many more examples can be drawn from the data in Table 3.6 but they all point to the results that have been found in practice: variable responses of crops to applied nitrogen. Sulphur responses have not been demonstrated in the field, but pot trials indicate that they are possible in old arable soils when the nitrogen supply is adequate (D. H. Parish, personal communication). Sulphur deficiency will always be secondary to nitrogen deficiency under natural conditions in the Namulonge soils, because sulphate is not as easily leached as nitrate from soils rich in sesquioxides.

Table 3.8. *Reserves of calcium, magnesium and potassium*

Ratio of total to exchangeable bases at different
soil depths; average of five profiles

Depth (cm)	Calcium	Magnesium	Potassium
0–30	1.05	1.60	2.83
30–60	1.42	1.21	3.76
60–90	1.39	1.49	4.98
90–120	1.38	1.11	5.02
120–150	1.18	1.26	5.20
150–180	1.03	1.56	5.02
180–210	1.11	2.04	4.62
210–240	1.06	1.43	5.12
240–270	1.24	1.24	3.82
270–300	1.23	1.20	3.42

Total = extracted after boiling for 2 hr in 5 N HCl.
Exchangeable = extracted after shaking for 2 hr with 0.1 N HCl.

Calcium, magnesium and potassium. The quantities of calcium, magnesium and potassium present on the exchange complex of the soil are considerably in excess of the requirements of crops (Table 3.7). There are no indications of what levels of calcium and magnesium in the soil would result in crop deficiencies, but there is a suggestion that the calcium:magnesium ratio may be too narrow. Adding magnesium sulphate to soil, for example, can reduce the yields of crops (Stephens 1961) and this may be caused by an induced deficiency of calcium. If this interpretation is correct, magnesium deficiency is unlikely to occur in these soils until the ratio of calcium to magnesium is much wider. Experience at Namulonge and country-wide in Uganda indicates that potassium deficiency can occur at soil contents of 180 to 200 ppm and that these levels of potassium develop after a number of years of arable cropping (Stephens 1969a; Foster 1972). It is of interest to consider what reserves of bases occur in these deeply weathered soils. Table 3.8 gives the ratio of total to exchangeable base-content and the data show that the reserves of calcium and magnesium are very small and even if they are easily released they could not replace any major losses. In contrast, the reserve of potassium is larger but only three to four times the amount on the exchange complex. It has been shown that the potassium reserves are not easily released (Kintukwonka 1972). The nature of the reserves has not been investigated, but it is known that they are present mainly in the clay fraction and that the amounts present are correlated with the total inorganic phosphorus in the soil. It is conceivable that the reserves are in the form of compounds such as taranakite which is known to be formed in soils rich in sesquioxides (Taylor & Gurney 1965). To summarize, the supply of bases to crops from these soils is generally adequate, but losses by crop removal cannot be replaced from within the soil, except possibly in the case of potassium.

Phosphorus. In work on soil productivity at Namulonge, phosphorus has attracted more attention than any other nutrient. It is considered to be a special problem on ferrallitic soils because of their high phosphate-fixing capacity. The results of many field experiments indicate that the soils are not consistently deficient in phosphorus (Le Mare 1953; Manning & ap Griffith 1949; Stephens 1966). Le Mare (1957) proposed that phosphate deficiency is related to the moisture status of the soil and is most likely to occur in seasons that are drier than average. Consequently, some of the factors already discussed in relation to nitrogen supply may also be relevant here. Over a period of from three to six months the average arable crop demands from 20 to 30 kg phosphorus per ha from the soil. For the supply of such a quantity, the Namulonge topsoil contains per ha about 1600 kg phosphorus and, in the immediate subsoil, a further 2800 kg; a total of 4400 kg phosphorus, 0.6 to 0.7 per cent of which is required by the average crop. The total phosphorus itself comprises from 40 to 50 per cent of organic phosphorus, the remainder being a complex mixture of compounds of iron, aluminium and other elements. This organic phosphorus represents from 1800 to 2200 kg phosphorus and if this is mineralized at the rate of 2.4 per cent per cropping season (the rate at which nitrogen is mineralized), 43 to 53 kg phosphorus will be released into the soil solution in any six-month period. In solution much of the phosphorus will be precipitated and become non-available, unless the crop roots are present to absorb it immediately. The distribution of roots in the soil is a characteristic of the individual crop but it can be greatly modified by the pattern of moisture availability. Thus the roots of young cotton will remain in the topsoil if there is adequate moisture present but, under dry conditions, they will exploit the deeper soil (see Chapter 4). It can be envisaged that, under conditions of adequate moisture and good root distribution, little or no response to phosphorus can be expected whereas, under dry conditions, a band of phosphatic fertilizer placed deep in the topsoil, in contact with the roots, could maintain the supply of the nutrient and result in a large crop response. Thus, part of the phosphorus story is very similar to that discussed under nitrogen, where the supply of nutrient is dependent on the soil water status, a factor that has been shown to exhibit extreme variability in Namulonge soil. Another aspect of the phosphorus story at Namulonge is the depression in crop yields following small applications (200–400 kg per ha) of phosphatic fertilizers (Le Mare 1968). This depression was observed in field experiments on arable land. Kabaara (1965) in pot experiments showed that it could be prevented by liming the soil to a pH in excess of 5.5. The mechanism responsible for the depression has not been fully elucidated, but is probably related both to the acid nature of the phosphatic fertilizer in solution and to manganese availability (P. H. Le Mare, personal communication). The practice developed at Namulonge was to lime all arable land in order to maintain a pH value of 5.8 to 6.2, as measured in 1:5 soil suspension in 0.01 M $CaCl_2$. This pH is high enough to prevent the development of the depression effect and probably represents the optimum value for phosphorus availability in these soils.

To summarize: in the presence of sufficient water the phosphorus supply in Namulonge soils is probably adequate for crops. If the root development of the crop is restricted either by adverse weather conditions or by soil structure, phosphorus supply can be inadequate. Small applications of phosphorus to strongly acid soils can depress crop yields.

Liming. The application of ground limestone to Namulonge soils has been shown to benefit cotton, beans and groundnuts. Under natural conditions pH values of the soils range from 4.6 to 6.2. The higher values are found under grass or bush and they generally decline under arable cropping. Where the topsoil is free from gravel, the pH under arable cropping will fall to between 5.0 and 5.5, but on areas with gravelly topsoil the values can fall to 4.6 or lower. Application of limestone was first tried on old arable land by Le Mare (1961), who obtained large and consistent increases in cotton yields. He also showed how pH controls the phosphorus supply in the soil, and demonstrated a close relationship between cotton yields and pH of the soil under natural conditions. Subsequent experiments have confirmed the beneficial effects of liming which last for at least five years, but do not prevent the natural fall in fertility (Jones 1972). Apart from the possible effects on phosphate supply, liming can also benefit soil productivity by widening the narrow calcium: magnesium ratio already noted, or by improving the supply of one or more micro-nutrients. Very little is known about the mechanisms responsible for the beneficial effects of liming, but there is no doubt that yields of cotton, beans and groundnuts can be raised by liming the soil to pH values between 5.8 and 6.2.

Micro-nutrients. There is very little information available on the content of micro-nutrients in Namulonge soils. Over a number of years individual nutrients such as molybdenum and boron and several mixtures of micro-nutrients have been tried in field experiments but the effects have always been very small and inconsistent (Le Mare 1963; Ogborn 1964c; Jones 1967; Stephens 1969b). Consideration of the mode of origin of the soils suggests that with increasing use of major nutrients, deficiencies in micro-nutrients must develop. There is an urgent need for basic information on the amounts and nature of the micro-nutrients in the red ferrallitic soils.

Soil physical conditions

An indication of some of the soil physical conditions encountered at Namulonge was presented in Table 3.2. Two of these physical attributes, the gravel content and the dense subsoil, may directly affect soil productivity. The gravel content of the soils is very variable and experience shows that where gravel occurs near the surface, the soils have naturally low productivity. Several factors may be responsible, such as lower water-holding capacity; liability to surface capping leading to increased losses of water through runoff; a smaller soil volume available for root exploration; and a tendency for the fine-earth fraction to be more acid with a lower content of organic matter and nutrients. However, experience has shown that gravel areas can produce heavy crop yields if they are limed, adequately supplied with nutrients as inorganic fertilizers and suitable cultural methods adopted so that runoff losses of rain water are reduced to a minimum. Areas with a gravelly topsoil form a significant proportion of the available arable land and it may be necessary to give them special management if they are to produce maximum yields.

The zone of high clay content below the topsoil can also be expected to affect productivity. As shown in Table 3.2 it has, in comparison with the topsoil, a higher apparent density and lower water-holding capacity. These two properties, combined with its naturally lower nutrient content (Table 3.3), make it a less desirable environment for the growth of roots which tend to be restricted to the topsoil. This heavy clay subsoil, when ploughed by tractor, is the ideal medium for the creation of a compact zone that restricts root development (plough pan). At Namulonge, all primary cultivations for twenty years were based on the use of wheeled tractors and disc or mould-board ploughs and there is now evidence that hard-pans were created in these soils.

Another factor that may contribute to the formation of a hard-pan, but on which there is no direct evidence, is the possibility that cultivating the topsoil speeds up the process of clay eluviation which, if deposited in the subsoil, would further reduce the natural pore space left after the passage of tractor wheels and plough soles. A discussion of the effect of hard-pans on the growth of cotton roots is presented in Chapter 4. Another possible effect of the hard-pan is the creation of temporary water logging in the topsoil after heavy rain. This has been observed in the field but no information is available on its effect, if any, on nutrient supply and root activity. Replacing the wheel-tractor and plough with other methods of primary cultivation could be expected to help in reducing hard-pan formation, if clay eluviation from the cultivated zone is not an important factor. These aspects of soil productivity require further investigation.

It has been shown (Table 3.5) that, in some seasons, the total rainfall does not exceed the requirement of arable crops for water. The requirements of permanent grass, however, are seldom met. Therefore it is essential that all the rainfall received enters the soil and as much as possible is stored for future use. This presents a major problem with a high rainfall intensity (up to 250 mm per hr) in a gently rolling topography like that at Namulonge, because the surface of the soils under cultivation is prone to capping, and soils under grass are com-pacted by grazing animals. The creation of graded terraces is the ultimate answer to this problem but much can be achieved by suitable field layout, cultivation and crop management (Chapter 9).

Few detailed studies have been carried out on the effects of soil physical conditions on the productivity of Namulonge soils. In the past it was difficult to separate the effects of nutrients from physical effects, but now that nutrient supply is better understood it is becoming possible to carry out more penetrating work in this field.

The crop rotation

In the previous section the chemical and physical factors of soil productivity at Namulonge and their possible interactions with climate were discussed. It is now convenient to consider the effects of these factors within a crop rotation. The traditional agriculture of the Namu-longe area has no well-defined rotation, because neither the sequence of crops nor the duration of cropping, nor the length of the period under rest vegetation are fixed. However, the recommended six-year rotation of three years' arable cropping followed by three years of rest has proved to be a useful standard for studying rotation effects, and this was the basis of a study at Namulonge during the years from 1963 to 1968.

Rotation effects without fertilizer inputs

The rotation studied was three years in Rhodes grass (*Chloris gayana*) or elephant grass (*Pennisetum purpureum*) followed by three years of arable cropping with beans grown in the first rains and cotton in the second. The field, which had just completed nine years in the 3:3 rotation, had not received any fertilizers. The aim was to measure the quantities of the major nutrients involved in the nutrient cycle (Figure 3.1). The technique used was to sample carefully both the crops and the soil (to a depth of 45 cm), before and after the three years' rest and three years' arable cropping. The results are summarized in Table 3.9.

Table 3.9. *Soil nutrient changes through a rotation*

Change in the nutrient content in kg per ha, organic carbon and pH of the 0–45 cm depth of soil following a resting phase and arable phase on the same area of land

	Resting phase, 1963–65			Arable phase, 1966–68		
	Elephant grass	Rhodes grass	S.E.	Elephant grass	Rhodes grass	S.E.
Total nitrogen	+768	+817	±210	−920	−899	±310
Total phosphorus	+85	+85	±53	−54	−88	±56
Total sulphur	+45	+41	±55	−86	−94	±35
Exchangeable potassium	+476	+243	±182	−460	−241	±233
Exchangeable calcium	+971	+1366	±507	−1896	−1850	±381
Exchangeable magnesium	+465	+471	±247	−404	−802	±118
Organic carbon	+17800	+15500	±4020	−19700	−17500	±1620
pH	+0.31	+0.19	±0.04	−0.31	−0.28	±0.08

The data show that large amounts of nutrients move in the nutrient cycle. It is noteworthy that the losses of nutrients during the arable period are of the same order as the gains after three years in grass, an observation that is in agreement with practical experience in that the 3:3 rotation does maintain soil fertility for limited periods. Another feature of the results is the similarity of the relative amounts of nutrients lost to those of nutrients gained. The theory proposed to account for the nutrient cycle within the profile states that nutrients are lost from the topsoil during the arable phase by leaching, and returned to the topsoil during the rest by root uptake. These two mechanisms are unlikely to result in the relative amounts of nutrients lost being similar to those of nutrients gained, which is likely to occur only if the same mechanism is responsible for the gains and losses, but acting in two directions. Of the two mechanisms proposed, only the leaching process is capable of reversal, and this would involve the mass movement of soil water upwards in response to a hydraulic gradient, during the rest phase. Theoretical calculations of the possible movement of water from subsoil to topsoil during three years under grass indicate that this mechanism can account for the bulk of the gains of calcium and magnesium recorded. Assumptions made in the calculations were

(i) that the soil solution at 50 per cent water-holding capacity contained from 60 to 80×10^{-6} g magnesium per cm^3 and from 120 to 140×10^{-6} g calcium per cm^3,

(ii) that the average hydraulic conductivity of the soil was 2×10^{-9} cm per sec.

The potential hydraulic gradient between soil depths of 30 and 100 cm was estimated to average 99 cm based on measurements of the electrical resistance of nylon units under grass for the three years. Other facts that suggest the mass movement of soil water as being important are (i) the gains of calcium and magnesium, which are larger than could conceivably be brought up by either of the two grasses, and (ii) that the two grasses with their very different growth habits result in very similar gains of nutrients. In view of the foregoing it is reasonable to think of a rhythm of fertility through the rotation in terms of the movement upwards or downwards of soil water carrying with it a proportion of the available crop nutrients.

Other factors studied during the rotation were organic carbon content and pH of the soil. The data for these are also given in Table 3.9. The change in the organic carbon content of the soil reflects the different equilibrium values for carbon under arable cropping and rest vegetation, but in addition it is an indication of the loss and gain of nutrients. Foster (1969) has shown that the organic carbon content of a wide range of Uganda soils is a good measure of their content of other nutrients. The increase in pH values supports the theory of mass movement of soil water, because the rise in pH under grass must reflect the addition of bases into the system other than through the growing plant (Nye & Greenland 1960).

To summarize: during the years from 1963 to 1965 land under grass gained considerable quantities of nutrients in the topsoil. During the years from 1966 to 1968 the same area of land under arable crops lost similar quantities of nutrients. The major mechanism indicated for the gains and losses is the mass movement of soil water up the profile in response to a hydraulic gradient, or down the profile in response to gravity. Because this mechanism is dependent on the weather pattern, it is important to consider how typical the weather was during the years of the study. For this purpose reference must be made to Tables 3.5 and 3.6.

Were the years of the rest study unusually dry? According to the data in Table 3.5 the years from 1963 to 1965 were drier than average in the first rains and wetter than average in the second. In terms of total rainfall they were average years, as judged by the twenty years of data available. Several three-year periods in the 1950s and late 1960s were drier. Thus it is unlikely that the results for 1963–65 were favoured by unusually dry weather.

Were the years of arable study unusually wet? It would seem from the data in Table 3.5 that the years from 1966 to 1968 were wetter than average. Table 3.6 gives an indication of the frequency of leaching that may have occurred: twenty-two occasions in the three-year period or an average of seven per year. Over the twenty-year period, eleven years experienced six or more leaching occurrences, the other nine years experienced fewer. It may be stated that the period from 1966 to 1968 was typical of half the years recorded at Namulonge.

It would appear that neither the period of study of the rest phase nor that of the arable phase of the rotation could be considered to be atypical of conditions at Namulonge since 1950. However, it is instructive to consider what results would have been expected during the arable phase if the study had been carried out in the years from 1952 to 1956. These five years were generally drier than later years and averaged less than three leaching occurrences in each year. Accordingly leaching losses would have been small and consequently an evaluation of the arable phase of the nutrient cycle during these years would probably not have given results comparable with those recorded from 1966 to 1968.

In the particular case of Namulonge, much of the land was cleared from five to ten years' rest in bush during the period from 1952 to 1956 and, as a result, was naturally high in fertility. Moreover, most of the exploratory fertilizer studies, designed to test the effects of applying different major nutrients, were carried out in this period by Le Mare (1953). His results were generally negative and indicated no consistent deficiencies of any nutrients. Thus, retrospective consideration of the weather pattern in the context of the nutrient cycle goes some way to explaining the anomalies of results from fertilizer trials obtained at Namulonge over the past twenty years.

As was stated at the beginning of this section the farmers in the Namulonge area have not developed any form of standard arable crop rotation, particularly with respect to the length of time in arable crops before resting the land. It is apparent from the arguments used above that the run-down in fertility during an arable period can vary according to the wetnesss or otherwise of the prevailing conditions. It may be presumed that centuries of experience has taught the Baganda farmers that the useful length of the arable period can vary considerably and cannnot be predicted on a time basis as it can in many parts of Africa. This gives rise to the impression of a completely haphazard farming system, which in fact reflects the farmers' response to the interplay of climate and soil fertility in his area.

Some effect of fertilizer inputs

From 1966 onwards, the yield levels of most arable crops grown at Namulonge increased dramatically (Chapter 9). This was achieved by a combination of better crop varieties, improved agronomic practices and the routine use of inorganic fertilizers. The rationale behind the fertilizer programme was a compromise between imitating the gain of nutrients by the topsoil following three years of grass and correcting

Table 3.10. *The fertilizer mixture*

The composition of a mixture of calcium ammonium nitrate (20.5 per cent nitrogen), single superphosphate and muriate of potash mixed in the proportions 2:2:1 by weight

	Nitrogen	Phosphorus	Sulphur	Potassium	Calcium
Composition (%)	8.2	3.5	5.0	10.0	12.5
Nutrient (kg) in 752 kg of mixture	62.0	26.0	37.6	75.2	94.0

the loss following three years of arable cropping. The quantities of nutrients supplied to each crop in the years 1966 to 1969 were those indicated by the actual gain and loss of nutrients by the 0–15 cm depth of soil after three years of grass and arable cropping during the 1962–64 period (Jones 1972). These gains and losses were similar to those shown in Table 3.9. The fertilizer mixture formulated to meet this requirement is given in Table 3.10. The mixture comprised the three fertilizers most readily available locally. A high content of phosphorus was included to try to counteract any deficiencies in this nutrient arising from fixation (Le Mare 1968). The nitrogen content of the mixture was low relative to the content of other nutrients, because the mixture was applied in the seedbed and extra nitrogen could be supplied to the crop as a top-dressing. A particular feature of the mixture was the high calcium content.

The effects of using the fertilizer mixture on the yields of maize, cotton and beans during the years 1965 to 1968 have been reported in detail elsewhere (Jones 1972), but are summarized in Table 3.11a. The yields of all three crops were raised by the use of the fertilizer mixture. In subsequent years even heavier yields were obtained (Chapter 9) as the management of crops under the higher levels of fertility became better understood. The results presented in Table 3.11b are of particular interest because they show that the routine use of the fertilizer mixture on a bean crop completely counteracted the fall in productivity that occurred during the same period without fertilizers. The data for the control treatment form an unusually good example of the fall in fertility under arable crops, for far too often this inferred steady decline in yields is upset by season-to-season variation in climate and attacks of pests and diseases.

After six applications of the fertilizer mixture, an attempt was made to measure the effects of its use on the nutrient status of the soil. Table 3.12 shows the nutrient content of soils after three years of arable cropping with and without fertilizer application. The results show no change in the organic carbon, total nitrogen and magnesium content of the soil; a suggestion that the phosphorus, sulphur, and pH values have

Table 3.11. *Fertilizer use and crop yields*

(*a*) The average yields in kg per ha of maize, cotton and beans with and without fertilizers from a number of trials carried out between 1966 and 1968.

Fertilizer mixture (kg per ha)	Maize[a]	Cotton	Beans
0	2904	1115	852
752	5060	1484	1272
Number of trials	3	5	4

(*b*) Yields of beans, in kg per ha, with and without fertilizers on land cleared from three years grass in 1965.

	1965	1966	1967	1968
No fertilizer	1304	1033	904	776
627 kg per ha fertilizer mixture	1368	1571	1404	1393
S.E.	±40	±80	±54	±119

[a] The maize crop received a top dressing of 102 kg nitrogen per ha at tasselling, as calcium ammonium nitrate.

Table 3.12. *Soil composition and fertilizer use*

Comparison of composition of the 0–20 cm depth of soil after three years of arable crops with and without fertilizers

	Organic carbon (%)	ppm total			ppm exchangeable			
		Nitrogen	Phos-phorus	Sulphur	Potas-sium	Calcium	Mag-nesium	pH
No fertilizers	1.74	1090	590	187	201	1497	386	5.89
With fertilizers	1.72	1060	604	199	302	1731	382	5.98
S.E.	±0.20	±130	±29	±6.0	±11	±281	±32	±0.10

The fertilized plots received six applications of a fertilizer mixture supplying a total of: 309 nitrogen, 132 phosphorus, 189 sulphur, 376 potassium and 471 calcium all in kg per ha.

risen; and definite increases in potassium and calcium. It has been shown elsewhere (Table 3.9) that three years of arable cropping without fertilizer will lower the nitrogen content of the soil and it would seem that the routine application of fertilizer nitrogen cannot prevent the loss of this nutrient. For phosphorus, sulphur, potassium and calcium some correction of the losses is achieved, the largest effect being found with potassium and calcium. From the practical point of view, it was concluded that the fertilizer mixture supplied excessive amounts of potassium, and after 1969 the amount of potassium applied was reduced to one half unless analysis showed the soil content to be

Figure 3.3. Taking a soil sample. Analysis of soil samples from all the farm fields was a routine feature of the work in soil chemistry.

less than 200 ppm (0.5 mequiv. per cent). This change was achieved by applying potassium only to the cereal crops in the rotation. It would seem that the supply of phosphorus, sulphur and calcium was generally adequate and their application rates were not changed. Accumulation of phosphorus is not surprising, since the amounts applied were deliberately heavy to counteract the possible fixation of this nutrient in the soil. It is assumed that sulphur accumulates as sulphate, held on the anion exchange complex and absorption surfaces of these soils rich in sesquioxides; its accumulation as organic sulphur is unlikely because of the unchanged values for organic carbon. The accumulation of calcium in the soil is desirable because, as was noted earlier, the supply of this nutrient relative to other nutrient bases is thought to be low. A consequence of the accumulation of calcium and potassium is the slightly higher pH value of the soils receiving fertilizers. This has many implications, such as a higher rate of mineralization of the soil organic matter, and greater availability of the large mineral phosphorus supply in the soil. The higher pH value is the result of using a fertilizer mixture with a high base content. The use of calcium ammonium nitrate as the source of nitrogen is particularly desirable because other nitrogen sources, such as sulphate of ammonia and ammonium sulphate nitrate, lower the soil pH values (Stephens 1969b) and will eventually enforce the application of limestone. Limestone is not readily available in Uganda, and if a fertilizer programme such as the one used at Namulonge can maintain pH values in the long-term, it will surely be worthwhile.

The failure of the fertilizer programme to prevent the lowering of the nitrogen content of the soil is of particular interest. The accumulation of nitrogen is closely associated with the accumulation of organic matter, which is the main mechanism available for holding it against losses by leaching. Nye & Greenland (1960) have discussed the concept of the content of organic matter being in equilibrium with the growing vegetation. Organic matter content will rise only if the initial value is below the equilibrium value for a particular soil and the vegetation growing in it. At Namulonge, the soil organic matter is greatest under permanent vegetation, where there are continual inputs of organic material, and where good protection is afforded to the soil against high temperature. Under arable crops, the equilibrium is much lower and, during any arable phase of a rotation, levels of organic matter fall. The final equilibrium value for organic matter will vary with different crop sequences and fertility inputs; the time taken to achieve this value will depend on the climate. Under Namulonge conditions the equilibrium value under arable cropping is not known, and it is unlikely to be achieved in three years. Therefore, for the results shown in Table 3.12 which represent the first three years of arable cropping after grass, it

Table 3.13. *Nitrate distribution in soil profiles*

The nitrate nitrogen content of soils at different depths
under different cultural conditions in ppm

Depth (cm)	Continuous arable[a] cropping		3:3 rotation without fertilizers[b]	
	Without fertilizers	With fertilizers	After arable	After grass
15	4.4	35.6	—	—
30	3.9	31.7	10.6	8.1
60	0.9	31.7	7.7	4.1
90	3.4	28.8	8.6	2.9
120	4.4	24.4	9.9	3.2
150	7.8	24.9	12.1	3.6
180	—	—	12.4	4.7
210	—	—	12.0	5.2
240	—	—	14.4	6.2
270	—	—	15.0	6.1

[a] Twenty-one years continuous arable cropping, fertilizers applied in the last seven years only.
[b] Mean of ten profiles.

is not surprising that fertilizers have not affected the organic carbon and nitrogen values.

What then is the fate of the extra nitrogen applied? Extra crop can only account for a proportion of it because all crop trash was returned to the soil. The results in Table 3.13 show the distribution of nitrate nitrogen in soil profiles under different farming conditions. Under continuous arable cropping, with no fertilizers, the nitrate content varies from 0.9 to 7.8 ppm compared with 24.4 to 35.6 ppm where fertilizers were applied. Considerable accumulation of nitrate has occurred in the soil receiving fertilizers. The high level of nitrate in the 15 and 30 cm depth is unusual and was caused by the exceptionally dry weather encountered during the twenty-four months prior to sampling. A more usual distribution is shown in the column headed '3:3 rotation', with an accumulation of nitrate below 120 cm. This accumulation disappears after three years in grass, a feature also noted by Mills (1953b) and Simpson (1961). These results indicate that much of the nitrate supplied in fertilizers, or already present in the soil in excess of crop requirements, can be stored at different depths in the profile depending on the prevailing climatic conditions. Where the excess nitrate remains within the rooting zone of arable crops it will be available to following crops. This helps to explain the very variable crop responses to applied nitrogen experienced at Namulonge. When the nitrate is leached beyond the rooting depth of arable crops it is lost to

Table 3.14. *Residual effect of fertilizers*

Yield in kg per ha of seed cotton (the last crop in the arable phase) and dry-matter
production of grass in the following grass phase, showing the effect of fertilizers
applied in the arable phase

Year ...	1966		1967		1968		Cotton 1968	Grass
Season ...	1st	2nd	1st	2nd	1st	2nd	2nd	1969–71
Control	0	0	0	0	0	0	1383	16400
R_1	1	1	1	1	0	0	1284	19800
R_2	2	2	2	2	0	0	1433	20500
F_1	1	1	1	1	1	1	1556	21100
F_2	2	2	2	2	2	2	1606	23300
S.E.							±109	±301

0 = no fertilizers.
1 = 376 kg per ha of a nitrogen/phosphorus/potassium compound 10:10:18 except
    1968 when the fertilizer mixture shown in Table 3.10 was used.
2 = 752 kg per ha as for 1.

them but may move up into the root zone under drier than average
conditions or when the land is planted to grass. Thus the full benefit
of using nitrogen and other fertilizers on these soils may not become
evident until the land has completed at least one cycle of rotation
involving arable crops and grass. Some results of an experiment at
Namulonge illustrate this point (Table 3.14). It can be seen that for
treatments R_1 and R_2 that twelve months (two crops) after the fertilizers
were last applied there was little residual effect in the cotton crop,
whereas the following grass crop showed a large residual effect of the
same fertilizers. These results confirm the statement that the full benefit
of fertilizer use can only be measured when the land has completed a
ley–arable rotation, and are in agreement with the theory of nutrient
re-cycling within a soil profile.

   The results presented in Table 3.13 and 3.14 show that agriculturally
significant amounts of nutrients can be stored in the soil profile, despite
the movement of water through a soil profile that must be occurring
to feed the extensive areas of swamps that are present in all the valleys.
Some idea of the amount of water moving through the profile can be
gained from the data in Table 3.5. In each year there is an average of
243 mm water in excess of the crop requirement, and the values range
from zero in 1952 and 1954 to 576 mm in 1961. These figures will vary
with the type of arable crop grown and will be smaller under grass or
any permanent vegetation. Information on the pathways of water
movement to the swamps would be of value in planning a programme
of fertilizer use.

   The storage of nitrates in the soil profile suggests that, under these

Table 3.15. *Recovery of applied nutrients*

Percentage recovery of nutrients supplied as fertilizer by six
arable crops and three years of grass, a 3:3 rotation

	Nitrogen		Phosphorus		Sulphur		Potassium		Calcium	
Level of application ...	1	2	1	2	1	2	1	2	1	2
Nutrients applied (kg per ha)	151	213	66	92	173	210	226	301	51	145
Nutrients recovered (kg per ha)	115	161	68	75	18	26	84	143	18	18
Per cent recovery	76	76	103	82	10	12	37	48	35	12

Nutrients recovered were calculated from the difference between control and fertilized plots in composition and yield of grass and the essential parts only of arable crops.

Arable crops were grown in 1966, 1967 and 1968 and grass cut at two-month intervals in 1969, 1970 and 1971.

conditions, recovery of applied nutrients through a ley–arable rotation could be substantial. The experiment reported in Table 3.14 was monitored for nutrient removal in crops and grass and afforded an opportunity for calculating the recovery of applied nutrients. The results presented in Table 3.15 show that, over the six-year period of the rotation, 76 per cent of the nitrogen applied in the first two years was recovered. In comparison with the temperate areas this extent of recovery is very high. The recovery of phosphorus was greater than that of nitrogen. The data for sulphur, potassium and calcium show much lower recoveries, which is partly related to the excessive quantities of these nutrients applied, relative to demand by the crops. The very high recovery of phosphorus militates against the argument used in formulating the fertilizer policy, that the soils have a high phosphate fixing capacity. Instead, it favours the conclusions of an earlier section (p. 61) that, generally, phosphate supply is adequate. The study of nutrient recovery reported here was not very thorough; no attempt was made to account for any extra nutrients remaining in the soil or in the material left after threshing maize cobs and bean pods. It is possible, therefore, that recoveries could be even greater. This suggests that nutrient supply in the Namulonge soils is a closed system with little losses outside those removed in crop harvests.

Method and timing of fertilizer application

During the period from 1952 to 1965, the methods used to apply fertilizers varied from broadcasting and mixing with the soil, to deep placement in the ridge. Trials were conducted to test different methods of application, but no consistent differences were found (Le Mare 1957). Time of application of nitrogen fertilizers was also studied with no consistent results (Ogborn 1965). After 1965, with the introduction of the fertilizer mixture (Table 3.10) it was argued that deep placement of the fertilizer was the best method because of the risk of dry conditions in the surface layers of the soil. Two attempts were made to test the benefit of deep placement over shallower placement, but the results were conflicting (Jones, E. 1969; Jones, M. A. H. 1969). However, when fertilizers were first applied to run-down arable land it was noted that plants growing near the placed fertilizers always had a better start than those 15 cm or more away from the fertilizer. After several years of routine fertilizer use, this effect disappeared from the fields, indicating that placement was no longer necessary. A similar situation was found specifically with nitrogen application. When the land was run-down, application of nitrogen in the seedbed improved the growth and appearance of the young crop, but after several years of fertilizer use the effect was less noticeable. Timing of the application of nitrogen fertilizer became more important on the fertile soils. Under these conditions it was assumed that the soil could supply 90 kg nitrogen per ha per season and that this was adequate to grow all crops through the vegetative stage of growth, but inadequate to meet the full demands of the period of reproductive growth. Nitrogen was applied to meet this demand. In some seasons, with dry conditions in the middle and towards the end, this approach would fail, but on the basis of results already discussed, the top-dressed nitrogen could be considered as stored for the next crop. Namulonge conditions present an excellent opportunity for basing the nitrogen fertilizer policy on a book-keeping principle of gains and losses.

Summary of Namulonge experience

This chapter started by noting the anomalous observations of no consistent fertilizer responses in the 1950s contrasted with large, consistent responses in the late 1960s. It has been shown that at least two factors were in action. First, in the 1950s most of the land was newly cleared from bush and was therefore very fertile, and second, the usual mechanism for lowering fertility under arable crops, leaching, rarely occurred. By the middle 1960s, the land had been farmed for fifteen or more years and, during the early 1960s, it had been subjected to considerable leaching. As a result, the general nutrient levels of the soils had been lowered by losses from removal in crops and livestock and possibly, to some extent, by leaching. Hence, when extra nutrients were applied, the crops responded. The extent of the response was, however, determined by other factors, such as improved crop varieties (Chapters 7 and 8) and management (Chapters 4, 5, 6 and 9), which raised the demand for soil nutrients.

The failure to obtain fertilizer responses in the 1950s led to the investigation of the principles governing soil fertility in the area. These principles can be summarized as follows:

Namulonge soils contain limited amounts of useful nutrients and under permanent vegetation these are concentrated in the surface layers, in association with organic matter.

Under arable cropping, nutrients move into the deeper soil beyond the root range of arable crops. These nutrients can be stored in the deeper soil, and returned to the surface layers during a period in grass.

Nutrients supplied in fertilizers behave in the same way as naturally occurring nutrients, and the full benefit of fertilizer-use, in terms of agricultural production, cannot be measured until the land completes a ley–arable type of rotation.

A programme of fertilizer use for a ley–arable rotation can be based on replacing the amounts of nutrients removed in agricultural produce, because irrevocable losses of nutrients out of the soil by leaching are likely to be small.

Maintaining the topsoil pH at values ranging from 5.8 to 6.2, as measured in a 1:5 suspension in 0.01 M $CaCl_2$, is beneficial to most crops, but the mechanism of the benefit is not known.

Implications for other areas in the tropics

The particular conditions of soil and climate found at Namulonge determine the soil productivity situation that has been described. Similar, red ferrallitic soils occupy 26 per cent of the land area of Africa south of the Sahara, and other related soils, described as having only a small content of weathering minerals, occupy a further 19 per cent (D'Hoore 1964). These soils contain limited amounts of nutrients concentrated in the topsoil and closely associated with the organic matter. Their behaviour when used for agriculture will vary with the prevailing climate, but even within one climatic type, their behaviour will vary according to the nature of the crops. Perennial crops as a general rule use more water than annual crops because they cover the ground for longer periods. They also have deeper roots than annual crops and, because nutrients are not so easily moved out of their root range, losses by leaching are smaller. It is axiomatic that only crops that suit the rainfall pattern should be grown in any area. If the interpretation of Namulonge experience is correct, the use of inorganic fertilizers should also be planned to fit in with the rainfall or, more precisely, the soil water status. Exceptions to this statement are soils that are intrinsically deficient in one or more nutrients, but when these deficiencies are corrected the same principles will apply.

A particular feature found at Namulonge is the absence of serious, irrecoverable losses of nutrients by leaching and this must be part of the explanation for the existence of the 'fertile crescent' in the area. In other parts of the tropical zone, with greater rainfall or similar annual rainfall concentrated into a single season, losses of nutrients by leaching must be greater as indicated by the widely differing ratios of arable cropping to rest recorded for shifting cultivation in different areas (Allan 1965). What is not known is whether the nutrients are lost out of the soil or stored at considerable depth where only a long period in forest will recover them. Information on the fate of rainwater entering the soil and nutrients in solution is largely lacking, but is essential for a greater understanding of the potential of these areas for agricultural production.

4

Crop physiology

A. B. HEARN

Introduction

This chapter is concerned with the way environment and genotype interact to determine how the crop grows and how, in turn, growth determines yield. The relationships among these factors, which include farming operations, form the system to be studied in this chapter (Figure 4.1). Farming operations affect and are affected by the environment and by the variety of a particular crop chosen. The subject matter, which we have come to regard as *crop* physiology, is thus both broader than plant physiology and deeper than crop agronomy.

The processes within individual plants or organs concern the plant physiologist. Because he is dealing with the whole system rather than part of it, the agronomist operates at a higher and more complex level of organization. He may necessarily limit his study to the first and last terms in the system: what operations a farmer can do, and the yields associated with them. The agronomist wants to know if one treatment is better than another, without necessarily wanting to know why. Agronomy may therefore be regarded as a technology concerned with developing recipes to grow crops. The relationships between the yield and the cultural operation that constitute the recipe, vary from year to year and place to place. Crop physiology is the science that tries to explain or predict these relationships, and their variation, in terms of the physiological processes within the plant, and the response of those processes to the physical environment. The crop physiologist not only

Figure 4.1. Diagram of a crop production system.

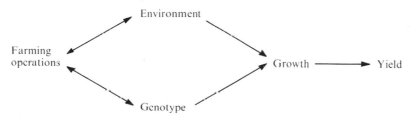

tries to explain why one treatment is superior, but also tries to see if the better treatment realizes the potential of the environment.

History of crop physiology at Namulonge

These ideas have been implicit in the work at Namulonge since its inception. In presenting the Namulonge programme, Hutchinson (1950) used the term 'crop physiology' to describe that part of the crop husbandry programme which would study how crop environment, especially water and nutrient supply, affected leaf area, plant development, and cropping capacity to determine yield. These studies were to correlate with pathological and entomological studies to assess the importance of pests, diseases and pathogens. From these ideas arose the communal experiments referred to in successive Namulonge *Progress Reports* up to about 1956, and from which Farbrother obtained much of his data.

Thus crop physiology began at Namulonge as an offshoot of crop husbandry. The field of interest soon broadened as Evenson joined Farbrother and deepened with the arrival of Dale until, by 1960, five scientists were involved. Their interests, which are reviewed in this chapter, ranged from the whole plant physiology and water relations of Farbrother, through Rijk's work on the physics of evapotranspiration, to Morris' detailed studies of boll development and the stomatal and hormone investigations of Dale and Milford.

Systems analysis, models and the cotton crop

In reviewing the crop physiological work at Namulonge, can we assemble sufficient facts in adequate order to explain or predict crop performance? Although this may appear ambitious, the extent to which we fail will reveal the gaps in our knowledge. The degree to which we succeed will determine our contribution to the solution of the technological problem, first posed in a more limited form by Fergus (1936), of either adapting the crop to suit the environment, or manipulating the environment to suit the crop. We are therefore concerned both with defining a system (i.e. assembling in adequate order) and describing its components (i.e. assembling sufficient facts). Ideally, defining the system should precede describing the components, but once started, both can go on side by side, each catalysing the other. In practice, describing components has tended to outstrip defining the system because the components are of scientific interest in their own right, without regard to their place in the crop production system as a whole, the understanding of which can help to solve agronomic problems. Many facts about the components have accumulated and obscured the structure of the system as a whole. Because it is the function of this chapter to bring order to the many and diverse facts, it is convenient to reverse the historical order and first consider the system as a whole in order to provide a framework to review what has been learnt at Namulonge about the components.

To apply systems analysis to crops and to construct models that describe and simulate real crop systems are much in vogue. These activities have as one of their aims just what has been set out in the preceding paragraph. Work at Namulonge testifies that such aims and their attainment are not new. Hutchinson *et al.* (1958*b*) constructed a model to synthesize, into a coherent story, all that had been learnt about water relations and growth of cotton at Namulonge up to 1957. What is relatively new is that computers can now use comprehensive and dynamic models to simulate complex real systems. The possibility of computer simulation has focussed attention on such models giving fresh impetus to the study of crops as systems, providing a methodology to build models of crops, and a terminology and conventions to communicate them to others. In this chapter the concepts are presented for a model that will provide a framework within which to review the crop physiological work done at Namulonge.

Purposes of a model

More specifically, the aims of constructing a model are:
- (*a*) to bring order to the diverse physiological data collected at Namulonge and relate them to those collected elsewhere,
- (*b*) to reveal the gaps in our knowledge both of the system as a whole and of the components of the system,
- (*c*) to explain or predict variation in crop yield,
- (*d*) to indicate how the farmer might manipulate growth to obtain the best economic results.

Before we elaborate the system set out in Figure 4.1 we must recognize that it is part of the larger system, shown in Figure 4.2, which includes the market and other farm enterprises.

The elements in Figure 4.2 are discrete entities or subsystems between any two of which are physical interfaces that help to define the part of the system in which we are interested. The arrows linking the elements indicate the flow of material, energy or information across the interfaces.

Figure 4.2. The economic and social context of the system.

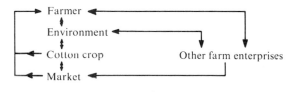

In the context of Uganda the term 'farmer' includes his advisers in the Government extension service who make decisions for him, such as the variety to sow, and the period in which the crop should be sown. The two-way link between the farmer and the environment indicates that the farmer not only alters the environment, but that the environment influences the farmer. For example, to decide about a particular operation a farmer will consider the current weather as well as his prediction of subsequent weather, determined either objectively by meteorological probabilities or subjectively by the farming lore of the area. His decisions will in turn influence the weather subsequently experienced by the crop, by determining the phasing of the pattern of crop growth with the actual pattern of weather after the particular operation is done. This aspect is discussed by Rijks (1967*a*) in relation to sowing date.

The arrow linking the cotton crop with the farmer represents the farmer's response to the appearance of the crop. Within one season, he may decide to spray with insecticide, top dress with fertilizer or to irrigate; on the basis of one year's experience, he may change his strategy for the next year. Such considerations do not always apply in a country like Uganda, however, because many farmers do not yet command the necessary resources.

The link between market and farmer indicates the cash returns to which the farmer responds. In any one season this may influence how much of the crop he picks, while from one season to the next it may affect the acreage and variety sown and the quantities of fertilizer and insecticide used. Links between other farm enterprises and the cotton crop are usually indirect. Other enterprises normally affect cotton either through the environment (effects of crop sequence and pests and disease) or through the farmer, directly by competition for farm resources or indirectly through the market returns.

Although Figure 4.2 is only part of a larger economic, ecological and sociological system it is adequate to define and show the context of the subject matter of this chapter. The rest of the chapter is concerned with the cotton crop and the physical components of the environment. First, actual models which have been developed or used at Namulonge will be considered. We shall then look at a comprehensive model which has been proposed to meet the needs of this chapter. The physiological work at Namulonge will be reviewed as components of this comprehensive model. Work done elsewhere will be drawn on to fill gaps, and to evaluate the Namulonge work. Finally, a working model will be discussed.

Models used at Namulonge

A model of water relations

Hutchinson *et al.* (1958*b*) constructed a model of crop water requirements, and incorporated this into a larger model to relate yield to crop
water supply. They showed that the yield of cotton in different seasons
after various sowing dates was statistically correlated with the extent
to which rainfall satisfied crop water requirements. As long as rainfall
satisfied requirements leaf area continued to expand exponentially,
showing that growth was uninterrupted. As soon as rainfall was
inadequate, exponential expansion of leaf area stopped, and apparently
could not be resumed. It was argued that this event determined the
potential of the crop and that rainfall subsequent to it determined how
much of the potential was realized. When analysed in this way crop
water supply accounted statistically for most of the effect of date of
sowing on yield, but a significant part (a negative effect) still remained
unidentified. In years when the rains are prolonged into December (e.g.
1953, 1967), late sowings do not benefit from the late rain (Farbrother
1954; Arnold 1968). The unidentified effect of date of sowing apparently
overrides any benefit that the late rain might confer on late sowing. It
is suspected that this unidentified effect is related to the cloudy
weather and heavy rainfall at the end of October and in November,
associated with the southward passage of the inter-tropical convergence
zone. Such weather is more likely to affect later-sown crops adversely
by causing boll shedding and rotting.

This model contributed to all the aims defined on p. 79 in that it (*a*)
put the data in order, (*b*) revealed gaps in knowledge (in particular the
unidentified sowing date effect and the physiological definition of crop
potential), (*c*) explained variation in crop performance, and (*d*) predicted the sowing date that would be optimum for yield.

A spacing model

Bleasdale (1966) and his colleagues have developed models that relate
yield to plant population and arrangement. The bases of the model are
(*a*) dry weight increases logistically, (*b*) as population density increases
yield of dry matter approaches an asymptote, and (*c*) the dry weight
of part of a plant is allometrically related to the total dry weight of the
plant as these vary with population density. The model implies that there
will be an optimum population (lighter yields being obtained at populations both denser and sparser) only if seed cotton as a fraction of total
dry matter decreases as population density increases. The model
further implies that the optimum population will be denser when either
plant dry weight, or the proportion of the dry matter in seed cotton,

decreases more slowly as the population density increases. Although these models were developed for vegetable crops not grown for their reproductive parts, the models have proved useful at Namulonge to interpret the response of cotton to spacing in terms of the crop's indeterminate habit and its production and partition of dry matter (Hearn 1972*a*). On more fertile sites the optimum population was denser because plant dry weight decreased less rapidly as population increased; the proportion of dry matter found in seed cotton was unaffected. Water shortage and insect infestation on the other hand did affect the partition of dry matter; less of the dry matter went into seed cotton at dense populations under such conditions. This was probably the result of proportionally greater loss of bolls from shedding and mummifying which would decrease the size of the sink formed by the bolls and increase the proportion of assimilates allocated to vegetative growth.

The recommended spacing for cotton in Uganda has become progressively closer over the years, from 1.5×1.2 m in 1911 to 0.6×0.3 m in certain areas now (Hearn 1972*a*). Assuming that the recommended spacing reflects the optimum spacing and that Bleasdale's model is valid, this change suggests that conditions for the crop have become more favourable, which the following observations support. Varieties have been bred that are better adapted to the environment (Arnold 1970*c*), particularly as regards to pest and disease resistance. Because the environment is better understood, particularly the factors concerned with sowing date, crops can be sown with the greater probability of receiving a favourable water supply (Rijks 1967*a*).

A model of production and distribution of carbohydrates

In order to relate yield to growth, Hearn (1969, 1972*b*) constructed a model to take into account the production and distribution of assimilates in the crop plants. The basic concepts are to estimate the sink capacity of one class of sinks, the bolls, and to allow that class to take priority in the competition between classes of sinks for the limited carbohydrate supply.

Production of assimilates can be estimated either by measuring increase in dry weight and analysing it classically into the size and efficiency of the photosynthetic systems (Watson 1952) or it can be estimated by a model, such as de Wit's (1965), of the photosynthesis of the canopy. The size of the sink formed by the bolls is calculated, in terms of daily carbohydrate requirement for growth, from crop development and boll growth data (the method is discussed later). In allowing the boll sink priority for assimilates, the model is consistent with the nutritional theory of boll setting and shedding. It is implied that any assimilate surplus to the requirements of the boll sink, is allocated to vegetative sinks. As flowering proceeds and boll numbers increase, the boll sink becomes large enough to take the whole of the assimilate supply; the model then ceases to set any additional bolls but sheds them, thus preventing further increase in sink size. Periods of water shortage in a rainfed crop temporarily close the stomata and curtail the assimilate supply; flowers opening at such times do not set bolls because assimilate supply is inadequate for their growth.

This model explained, in some circumstances, differences among varieties, sowing dates, effects of water shortage and the start of boll shedding. In other situations heavy shedding started which the model did not explain. The model suggested that the pattern of assimilation and the pattern of sink development are inevitably out of phase. The phase lag arises because at each node on a fruiting branch the leaf expands before the bud flowers, and senesces before the boll is mature. As a result assimilation declines as the boll sink expands.

The crop forecast model

The models used to forecast the Uganda cotton crop (Manning 1952; Kibukamusoke 1958; Arnold 1969a) were dealt with in detail by Rijks in Chapter 2 but are mentioned here for completeness. Regression models predict cotton production in each cotton-growing area as a function of rainfall and estimated sowing date and acreage. The models show the relative importance of these factors in each area. Sowing date is more important in some areas than others. Negative quadratic terms show that it is possible to have too much rainfall as well as too little. Differences in the regression coefficients for rain at different times show that the relative importance of water supply varies with the stage of growth.

An extended water-balance model

The model of crop water requirements that Hutchinson *et al.* (1958b) used has been developed by Jones for a continuous maize–cotton sequence in Chapter 3. Each crop has its own factor which increases through time to account for the increase in the rate of crop water use as leaf area expands and the crop covers more ground. The model is used to estimate the effect of soil water on the availability and leaching of mineral nutrients. The model does not recognize that water shortage can cause leaf area to stop expanding, and consequently conserve water.

A comprehensive model

Concepts for the model

Each model presented so far is limited to a few aspects of the crop. Figure 4.3 combines their features to provide a conceptual framework into which to fit all the physiological work done at Namulonge, which can then be reviewed as part of a single coherent story. Each of the inputs is the subject of another chapter. In this chapter we consider how they interrelate to affect growth and yield.

The core of the model is the production and distribution of carbohydrates. By its orderly morphological development, the crop generates both a photosynthesizing surface and a succession of growing organs to use the products of photosynthesis. The plant's morphology is such that the production of each flower bud is inevitably accompanied by the production of a leaf. Crop photosynthesis depends on the attributes of the crop canopy reacting with the relevant environmental inputs, allowing the daily production of carbohydrate to be predicted. Respiration depends on dry weight of crop tissues and is the first requirement met from the carbohydrate supply. Remaining carbohydrate is available for growth and is partitioned among four classes of sinks; bolls, leaf, stem and root. The capacity of the boll sink can be estimated from the rate of initiation of flower buds and the rate of growth of a boll. The amount of carbohydrate that accumulates in one particular sink, the bolls, determines yield. Accumulation of carbohydrates in the sink controls the system internally. As the tissues increase in weight, they feed back negatively by respiration to decrease the proportion of carbohydrates available for growth. Until full cover is achieved, the increasing leaf tissue feeds back, both positively through a larger photosynthetic system, and negatively by greater water consumption leading to earlier water shortage. Carbohydrate accumulating in the bolls feeds back negatively in three ways: less is available for new leaf, less for new flower buds, and less for new bolls to set. As flowering proceeds, the number of bolls eventually becomes sufficient to use all the carbohydrate supply whereupon production of new leaf and flower buds, and setting of new bolls stop.

The weather inputs for the model are evaporation, rainfall, radiation and temperature. They control the model externally, can slow down the rates of the various processes, or stop the system completely. Radiation ultimately drives the whole system through the photosynthetic apparatus. Temperature governs respiration rates and the rates at which organs are initiated and grow, thus influencing sink capacity. Rainfall and evaporation operate through crop water relations. In the present model the influence of the soil is limited to that exerted by the soil water constants and rooting depth. In respect to crop culture, the model only recognizes spacing and sowing date. There are no data yet that permit us to include farming operations such as weeding and land preparation. Sowing date has two major roles; it determines the weather at establishment, which affects plant population, and it determines how the pattern of subsequent weather matches the pattern of growth. Plant population affects the system through leaf area and sink capacity on a unit area basis.

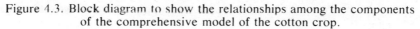

Figure 4.3. Block diagram to show the relationships among the components of the comprehensive model of the cotton crop.

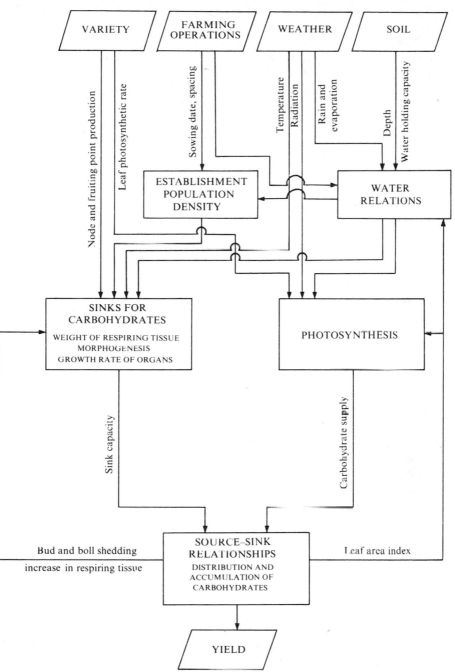

Varietal characteristics affect both the supply of carbohydrates and the capacity of sinks for carbohydrates, the former through leaf photosynthetic rate and leaf display, the latter through the rates at which plants initiate organs and at which organs subsequently grow. Varieties may differ in the relative strengths of, or order of priority among, different classes of sinks, but there are no data yet to include this feature.

In presenting the concepts for this model, the major components, 'farming operations', 'environment' and 'growth' in Figure 4.1, have been divided in Figure 4.3 into smaller units, each of which forms a subsystem that will be examined subsequently in this chapter. The relevant data will be reviewed and the concepts further developed, resulting in further subdivision of the units. These subdivisions will still be 'black boxes', whose inner workings remain unknown. Such ignorance does not constitute a gap in our knowledge as defined earlier provided the input and output for a box can be defined precisely, and related quantitatively. Gaps in the sense defined, only occur when no relationship can be found to put into the box. For example the model omits pests and diseases and mineral nutrition.

There are yet no data to predict pest populations or intensity of disease infection, and even if such data were available it would not yet be possible to relate the intensity of attack to the reduction in bud and boll numbers and leaf area.

Mineral nutrition has also been omitted because data are lacking. What is required is:

(*a*) to predict the supply in the soil,

(*b*) to predict the rate at which nutrients are taken up from the soil into the plant,

(*c*) knowledge of the effect of the supply in the plant on the carbohydrate economy.

Nitrogen, for example, is likely to affect the partition of the carbohydrate supply between leaf and root, and also the initiation of organs, and thus the sinks for carbohydrates.

Establishment

Prentice (1957) sowed seed every day for a three-month period that spanned the normal sowing season. In a particularly dry year the mean emergence was 35 per cent and the worst 22 per cent. Ndegwe (1968) found too much water as well as too little, decreased emergence. Sowing on the day rain fell, or the day before or after was less effective than two or three days after rain. Emergence has not yet been quantitatively related to soil water supply. With such a functional relationship the probability of emergence could be predicted from rainfall probability. The effects of sowing date on the likelihood of obtaining a given stand could then be assessed.

Ndegwe (1968) also investigated the effects of seed quality on germination and emergence. Harvesting under wet conditions, followed by poor storage that hindered the drying of seed cotton, decreased germination. Viability was not affected. Such seeds had a more permeable seed coat and absorbed water faster and excessively, suggesting that the germination processes were impaired through lack of oxygen. Gubanova & Gubanov (1961) and Christiansen & Justus (1963) found that picking and storing seed cotton under conditions similar to those described prevented adequate seed coat lignification, and so increased seed coat permeability. Their findings explain Ndegwe's observations and stress the need for good picking and storage conditions for seed cotton in order to obtain seed of good quality.

Water relations

The box labelled 'water relations' in Figure 4.3 can be elaborated as shown in Figure 4.5. Work at Namulonge has been concerned with the elements in the system and some of the relationships among them.

Soil water status. The usual approach has been to estimate soil water status by means of a water balance based on meteorological data and to check it by measuring soil water (Hutchinson *et al.* 1958*b*; Rijks & Owen 1965; Hearn 1972*b*; Chapter 3). The data needed are: (*a*) the water storage capacity of the soil, (*b*) the gains of water by the system, (*c*) losses of water from the system, and (*d*) soil water content. Gains comprise rainfall and, in some experiments, irrigation, and require no further comment.

The storage capacity of the soil within the system depends on the upper and lower limit of availability of soil water and on rooting depth. Although field capacity and wilting point have been useful concepts at Namulonge to indicate the upper and lower limits, they are no longer accepted as precise physical attributes. This is evidenced by the difficulty of measuring them, as a result of which estimates of available water range from 16.1 to 11.9 per cent on a volumetric basis (Manning & Farbrother 1953; Farbrother 1958; Hearn 1970). Part of this variation arises from difficulty in measuring bulk density, and part from the criteria used to indicate wilting point and field capacity.

Hutchinson *et al.* (1958*b*), following Prentice's (1957) observation that most roots were confined to the upper 30 cm of soil, used changes in this zone to characterize soil water status. However, Farbrother (1954, 1956) found that the crop did not extract water from below 150 cm. Root systems excavated between 1966 and 1968 confirmed that roots seldom penetrated beyond 120 cm. Farbrother (1960*c*, 1961*b*) found differences among varieties in the way they extracted water. An Albar variety consistently extracted more water from the second and third 30 cm layers of soil than a BP52 selection. A. B. Hearn (unpub-

(a)

(b)

Figure 4.4. Effect of spacing on root penetration. Photographs from pits at two sites described by Hearn (1972a). (a) and (b) show opposite sides of a pit dug between two spacing treatments on site A; (c) and (d) show a similar pit on site D.

(c)

(d)

lished) confirmed this varietal difference but found no consistent differences between the root systems of the two varieties.

Later work (Hearn 1968 *et seq.*) indicated that plant population density could affect rooting depth. Excavation and washing of root systems revealed two layers in the profile that appeared to impede the roots: a cultivation pan of denser soil at 15–20 cm and the clearly demarcated interface at about 30 cm between the brown topsoil and more acid, red subsoil. Roots of plants in dense populations more frequently penetrated these layers (Figure 4.4). This phenomenon did not always occur and the study of how frequently it did occur was prevented by the heavy labour requirement for digging pits and washing root systems, which necessarily limited sampling. Harris & Farazdaghi (1969) reported a similar observation with sugar beet.

Water is lost from the system through deep drainage, surface runoff, and by evapotranspiration. Rainfall in excess of the amount needed to restore the soil to field capacity is assumed to run off or drain deep. Evapotranspiration is discussed in detail by Rijks in Chapter 2. Briefly, potential evapotranspiration can be estimated by the Penman formula with locally validated constants. Actual evapotranspiration is derived from potential evapotranspiration by applying two factors, one to account for the extent of crop cover and the other to account for the effect of shortage of water in the system. The distinctive and novel feature of the water balance model proposed by Hutchinson *et al.* (1958*b*) was the increase in the crop factor as crop cover increased. They recognized two components in the crop factor: the transpiration component arithmetically equal to the leaf area index (LAI) up to a maximum of 1.8, and the soil evaporation component which diminished from 0.3 at sowing to zero when LAI reached 1. It was subsequently suggested (H. L. Penman, unpublished) that potential evapotranspiration had been underestimated and actual evapotranspiration overestimated, so that the crop factor is not arithmetically equal to LAI. Nevertheless the principle that the crop factor, and hence the water used, depends on the stage of growth, still stands as a significant contribution to the understanding of the water relations of annual crops, which has been recognized or confirmed in much subsequent work (e.g. Slatyer 1960; Fitzpatrick & Nix 1969).

Hutchinson *et al.* (1958*b*) implied that the crop factor and water use reached their maxima when LAI reached 1.8. This may be compared with a value of 2.0 or more found by D. A. Rijks (personal communication) in the Sudan and 2.7 by Ritchie & Burnett (1971) in America. Stern (1967) was not able to find a consistent relationship between water use and LAI.

The water balance can be expressed as the amount of water stored in a given depth, or as the water deficit, i.e. the amount of water required to restore the soil profile to field capacity. The latter is preferable if the storage capacity is not known previously, as at Namulonge, provided the deficit is less than the imprecise estimate of storage and the profile is at field capacity when the balance is started.

Farbrother (1957) adapted and developed for Namulonge conditions Bouyoucos' (1949) electrical resistance technique to measure soil water status. Farbrother (1964) concludes that the resistance is best used to indicate whether the soil is at field capacity or wilting point. When the calculated balance shows the profile is at field capacity or wilting point, the resistance of the units can provide a check. The units also indicate when, and from what depth water is extracted.

Plant water status. Relative water content (RWC) (Weatherley 1950) has been used at Namulonge to indicate plant water status. Farbrother (1956) showed that soil water status was closely related to RWC at dawn. When the upper 30 cm of soil was at wilting point, dawn RWC was 90 per cent. Dale (1961) showed that RWC in the field could decrease to 70 per cent by the middle of the day.

Hearn (1972*b*) constructed a regression model to relate RWC to the calculated soil water deficit, the age of the crop, and the evapotranspiration (*ET*) the previous day and (for 2.00 p.m. observations) the *ET* the same day. These factors interacted. The effect of a given soil deficit was less on older crops and when *ET* was small. These results are consistent with recent thinking (Slayter 1967); plant water status depends on the relative rates of water uptake and loss, and indicates the response of the plants to the interaction of environmental supply and demand for water, shown in this study by past weather (affecting the soil deficit) and current weather (affecting *ET*) respectively. The response decreases with age of crop, the result probably of deeper roots.

Stomatal response to water shortage. Dale (1961) studied the effect of plant water status, as one of several factors that affect stomatal resistance. Stomata started to close when RWC had decreased to 83 per cent and were completely closed at 70 per cent. Troughton (1969) found that leaf resistance to carbon dioxide exchange started to increase when RWC was about 83 per cent. The functional relationship of RWC to past and current weather (Hearn 1972*b*) with Dale's and Troughton's observations enables stomatal behaviour to be predicted from meteorological data. RWC of 83 per cent at dawn, would prevent the subsequent opening of the stomata. At 100 days from sowing with *ET* at 4 mm per day in order to obtain a dawn value of 83 per cent for RWC a soil water deficit of 260 mm would be required. Such a deficit exceeds the storage capacity of soil in the root range. It is therefore concluded that as long as there is some water in the root range, shortage of water will not prevent opening of stomata in the morning; some photosynthesis will occur even if it is likely to be curtailed early in the day when evaporative demand increases and stomata close. At 2.00 p.m., RWC of 83 per cent would not occur and stomata would not start to close until a deficit of 40 mm is reached, usually after about eight to ten days without rain. Stomata will be fully closed when the deficit reaches 160 mm. Because this deficit is equal to the storage capacity of the soil in the root zone, it is concluded that stomata will not be completely closed until all available water is consumed. Carbon dioxide exchange will still be one-fifth to one-quarter of the rate when the stomata are fully open (Troughton 1969) which will be adequate for respiratory requirements.

Growth response to water storage. Farbrother (1956) concluded that when the top 30 cm of soil had dried to wilting point and RWC at dawn was 90 per cent, plant dry weight ceased to increase, flowers ceased to set bolls, new fruiting points ceased to appear and leaf area ceased to expand exponentially. Hearn (1972*b*) using a different approach, also found that the plant behaved as though growth stopped at a certain critical soil water deficit, which increased with age. The RWC associated with this deficit did not change with crop age and was estimated to be 92 at dawn and 82 at 2.00 p.m. In contrast to Far-brother's findings and consistent with those of Gates (1968), growth was freely resumed when the deficit was removed. The difference between the earlier and the more recent results may be the result of the change in variety and the use of fertilizer.

Conclusions on water relations. Results at Namulonge are consistent with work elsewhere (Slatyer 1955, 1967; Namken 1965; Troughton 1969) in that water shortage does not affect growth until RWC during the middle of the day declines to 83 per cent, though actual effects observed vary among the different environments.

Possible mechanisms by which the effects of water shortage on growth influence yield are suggested in Figure 4.5: (*a*) through the photosynthetic system, (*b*) through morphogenesis, and (*c*) through a direct effect on bud and boll shedding. These will be discussed in detail later.

Whatever the mechanism involved, water supply is clearly the major environmental factor affecting yield at Namulonge (Hutchinson *et al.* 1958*b*). It determines the start and finish of the season, and to a lesser extent the activity during the season. The cotton crop would benefit from any measure that would decrease its sensitivity to water shortage. Three such measures suggest themselves, and have been tried as part

Figure 4.5. Block diagram to show the structure of the water relations subsystem.

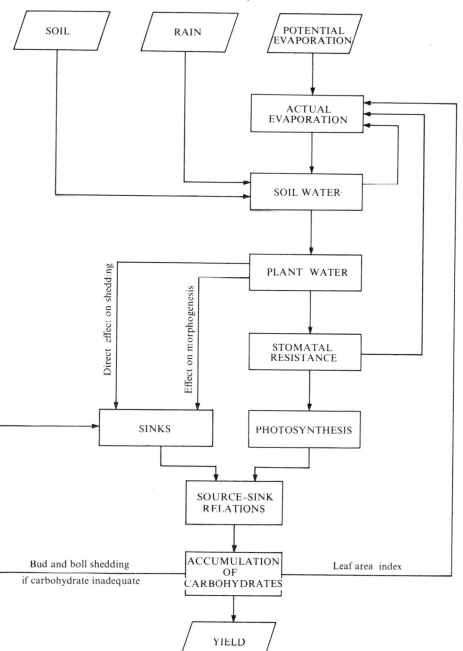

of an overall plan to intensify cotton growing on the research farm; (a) a variety such as Albar that extracts water more effectively, (b) closer spacing to give better root penetration, and (c) deeper cultivations to break up the pan. The opinion of all concerned (Arnold 1970a) that the crop is now less sensitive to water shortage was strikingly endorsed by J. B. Hutchinson (personal communication) when he visited the station after twelve years' absence. In the severe drought in October 1969 the crop did not appear to him to be severely droughted by the standard of two decades earlier and he was surprised to learn that the crop had been without rain for a month. It is not that it now needs a greater water shortage to affect the crop, but that the effects are less severe, as evidenced by the resumption of growth in Hearn's experiments compared with Farbrother's.

Photosynthesis

The photosynthetic system of the cotton crop was studied in the early years at Namulonge by classical growth analysis. Increases in dry weight were analysed into the conventional components to indicate the size and efficiency of the system. Such data were collected between 1951 and 1958 as part of the broader study, discussed by Farbrother (1958). in which he collected as comprehensive a set of data as possible to study the effect of the environment on the crop. The results revealed the dominant role of water supply, which has already been discussed. However, effects of water shortage and other factors on the rate at which the crop increased in dry weight, could not be directly related in any simple way to differences in yield. Measurements of crop dry weight were therefore discontinued after 1958.

When Hearn (1969) showed that the metabolic sink formed by the bolls provides a link to relate photosynthesis to yield, the study of the photosynthetic system was resumed at Namulonge. Workers elsewhere have constructed models to simulate the photosynthesis of crop canopies (e.g. de Wit 1965; Monteith 1965). All depend on estimating how much light the canopy intercepts and what the canopy does with that light.

The crop canopy. The amount of light intercepted depends on the size of the canopy, measured by its leaf area index (LAI) and how the leaves are displayed. Hearn (1970) measured the angles at which the crop displayed its leaves; cotton proved to have its leaves closer to the horizontal than other crops. This means fewer leaves intercept more light, which enhances photosynthesis in sparse canopies but uses more water.

Light interception measured with a photoelectric cell, was similar in the field at Namulonge (Hearn 1972b) to that in artificial communities of cotton plants under controlled conditions (Ludwig, Saeki & Evans 1965). In both situations the extinction coefficient was 1.1. De Wit's model predicted an extinction coefficient of 0.86. However, when LAI was between 1 and 2 there was little difference between the predicted and measured interception.

It is generally accepted that crops will photosynthesize fastest when their lowest leaves are at the light compensation point, which is approximately 5 per cent full day-light (Huxley 1964). Therefore, most dry matter can be expected from a canopy that intercepts 95 per cent of the light.

Up to 1956, only one of the Namulonge crops in which LAI was measured had a maximum LAI greater than 1 (Farbrother 1952 and subsequent *Progress Reports*). De Wit's model and actual measurements agree in this range and indicate that at the best only 60 per cent of the light received was intercepted, and usually considerably less. Such poor interception could only be justified if it were balanced by an increase in the proportion of assimilate accumulating in the boll sink, or an increase in the period during which accumulation took place. It was thought (Farbrother 1960*b*; Manning & Kibukamusoke 1960) that 90×60 cm spacing achieved just that, by deferring or decreasing water shortage. In recent years, with more intensive methods of husbandry, measured values of LAI have been between 2 and 3 and such crops intercept between 85 and 90 per cent of the light received. The heavier yields obtained suggest that the advantages of intercepting more light outweigh the disadvantages of greater risk of water shortage, which might shorten the growing period or decrease the proportion of assimilate accumulating in the boll.

The rate of canopy photosynthesis. The de Wit model predicts the daily photosynthesis of the crop canopy. It requires the curve that relates the rate of photosynthesis of a single leaf to the light intensity. The many such curves that have been published (Boehring & Burnside 1956; El Sharkaway & Hesketh 1965; Bierhuizen & Slatyer 1964; Ludwig *et al.* 1965; El Sharkaway, Hesketh & Muramoto 1965) fall into two groups. those which show light saturation at about 0.25 ly per min visible radiation with photosynthesis at 95 mm^3 carbon dioxide per cm^2 per hr and others in which photosynthesis increases to over 200 mm^3 carbon dioxide per cm^2 per hr in full sunlight at 0.75 ly per min. El Sharkaway *et al.* (1965) found that plants raised under low light intensity had shade leaves that were light saturated at low intensities. Obviously curves of the first group cannot be used to predict the field situation and a curve of the second group from plants raised in full sunlight is required for use in the de Wit model.

Shortage of water will affect the photosynthetic system through the stomata as already discussed. Troughton's work suggests no other effects until RWC drops to 75 per cent, which at Namulonge would be reached at 2.00 p.m. with a soil deficit of 110 mm.

Radiation operates on the system in two ways: by varying the energy input into the photosynthetic processes, and through stomatal closure which Dale (1961) studied. Both effects are evaluated in the leaf photosynthesis function in the de Wit model.

Temperature affects photosynthesis in at least three ways: (*a*) through its effect on respiration, already mentioned; (*b*) through its effect on the stomata (Dale 1961) which close at lower tempertures; and (*c*) through its effects on the number and capacity of the metabolic sinks, to be discussed later.

In order to estimate from the photosynthesis of the crop canopy the amount of carbohydrates available for growth of the aerial parts of the plant, it is necessary to subtract that required for respiration and root growth. Baker & Hesketh (1969), Hesketh, Baker & Duncan (1971) and Baker, Hesketh & Duncan (1972) have thoroughly investigated that required for respiration of the plant. The proportion of assimilate which accumulates in the root is more difficult to assess, but an estimate of 18 per cent may be made on the basis of work done by Huxley (1964) and Stern (1965) and pot experiments at Namulonge. When roots and respiration are taken into account, measured increases in dry weight agree with those predicted by the de Wit model.

Metabolic sinks for assimilates

For present purposes we can recognize two classes of sink: reproductive and vegetative. The former comprises the bolls, and the latter, those sinks formed when newly initiated organs expand into stem, leaf and root. Two important attributes of sinks are evident: their influence on the rate of photosynthesis (Humphries 1967) and on the distribution of assimilates. The former effect is important in determinate crop plants like wheat (King, Wardlaw & Evans 1967) and tobacco (Hackett 1973) but may be relatively insignificant in an indeterminate crop like cotton which initiates more sinks, both vegetative and reproductive, than it can supply, so that paucity in their number is unlikely to limit photosynthesis. Our main concern is with the influence of the reproductive sink on the distribution of assimilates, and thus on the control of growth. Competition between vegetative and reproductive sinks has long been recognized in indeterminate plants including cotton (Leonard 1962).

As long as the mainstem of a cotton plant continues to grow, the number of fruiting branches increases, each of which can continue to initiate leaves and flower buds at a constant rate. Consequently the rate of initiation of leaves and flower buds on the plant as a whole increases, and the number of growing leaves and bolls also increases. The potential capacity of the sink formed by these organs increases in direct proportion to their number. By contrast, the carbohydrate supply does not increase proportionately to the number of leaves, but approaches an asymptote set by the radiation received. Thus, provided the plant continues to grow, it is inevitable that the capacity of the sinks will eventually exceed the carbohydrate supply. Sinks will therefore compete with each other for carbohydrate both within and between classes.

At Namulonge, Farbrother's (1954) and Dale's (1959) disbudding experiments confirm the conclusions of earlier workers that bolls normally compete successfully for assimilates and when removed permit extra vegetative growth. It is further concluded that when bolls are retained on the plant, and their number increases as flowering proceeds, they form a progressively larger sink.

As the crop continues to grow and the increasing demand of the sink formed by the bolls is satisfied, the quantity of assimilates for vegetative growth progressively decreases. Farbrother (1954) identified as critical 'the date when internal factors, stimulated by weight of crop, have begun to limit the development of the fruiting structure of the plant'. This concept will now be developed.

It is not clear whether metabolic sinks are active or passive, i.e. whether assimilates are 'pumped in' or are 'sucked in'. Are we to think of sinks in terms of a capacity to receive or a demand to be met? The latter is preferable where we are concerned with competition between sinks, and is consistent with the popular notion of the fruit or boll load of a crop.

To analyse the source–sink relations further, we need to know the amount of assimilate which a sink can accept, its capacity or demand. This capacity has been estimated for the sink formed by the bolls, but not for vegetative sinks. One solution to the problem of the relative strengths of competing sinks to attract assimilates, is to assume that one sink takes priority for a limited supply of carbohydrates; its capacity is satisfied first and any surplus is then available for other sinks. This is an oversimplification and is not always true, but will serve as a start to explore the source sink relations of cotton.

Size of the boll sink. At any time, the capacity of the sink formed by the bolls is determined by the number of bolls and the growth rate of individual bolls.

Number of bolls. The regular and orderly development of the fruiting structure has been well established since the work of Balls (1912). The construction and use of composite plant diagrams (Farbrother 1961*b*; Munro & Farbrother 1969) which were developed at Namulonge, depend on this feature of the cotton plant.

Development of the fruiting structure is regular and orderly because:

(*a*) after the first fruiting branch (sympodium) has appeared at a node on the mainstem, an additional fruiting branch usually appears at each higher mainstem node shortly after that node itself appears, and (*b*) the ratio between the rate at which flower buds appear along a fruiting branch and the rate at which mainstem nodes appear, is a varietal characteristic, constant over a wide range of environments (Farbrother 1961*b*; Hearn 1969). Thus, the varietal ratio and the rate of mainstem node production determine the basic rate of flower bud production. The rate is modified by the occurrence of vegetative branches (monopodia) and by the cessation of growth of the sympodium.

Mainstem node production is easily recorded and is a valuable parameter of the morphological development of the reproductive sink. Variety, temperature and water supply affect mainstem node production (Hearn 1969, 1972*b*). Temperature effects are described by accumulated day degrees above a threshold of 12 °C. Water supply acts like a switch; when a critical deficit is reached, node and flower bud production stop, and resume when water is available again.

The modification of the basic rate of flower bud production, caused by vegetative branches and by fruiting branches becoming inactive, is described empirically when the accumulated number of flower buds is plotted against mainstem nodes (Hearn 1969). After an initial phase in which the rate of flower bud appearance increases as the number of sympodia increases, a stage is reached during which flower buds appear at a constant rate relative to mainstem nodes. The plants are behaving as though, when each new sympodium appears at the top of the plant, the oldest active one stops producing flower buds. As a result, the number of active sympodia is constant, and depends on variety and spacing. During this stage the linear regression of cumulative flower buds on mainstem nodes describes plant development completely. The regression coefficient indicates the constant rate of appearance of flower buds and the intercept depends on the position of the first sympodium. The regression coefficient thus takes into account varietal differences in the ratio of flower bud production to node production, as well as density of plant population which affects the number of active sympodia. Upland varieties at Namulonge, on average, bear the first sympodium at the fifth node, though some such as BP52, show more variation about the mean than others, such as BPA.

The number of flower buds can therefore be predicted from meteorological data with the aid of (*a*) a function to relate the number of mainstem nodes produced to the number of elapsed day degrees; (*b*) a critical water deficit at which node production stops; and (*c*) a function to relate flower buds to mainstem nodes. In the functions (*a*) and (*c*) the constants differ among varieties but work so far has shown no evidence of differences in (*b*).

The potential number of bolls and thus the size of the boll sink can be predicted from the number of flower buds by allowing for the interval between bud initiation and anthesis. This interval is thirty days at Namulonge, but varies with temperature (e.g. Hesketh & Low 1968).

Sink capacity of individual bolls. The growth rate of an individual boll, when not limited by the carbohydrate supply, will indicate its sink capacity. In order to determine this rate, first-formed bolls are studied. Such bolls develop when competition among bolls for carbohydrates is minimal. Evenson's (1960) data on boll development at Namulonge can be used to show that the increase in the dry weight of a boll, including the husk, is essentially linear from five to six days after anthesis until five or six days before dehiscence. By dividing the final dry weight of the boll, inclusive of husk, by 80 per cent of its maturation period, the daily growth rate of a boll is estimated.

Varieties differ markedly in boll weight, but differences in maturation period are less clear. Morris (1964*a*) found bolls of an inbred line of BP52 matured a few days earlier than other varieties but this was not true of a multi-line seed issue of BP52 (Hearn 1972*b*). Varieties of diverse origins examined by Hesketh & Low (1968) did not differ in maturation period but Munro (1971) suggested that this attribute could be changed by breeding. Whatever the extent of variations the sink capacity of an individual boll is clearly a varietal character.

Temperature affects both boll weight and maturation period. Morris (1964*a*) presents some evidence, but Hesketh & Low (1968) provide the best set of data, from which it can be calculated that the sink capacity of an individual boll is greatest at a mean temperature of 28 °C.

Boll weight also varies with soil fertility and mineral nutrition of the crop (Hearn 1972*a, b*). The number of seeds, in particular, is increased by nitrogen fertilizers (e.g. Hughes 1966). Thus the sink capacity of a boll, and the development of the boll sink, will depend on the mineral nutrition of the crop.

Morris' (1964*a*) conclusion that water relations do not affect the maturation period is open to modification because he only varied water supply, and did not compare different evaporative demands. Evaporative demand is more likely than water supply to affect dehydration of the boll after the xylem vessels become blocked with tyloses during the maturation process (Morris, 1964*b*). However, the effect of water relations on boll dehiscence will not affect the sink capacity of the boll, because it does not operate until the boll is physiologically mature and has ceased to increase in dry weight.

Morris (1965) concluded, from experiments in which he shaded bolls and removed bracteoles, that boll walls and bracteoles may photosynthesize to contribute part of the carbohydrates used by the boll to grow.

Thus, part of the sink capacity of a boll may be filled by the boll's own photosynthesis and decrease its capacity for imported assimilates. However, shading and surgical experiments are difficult to interpret (e.g. Thorne & Evans 1964) because such treatments may lead to increased photosynthesis from other sources to compensate for the organs removed or shaded. The contribution of the boll's own photosynthesis may thus be underestimated but this is less likely if, as suggested earlier, cotton initiates a surplus of sinks. On the other hand, bolls deep in the canopy may not receive enough light to photosynthesize. Brown (1968) found few stomata in the boll wall and little uptake of external carbon dioxide by the boll. He concluded that the fruit may re-assimilate respiratory carbon dioxide as do other species (Kriedeman 1966). Photosynthesis of the boll can augment but not replace photosynthesis by the leaf canopy. The problem of predicting the contribution of the boll's photosynthesis in any situation is unresolved, and will be difficult to resolve because of the variable exposure of the bolls to light.

Sink and source relationships

Supply and demand for assimilates. A simple model, developed at Namulonge by Hearn (1969) to relate assimilate supply to the demand of the boll sink in order to show how yield might depend on growth, has been mentioned earlier in the chapter. The model is based on the nutritional theory of boll shedding (Mason 1922) and compares the supply of assimilates with demand. As long as demand is less than supply, each flower sets a boll.

Radiation sets the upper limit to the assimilate supply, which in turn limits the number of bolls whose assimilate requirements can be met. This number is the maximum boll load. As flowering proceeds, and bolls are set, the demand of the boll sink for assimilates increases. The crop normally initiates enough flower buds to provide a sink whose demand, should all buds set bolls, would be great enough to exceed the supply of assimilates by a large margin. In most crops the demand will therefore eventually equal supply. The crop then has a full boll load. The model predicts that additional bolls will be shed and vegetative growth will stop. Observations confirm these predictions in some situations (Hearn 1969, 1972b).

Translocation pathways for assimilates. The model assumes that there is a common pool of assimilates in a plant, to which all leaves contribute and from which all bolls can draw. This assumption conflicts with the work of Brown (1968) who showed that each boll can receive assimilates only from certain leaves, and conversely a leaf can only supply certain bolls. A boll receives assimilates from its subtending leaf, the leaf subtending the sympodium and from some higher leaves on the mainstem. There is a well-defined pattern of distribution of assimilates from each half depending on particular phloem strands. Bolls on sympodia inserted near these strands can draw assimilates from them. A much more complex model would be required to take into account Brown's work. First, the model would generate a plant structure consisting of mainstem and sympodia, whose leaves expand and buds flower in order. Each leaf would be examined in turn, with regard to its position, its photosynthesis computed, and carbohydrates allocated first to the nearest available boll according to Brown's pattern, then the next nearest and so on until there is none left. The model would then examine each boll in turn, and sum all assimilates from all sources. If receipts do not meet requirements, the boll would be shed.

Although the more complex model may eventually be developed and apply to a wider range of situations, the simpler model can be developed more readily, and does not necessarily conflict with Brown's pattern of assimilate distribution. His work suggests that a plant can be regarded as a set of parallel pools (the phloem strands) of assimilates instead of a common pool. A leaf can contribute to two or three adjacent pools and likewise a boll, if not satisfied by 'local' leaves, can draw from two or three pools adjacent to the point of insertion of its sympodium. If all pools are equally well supplied, a symmetrically developed plant will behave as though there is a common pool.

Stress shedding. Shedding that is accounted for by the nutritional theory of boll shedding can be termed stress shedding and distinguished from non-stress shedding to be described in the next section. When demand by the boll sink for asssimilate exceeds the supply, stress shedding occurs.

When a full boll load is attained there will be stress shedding of surplus bolls. Stress shedding also occurs when, for example, cloudy weather or shortage of water temporarily decreases the supply of assimilates. Demand may then temporarily exceed supply and the model predicts that bolls will be shed, as is commonly observed. A special case is 'the onset of water stress', described by Hutchinson *et al.* (1958*b*) and referred to earlier, which can be interpreted in terms of the model. The onset usually occurred between 90 and 120 days, as the demand of the bolls for carbohydrates was approaching the supply. The water shortage would have decreased the supply so that it was less than the demand. This event would cause shedding of surplus bolls and check the vegetative growth because of shortage of assimilates. The check was apparently permanent and, because there was no further vegetative growth, the potential of the crop was determined.

Non-stress shedding. Undamaged bolls are sometimes shed even when there is adequate water and radiation, and before the demand for carbohydrates exceeds supply. Both the simple and the complex models are inadequate to account for this situation, in which the bolls do not take priority over the vegetative sinks. Such shedding can be distinguished from that caused by metabolic stress and referred to as non-stress shedding. It is realized, however, that further research may reveal that as yet unidentified stress causes what is now termed non-stress shedding.

A result of non-stress shedding is that the number of bolls increases more slowly which defers the day when the demand of the boll sink for carbohydrates exceeds the supply. In contrast to the situation with stress shedding, vegetative growth continues because the supply of water and radiation are adequate, and the supply of carbohydrates still exceeds the demand of the bolls. With a long growing season this can be an advantage. Because the crop continues to flower, new bolls can be set to replace those that are ripening, thus maintaining the boll load. Further, the crop is also continuing to produce new leaves to maintain the carbohydrate supply. Thus the boll load, sink size and assimilate supply can be maintained until the end of the season. This is the basis of a 'top crop'.

Continued vegetative growth has also been regarded as a particular advantage when pests and diseases or bad weather are likely to interfere with the setting, maturing or picking of bolls at some time during the season. Prolonged flowering increases the chance that at least part of the crop will encounter good conditions. This attribute has been a recognized virtue of varieties selected for tropical African conditions, which led to the emergence of African upland varieties (Arnold 1970*b*, *c*).

Where the growing season is short, non-stress shedding is a disadvantage because there is no time to set and mature any more bolls after the first have ripened. It is probably better to enlarge the boll sink as rapidly as possible so that it can accept the whole of the assimilate supply and form a full boll load as soon as possible. This distinction between behaviour suited to long and to short growing-seasons introduces the concept of an apparent balance between vegetative and reproductive growth in some varieties, and its absence in others, commonly referred to as 'indeterminate' and 'determinate' varieties respectively.

Non-stress shedding can be excessive in warm fertile conditions, with plenty of water, and is associated with rank vegetative growth. Discussion often centres on which comes first, the shedding or rank growth. A factor that influenced the relations between sink and supply would cause both.

Non-stress shedding is an area of crop physiological research that urgently requires investigation in terms of the relations of vegetative and reproductive sinks and assimilate supply. If non-stress shedding could be quantitatively related to the environment or the carbohydrate economy of the plant, it could then be incorporated into the models. When understood, it could be controlled or manipulated to achieve agronomic ends.

Study of shedding at Namulonge. Although shedding has already been discussed as an aspect of the relations between sinks and sources, it is frequently treated as a special subject (e.g. Nyahoza 1966b, and papers reviewed by Cognée 1968). There is then a danger of overlooking the relationship of shedding to the physiology of the whole crop, as discussed in the previous section. To quote shedding percentages does not explain anything; it merely describes superficially what has happened. For example, the shedding percentages of varieties may differ but this does not necessarily mean direct genetic control of shedding. Differences could arise either from variation in boll size so that fewer bolls are needed to make a boll sink large enough to equal the assimilate supply, or because of a different degree of non-stress shedding, resulting from genetic control of the partition of assimilates between vegetative and reproductive growth.

At Namulonge, Nyahoza (1966b) studied a number of factors that cause shedding. Most were stress shedding. His results confirmed findings elsewhere, and can be explained qualitatively in terms of the simple model of source–sink relations.

Farbrother (1953, 1954) recognized that shedding of small flower buds (match-head sheds) was important at Namulonge. He found that such shedding was sometimes associated with water shortage, and at other times not. Dale (1957, 1958) studied the phenomenon with inconclusive results.

Water shortage has been cited at Namulonge, as elsewhere, as a cause of bud and boll shedding. The model illustrates how water shortage could operate through stomatal resistance on the carbohydrate supply to cause shedding. There is no conclusive evidence yet on whether this is the only mechanism involved or whether water shortage can cause shedding independently of the carbohydrate economy. Because the first consistent effect of water shortage is stomatal closure, it is most likely that water shortage initially works through the carbohydrate economy.

If the increase in the size of the boll sink is prevented by insect damage, or artificially by disbudding, vegetative growth continues unchecked. The continued vegetative growth may eventually lead to heavier yields in some varieties. Dale (1962) reviewed such work and Brown (1965) measured a varietal difference in this effect in Tanzania. It is not certain whether the heavier yield comes from a larger photosynthetic system, a faster increase in flower bud numbers giving a more rapid development of the boll sink, or both.

Hormonal control of shedding. Dale & Milford (1965) started to study the hormonal control of shedding at Namulonge. They extracted two growth substances from young bolls. By applying these externally to young bolls, they were unable to show that either extract controlled shedding, in contrast to American work (Carns, McMeans & Addicott 1959). Unfortunately the Namulonge work was not continued long enough to achieve the success attained by other research groups (Addicott 1970). Dale (1965) showed that pedicels from which bolls had been removed abscise, and that their abscission was delayed by the presence of other bolls on the plant. He therefore concluded that growth substances from within a boll are not necessary for its abscission layer to develop, and substances produced elsewhere in the plant promote its development. The vegetative meristem in the growing point of the mainstem is a possible site for the production of such a promoter. When there are bolls on the plant, they compete for assimilates with the vegetative growing point which becomes less active and produces less of the abscission promoter. Thus in Dale's (1965) experiments, when other bolls were present the vegetative growing point would have been less active and produced less of the abscission promoter, with a result that pedicels would abscise more slowly, as he observed. When the activity of the vegetative growing points is stimulated by, for example, high temperatures and a liberal supply of water and nitrogen, production of promoter would increase and cause the non-stress shedding that is associated with rank growth.

Morris (1964b) obtained an extract from growing seeds that delayed the abscission of cotton petioles. He also showed that the abscission layer in the pedicels developed as a normal prelude to the dehiscence of bolls. This observation is significant and emphasizes that the development of the abscission layer is an inevitable and essential step in the development of the boll, rather than an unfortunate accident or a pathological condition: shedding is merely the result of the premature development of the layer.

Morris (1964b) concluded that the seeds in growing bolls produce a substance that inhibits the development of the abscission layer. It is postulated that when seeds mature, or if they receive insufficient assimilate, they cease to produce the inhibitor, the abscission layer develops and shedding or dehiscence follows. Johnson & Addicott (1967) note that when the crop is flowering profusely so that many bolls are available to set, the bolls retained are those with most seeds. They suggest that such bolls, because they have more seeds, produce more of a hormone that enables them to attract more assimilate. Evenson's (1960) data are relevant to this view: bolls that developed from flowers at peak flowering when competition among bolls was most severe, had more seeds than earlier or later bolls.

In a recent review of hormonal control of abscission, Addicott (1970) describes abscission as a correlation phenomenon which involves the complex interplay of at least five hormones as well as other physiological factors, such as competition among plant parts for nutrients. Although this area of research has expanded rapidly and theories are frequently revised and become more complex, it is possible to summarize the Namulonge findings in simple terms consistent with existing theory (Addicott 1970). A hormone that promotes the development of the abscission layer is produced in the plant, probably in the vegetative growing point in quantities proportional to the activity of the growing point, and therefore varying with its carbohydrate supply. During the growth and development of a boll, the seeds produce a hormone to inhibit the action of the abscission promoter. When the growth of the boll stops, either at maturity or prematurely because nutrients are inadequate, the inhibitor is not produced allowing the abscission layer to develop, possibly accelerated by an abscission promoter produced within the boll.

Therefore, although hormones are undoubtedly involved in the control of shedding, this is not inconsistent with the view that shedding is a result of the functioning of the carbohydrate economy of the plant. Growth substances can be regarded as the means to convey information about the state of the plant's economy to sites where action is required to control or modify it: shedding is one of the most important control actions. Viewed in this way, detailed knowledge of the chemical nature or mode of action of hormones is not required in order to construct models to relate yield to growth.

Discussion of the model

Dry-matter production

The seed cotton that the farmer sells is part of the total dry matter produced by the crop. The model attempts to predict the production, distribution and accumulation of dry matter by the crop. Figure 4.6 illustrates diagramatically how the model brings together, in varying degrees of complexity, the components of the system in order to predict yield.

In each diagram the line *ABCD* is the rate of dry-matter production; *AB* is the period when leaf area is increasing and the amount of light intercepted limits growth; *BC* is the period when the canopy is intercepting 95 per cent or more of the light, and the light available limits growth. The area under *ABCD* is the total dry matter produced. The line *EX* is flower bud production, which when retarded by *EF*, an appropriate period to allow the bud to develop to anthesis, gives *FY* to indicate cumulative numbers of flowers. The line *FY* can be read

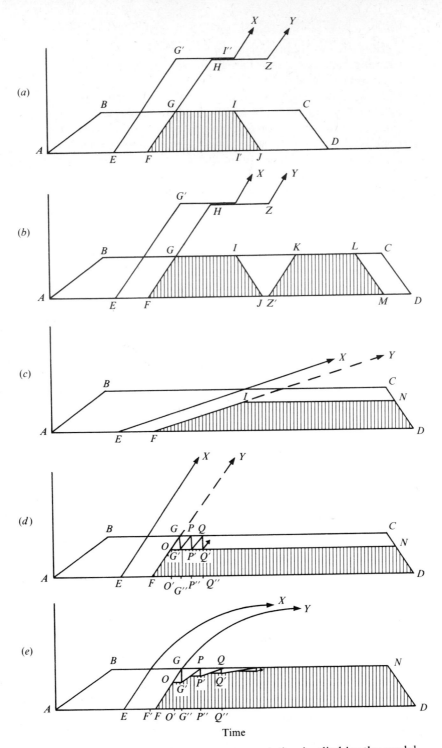

Figure 4.6. Patterns of dry matter accumulation implied by the model, with various assumptions.

either as numbers of bolls or as their combined rate of increase in dry weight (at Namulonge on the basis of 0.14 g per boll per day for the variety BPA). In each diagram the shaded areas represent the dry matter that accumulates in the bolls, and are the sums of the daily increments in dry weight of all the bolls that contribute to the harvest. Only 70 per cent of the shaded area is seed cotton, the rest is husk.

Figure 4.6*a* represents a crop with 'determinate' growth. Initially a boll sets from each flower so that the actual number of bolls follows the course of the potential numbers along *FY* until at point *G* there are enough bolls to use all the assimilate supply. Vegetative growth, and hence bud production, then stops giving the horizontal step *G'I''* in *EX*, with the result that at point *H*, thirty days after *G*, flower production stops. *FI'* is the time for the first boll to mature. At *I* the boll load starts to decrease, the crop can regrow and produce buds again at *I''* so that thirty days later the crop could start to flower again along *ZY* to begin a second fruiting cycle. Area *JICD* is the dry matter that accumulates during regrowth after the crop matures. The seed cotton yield is 70 per cent of the dry matter represented by area *FGIJ*.

Figure 4.6*b–e* is derived from 4.5*a*. Figure 4.6*b* represents the situation where temperature, radiation and water are adequate for the crop to complete a second cycle of fruiting. The area of regrowth *JICD* has become large enough to contain area *Z'KLM* which repeats *FGIJ*.

In Figure 4.6*c* flower buds are produced so slowly that the first bolls mature at *I* before there are enough to form a sink with sufficient capacity to use all the carbohydrate supply. Subsequently, bolls mature at the same rate as the crop flowers with the result that although the total number of bolls increases along *IY*, the number growing is constant, and less than a full load. In this way part of the carbohydrate supply, between *IN* and *BC*, is always available to produce new flower buds and leaves. The environment (temperature and water supply) or the farmer (by terminating the crop artificially) determines the position of *CD* in time. The yield is 70 per cent of *FIND*.

In Figure 4.6*d* the boll load increases rapidly as in Figure 4.6*a*. When a full load is reached at point *G*, not only are new bolls shed but also existing bolls up to an age equal to *O'G''* on the time scale. The number of bolls then drops to *O*. Meanwhile flowering continues along *GY* so that boll numbers increase along *G'P*. At *P* the events are repeated. In this way the number of bolls oscillates just below a full load, leaving part of the carbohydrate supply to continue vegetative growth.

Figure 4.6*e* is derived from Figure 4.6*d*. As bolls set along *FY*, progressively less carbohydrate is available to continue bud (and therefore leaf) production along *EX*; thus the slope of *EX* starts to decrease at *F'*. Therefore the slope of *G'P* is less than that of *OG*, and more time is needed for *G'P* to reach *BC*. Therefore the earliest bolls set along *GP* are too old to be vulnerable when shedding starts again at *P* and boll numbers do not fall as low as *G'*. *P'Q* is of less slope than *G'P* and the process is repeated. Thus the number of bolls retained gradually increases and approaches *BC* asymptotically.

In Figure 4.6*d,e* variation in the age at which bolls cease to be vulnerable would determine the rapidity with which a crop of bolls is set and the degree of determinacy of the crop. Such variation might arise because either genotype or environment affects the activity of the mainstem growing point and thus its production of an abscission-promoting hormone (see p. 105). Sensitivity of the abscission layer to hormones decreases with age so that the older the organ the greater the amount of hormone needed to cause abscission.

In the crop illustrated in Figure 4.6, all shedding is stress shedding in response to internal shortage of carbohydrates. In Figure 4.6*c–e* vegetative growth is continued in an indeterminate manner without non-stress shedding. In these situations the priority of the bolls is not absolute, but is limited to bolls over a certain age. The priority of young bolls is conditional upon supply; if supply is short they shed in favour of continuing leaf growth. When carbohydrates are adequate the flower line *FY* is parallel to the bud line *EX*. A new dimension is added if non-stress shedding were included in Figure 4.6; each situation would be repeated with *FY* at a lesser slope to *EX*.

Potential of the environment for cotton

Each flower bud has the potential to develop into boll. Nevertheless the total number of flower buds produced by a crop cannot be regarded as the potential of the crop (although such is often implied) because it is physiologically impossible for every flower bud to mature a boll. For example, at Namulonge a crop usually produces 200 to 300 flower buds per m^2, and sometimes many more to compensate for loss by bollworm infestation or disbudding. If 300 buds per m^2 set bolls, at 5.5 g seed cotton per boll, 16500 kg per ha of seed cotton would be harvested. Three hundred buds per m^2 would be produced in about 60 days so that for a short period all bolls would be growing simultaneously. The combined growth rate of 300 bolls per m^2 at 0.14 g per day per boll, would be 42 g per m^2 per day. The maximum rate at which the crop can produce dry matter in this environment is 17.6 g per m^2 per day, represented by *BC* in Figure 4.6. Clearly flower bud production does not indicate the potential of the crop in this environment. As Dale (1962) concluded, 'a certain amount of bud and boll shedding is the inevitable adjunct of the indeterminate habit of cotton'.

The preceding paragraph suggests that the rate of photosynthesis of the crop canopy in a particular environment provides a better basis from which to derive the potential of that environment. Growth studies

(Hearn 1972*b*) indicate that the maximum growth rate at Namulonge during September, October and November with average radiation is 17.6 g per m² per day, which is equivalent to 128 growing bolls per m². At 5.5 g seed cotton per boll, 128 bolls per m² would yield 7028 kg per ha. Figure 4.6 predicts such a yield if *BC* is set at 17.6 g per m² day, if the period of dry weight increase of a boll is 57 days and if 70 per cent of the final dry weight of a boll is seed cotton.

However, at point *G* in Figure 4.6*a* new leaves cease to appear, and existing leaves would cease to expand. The rate of photosynthesis of the canopy would then decline because LAI would diminish and the efficiency of the individual leaves would decrease as they aged. If the supply of carbohydrates falls linearly from *G* to *J* and bolls younger than 10 days shed in response to carbohydrate shortage, the potential yield of the extreme determinate type of crop in Figure 4.6 becomes 5186 kg per ha. The duration of the period *F* to *J* would be 84 days, 27 days to reach full boll load at five flowers per m² per day plus 57 days for the last boll to mature. At Namulonge for crops sown in mid-June, the water supply usually permits the period *F* to *D* to extend for 122 days from 1 September to 31 December with intermittent periods of water shortage. The determinate crop of Figure 4.6*a* does not fully use this period. To complete the second cycle in Figure 4.6*b* requires 171 days and *F* to *D* extends to mid-February. In most years the weather would be too dry to complete the second cycle, but in those years when it is possible, the potential yield would be twice that of Figure 4.6*a*, 11 200 kg per ha.

In order to estimate the potential of the indeterminate crops in Figure 4.6*c,d*, it is necessary to estimate the fraction of the carbohydrate supply needed to continue leaf and flower-bud production. An LAI of 3 will intercept 95 per cent of the light received. The lowest leaf will be at the compensation point for light and crop growth rate will be maximum. In order to maintain an LAI of 3 from *B* through to *C* in Figure 4.6*c,d*, new leaves must be produced as fast as old leaves die. Leaf life is 56 days (Farbrother 1956). Production of leaves equivalent to an LAI of 3 every 56 days, with an area to weight ratio of 150 cm² per g, and a leaf to stem ratio of 3 to 2, requires 6.26 g of dry matter per m² per day. Thus an LAI of 3 which produces 17.59 g of dry matter per m² per day, could provide 6.25 g to maintain leaf tissue and 3.86 g for root growth, leaving 11.34 g for boll growth. For growing conditions in which *F* to *J* extends from 1 September to 31 December (122 days), the potential yield of Figure 4.6*c* is 7425 kg per ha and that of Figure 4.6*d* is 8970 kg per ha.

The slope of *EX* (5 buds per m² per day) and the rate of leaf area production (equivalent to an LAI of 3 per 56 days) used to estimate the potential yield of Figure 4.6*d* at Namulonge, imply that the produc-

Table 4.1. *Potential and actual yields* (*kg per ha*)

	Water not limiting	Water limiting		
		Wettest year	Mean all years	Driest year
(a) Potential yields				
Determinate crop Fig. 4.6a	5186	5072	2858	709
Second-cycle crop Fig. 4.6b	10372	—	—	—
Indeterminate crop (slow flowering) Fig. 4.6c	7425	6608	3565	1034
Indeterminate crop (young bolls shed) Fig. 4.6d	8970	7982	4306	1250
(b) Actual yields				
Namulonge (Hearn 1968)			3120	
Serere (Arnold 1971)			3900	
Mubuku (Arnold 1971) irrigated			8966	

tion of each bud is accompanied by the production of 96 cm^2 leaf, which is a reasonable area for a leaf subtended by a flower bud. (The morphological association of leaf and bud production has been noted earlier.) The corresponding value for Figure 4.6c is 320 cm^2 which is not realistic and suggests that this crop could only develop with some non-stress shedding. A crop could only maintain a boll growth rate of 11.34 g per m^2 per day from B through to C in Figure 4.6c,d if the respiration of stem tissues remains constant in spite of the continued increase in stem tissue. (Respiring leaf tissue does not increase because old leaves abscise.) The amount of active (and therefore respiring) stem tissue needed to support an LAI of 3 is constant, and could be achieved if the increase in new stems were balanced by lignification of old stem tissue.

These potential yields have been estimated for cotton crops of a variety with a photosynthetic rate and display of leaves as defined, limited only by the mean radiation and temperature normally experienced at Namulonge in September, October and November (not by water, nutrients or damage from pests and diseases).

Because water supply is the major environmental variable affecting yield, it is important to consider the potential of the Namulonge environment with a restricted water supply. Shortage of water, determined by calculating the soil water deficit and noting days on which the critical deficit is exceeded, will affect potential yields estimated from the model in two ways: (a) no bolls will be set when water supply is inadequate, and (b) the photosynthetic rate will be halved so that there is less carbohydrate available for the growth of bolls already set. For the determinate crop of Figure 4.6a sown on 23 June, water balances calculated from twenty years' data at Namulonge show that in the 84 day period from *F* to *J* the number of days when water was inadequate ranged from 1 to 56, and averaged 22, to give corresponding yields of 5072, 709 and 2858 kg per ha respectively. For the indeterminate crop in Figure 4.6*d* the range was 2 to 83 days of water shortage with an average of 38 for a 122-day period *F* to *J*. Corresponding yields were 7982, 1250 and 4306 kg per ha respectively.

The heaviest experimental yields at Namulonge are of the order of 3000 kg per ha. Elsewhere in Uganda 3900 kg per ha were recorded at Serere and 8996 kg per ha at Mubuku (Arnold 1971). Potential and actual yields are summarized in Table 4.1 which emphasizes that a major challenge facing research workers is how to realize more of the potential crop more frequently. With average farm yields in Uganda of the order of 300 kg per ha an even greater challenge faces the extension worker.

Growing more profitable crops

The practical aim of research at Namulonge has been to increase the profit of the cotton crop to the farmer and to the nation. Where yields are so much less than the potential, the first step is to see if yields can be increased at an acceptable cost. This chapter only considers increasing yields, though it is recognized that the final evaluation depends on profitability. Figure 4.6 offers several strategies for increasing yields; this means increasing the shaded areas by manipulating the patterns of assimilate supply (*ABCD*) and sink development (*EX*) and the interaction between them (any effect the course of *EX* might have on *ABCD*).

When a crop is sown during the optimum period, its water requirements are most likely to fit the rainfall pattern, and the crop has the best chance to produce more dry matter than at earlier or later sowing, thus maximizing the area *ABCD*. The problem is how to maximize the proportion of this dry matter in the bolls.

The environmental and genotypic factors that affect the shape of *ABCD* and *EX* have been examined in detail earlier. Two things are postulated to affect the relative importance of genotype and environment: fluctuations in environment tend to affect carbohydrate supply before they affect sink capacity; selection pressure has been greater on sink capacity than on assimilate supply. These hypotheses will be examined in detail.

Changes in the weather affect supply of carbohydrates before the capacity of the sinks because there is a time lag between the initiation of a flower bud, and its bearing a boll. Thus a period of water shortage or cold weather could temporarily close the stomata and simultaneously stop initiation of flower buds; the effect on supply would be immediate but the effect on the capacity of the sink formed by the bolls would be delayed by four to six weeks.

It is difficult to detect physiologically significant differences in supply.

As Austin (1963) pointed out, to increase dry matter by 10 per cent over a season would only require 1 per cent increase in net assimilation rate. Likewise to increase leaf area would require a similarly small increase in the fraction of dry matter re-invested in new leaf. On the other hand it is relatively easier to detect physiologically significant differences in the development and capacity of the boll sink. In the early days, plants with larger bolls, or earlier flowers would catch a grower's eye, while more recently such attributes would be more easily detected by plant breeders in the conventional analysis of yield components, and the recording of plant structure and phenological data. Lint per seed, seeds per boll and position of the first fruiting branch are associated with sink capacity development and are strongly inherited. The number of bolls per plant is the only component which reflects variation in carbohydrate supply rather than in sink capacity, and is greatly influenced by the environment, both because carbohydrate supply itself is so influenced, and because of the major influence of insect pests. Thus heritability ratios of bolls per plant are usually small. In selection for yield, the attributes associated with sink capacity development, because of their large heritabilities, have been more heavily weighted than those associated with carbohydrate supply. There has been greater selection pressure therefore to change the pattern of sink capacity development than to change carbohydrate supply; in terms of Figure 4.6, in attempting to maximize the shaded area, breeding has changed the shape of *FY*, rather than *ABCD*. In physiological terms, it is postulated that breeders have been attempting to select patterns of sink capacity development to fit a pattern of carbohydrate supply imposed by the environment.

Manning (1956*b*, 1963) reported that increases in yield of the variety BP52 in successive modal bulks and in successive generations selected by his selection index were associated with an earlier crop and larger bolls. BPA which has outyielded and replaced BP52 has even larger bolls (Arnold, Costelloe & Church 1968) but is slightly later (Chapter 8). The evidence on earliness is conflicting, partly because it is difficult to define and measure (Munro 1971) and partly because it is more deeply involved in the genotype–environment interaction (Arnold 1972).

Breeding for heavier yields. Figure 4.6 well illustrates the magnitude of the breeder's task. Although sowing during the optimum period gives the greatest probability of maximizing the area *ABCD*, the weather that a crop experiences in any year or at any site will alter the size and shape of *ABCD*, whether or not the crop was sown during the optimum period. For each combination of circumstances the pattern of *ABCD* will be unique. Therefore the best strategy to accumulate dry matter in the bolls will vary from year to year, from site to site and among sowing dates within a year. The type of crop that will accumulate most dry matter in the bolls will vary among the four types illustrated in Figure 4.6*a–d*, and in the most suitable type the position of *EX* to maximize the shaded area will vary according to the shape of *ABCD* in the particular year. That the best crop type and the best position of *EX* will vary from year to year (and among dates and sites) is postulated as the physiological basis of the genotype–environment interaction that has plagued plant breeders in East Africa. A genotype selected with *EX* to maximize the shaded area for a particular pattern of *ABCD* in one year at one sowing date and site, may not be the best for other years, dates and sites. Arnold (1972) noted that the difference between yield of a modal bulk and unselected BP52 fluctuated over years and sites. He explained this observation in terms of the earliness of the modal material, i.e. *EX* shifted to the left in Figure 4.6. Years when the rains finished early favoured the modal bulk, whereas years of prolonged rains favoured the unselected material.

If physiological criteria are to be used to select for heavy yields, breeders are more likely to succeed if they continue what they have unconsciously been doing, and try to change the pattern of boll sink development, rather than improve the photosynthetic system. The question is what particular pattern of sink development to select for. If the crop physiologist is to answer that question he must refine and validate a model such as the one described. The model could then indicate what pattern of sink development is most likely, on average, to maximize the amount of dry matter that accumulates in the bolls. Plant attributes could then be chosen with high heritability that would contribute to such a pattern, which would constitute an ideotype (Donald 1968) appropriate to the locality. An ideotype would permit consistent selection for a constant physiological aim, in contrast to selecting in a succession of unique situations, each of which in turn would favour a different genotype. For this reason and because only strongly inherited characters would be specified in the ideotype, yield should advance more rapidly than by direct selection for yield.

The agronomic implications of the model. If the farmer is to benefit from the science done in his name, that science must provide a theoretical foundation for agronomic investigations. Penman's (1948) work provided the foundation for the synthesis by Hutchinson *et al.* (1958*b*); but a more comprehensive structure is needed to include other environmental and plant variables. The model presented in this chapter is offered as a contribution towards that structure. It is in respect of interactions between environment and genotype, and among the components of the environment, that the structure is particularly weak and where the model can contribute. Without such a structure, agronomic investigations are limited to field experiments with treatments chosen

Table 4.2. *Simulated photosynthesis and dry-matter accumulation*[a]
by model crops which maintain their canopies

LAI	Photo-synthesis (P)	Respira-tion[b] (R)	Increase in dry weight (C = P−R)	To maintain canopy[c]	Root growth (0.18 C)	Available for boll growth
(a) Photosynthesis not limited by water shortage						
1	13.0	1.55	11.45	2.08	2.06	7.31
1.5	18.0	2.33	15.67	3.12	2.82	9.73
2	21.7	3.10	18.60	4.16	3.35	11.09
2.5	24.4	3.88	20.52	5.20	3.69	11.63
3	26.1	4.65	21.45	6.25	3.86	11.34
3.5	26.8	5.43	21.37	7.28	3.85	10.24
4	27.0	6.20	20.80	8.32	3.74	8.74
(b) Photosynthesis limited by water shortage						
1	6.5	1.55	4.95	2.08	0.89	1.98
1.5	9.0	2.33	6.67	3.12	1.20	2.35
2	10.8	3.10	7.70	4.16	1.39	2.15
2.5	12.2	3.88	8.35	5.20	1.50	1.65
3	13.1	4.65	8.45	6.25	1.52	0.68
3.5	13.4	5.43	7.97	7.28	1.43	−0.74
4	13.5	6.20	7.30	8.32	1.31	−2.33

[a] All units, except LAI, are g per m^2 per day.
[b] 1.55 g per day per unit LAI.
[c] 2.08 g per day per unit LAI.

at the best intuitively, and otherwise arbitrarily, and which, however impeccable statistically, are no substitute for a sound theoretical foundation. The model may be regarded as an attempt to express explicitly the intuitive knowledge in the agronomist's mind.

The model incorporates the concepts concerning date of sowing formulated by Hutchinson *et al.* (1958*b*) and Rijks (1967*a*). Sowing during the optimum period gives the crop the best chance of maximizing the area *ABCD* in Figure 4.6. Hutchinson's unidentified negative date of sowing effect may be caused by non-stress shedding, which lessens the slope of *FY* relative to *EX*.

When the environment–genotype interaction was discussed earlier, it was noted that fluctuations in the shape of *ABCD* caused by the environment made the breeder's task more difficult. Any agronomic practices that dampen the fluctuations would assist the breeder. Crops less sensitive to water shortage would help; measures to achieve this have been discussed and are being implemented. Another possibility would be to select as optimum the sowing date that gives least variation in the shape of *ABCD*, rather than the one that maximizes its area.

It is of interest to repeat for a range of LAI the calculations made earlier (p. 109) to estimate the carbohydrate required to replace the canopy and that left over for boll growth. The results (Table 4.2) show that the canopy that can maintain itself most productively (i.e. has most carbohydrate left for boll growth) has an LAI of a little less than 3. In Table 4.2 the calculations have been repeated with the photosynthetic rate arbitrarily halved to simulate the effect of a degree of water shortage on stomatal resistance. The most productive canopy now has an LAI of 1.6. These results suggest that an indeterminate crop having a canopy with an LAI of 3 will intercept enough light to maximize the growth rate, and will also replace itself efficiently when growing conditions are favourable (i.e. adequate water and radiation). If, however, the potential for growth diminishes, because of drought or cloud, for example, most efficient replacement of the canopy will result with a smaller LAI. Measurements by Farbrother (1952, and in subsequent *Progress Reports*) of LAI of widely spaced crops grown without fertilizer rarely exceeded 1. Use of fertilizer and closer spacing has seldom increased LAI beyond 2. The impossibility of consistently obtaining an LAI of 3 suggests that there is a mechanism that operates under adverse conditions to ensure a canopy with a small LAI which is more productive than a larger one under such conditions. Nevertheless increasing LAI to 2 from less than 1, by use of fertilizers and closer spacing, has contributed to the recent large increase in the yields at Namulonge (Chapter 9).

Although Huxley's (1964) conclusion that the cotton crop in Uganda does not use all the light available to it is true, and although there is some scope to increase interception in order to raise yields, the crop at Namulonge appears to be programmed to produce and maintain a canopy that will intercept less than the 95 per cent of available light conventionally accepted as necessary for maximum growth rates. In terms of Figure 4.6, *BC* is kept below its maximum level in order that the shaded area can be greater. Thus the advantage of a small LAI (Farbrother 1960*b*) may lie as much in its ability to replace itself productively, as in its reduced crop water requirements serving to conserve a meagre supply.

The estimate of the potential yield of an indeterminate crop (Figure 4.6*d*) with a restricted water supply (see p. 111) was made for a canopy with LAI of 3 which is not the most productive size for such a situation (Table 4.2). If water was short every day, an LAI of 1.6 would be most productive. For a particular situation, the most productive size of canopy would depend on the proportion of days with a restricted supply and could be assessed from meteorological probability data for any site.

Work elsewhere (e.g. Ashley, Doss & Bennett 1965) in which yield was statistically correlated with LAI might be interpreted to indicate that LAI of greater than 3 is desirable. However, this relationship was not analysed physiologically and was unlikely to be a simple cause and effect. Alternative explanations, supported by correlations published in the same paper, could be (*a*) that an LAI of 5 is necessary in midseason to ensure an LAI of 3 at the end of the season or (*b*) that the cause and effect is between sink size and yield, and that sink size depends on number of flower buds which is inevitably correlated with number of leaves, and hence LAI.

In experiments at Namulonge (Hearn 1972*b*) the closest spacing gave the largest LAI but not the heaviest yield because the peak LAI was out of phase with the increase in capacity of the boll sink. In terms of Figure 4.6, *BC* started to decrease while *FY* was increasing, because, as already postulated, new leaves appear more slowly owing to competition from growing fruit, and because the older leaves are senescing. A gap in our knowledge concerns the factors controlling senescence of leaves, particularly the role of nitrogen.

When plant population density varies, the patterns of both sink development and assimilate supply change. With denser populations both increase faster or, in terms of Figure 4.6, the slopes of *AB* and *EX* become steeper. A result of this feature is that over a wide range of densities, yield is constant (Hearn 1972*a*). As plant population becomes denser, the proportion of dry matter in the bolls decreases so that ultimately yield declines at very dense populations. When the population is too sparse the assimilate supply does not reach its maximum level because the canopy does not intercept all the light.

It has long been thought that late-sown cotton should benefit from denser plant populations. The experimental evidence is conflicting (Farbrother & Munro 1970; Bowden & Thomas 1970). Arguments similar to those applied to the genotype–environment interaction also apply to this interaction. Population density affects the shape of both *ABCD* and *EX* in Figure 4.6. The basic shape of *ABCD* set by the population density is greatly modified by the environment and will be unique for each year and site. Therefore, the optimum position of *EX*, and hence the optimum density, will vary from year to year and from site to site.

The model, as presented so far, does not tell us how population density and date interact to affect yield, only that they will interact. When the model has been developed to simulate the growth of crops, it can be used to predict the interaction of population with sowing date for different sites and years. If the experimental results available verify the model, the best population can be predicted from the model when averaged over all years and sites for which there is meteorological data.

The nitrogen economy of the crop has been excluded from the model owing to lack of local data but constitutes a major gap in our knowledge.

PLATE 1

(*a*) Growth of crops on topsoil and subsoil. Maize and beans growing on
normal soil (on the right of the picture) and on a plot with the topsoil removed
(to the left of the picture). Crops on the right and left were sown at the same
time from the same seed lots and neither received any manure or fertilizer.
(*b*) A fertilizer experiment. The plots are clearly demarcated by the size and
colour of the maize plants.

PLATE 2

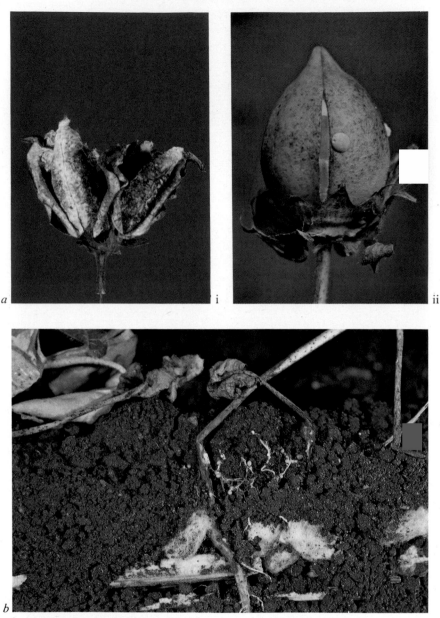

(*a*) Boll rotting. (i) Teleuto spores of a species of *Alternaria* appear as blackened areas on the rotting lint. (ii) Fungal spores of a species of *Molilinia* show as bright yellow and orange spots on this boll, which was completely rotten inside. (*b*) Damping-off caused by *Sclerotium rolfsii*. Infection can clearly be seen in this photograph of seedlings growing against glass.

PLATE 3

Resistance to bacterial blight. (*a*) Inoculating the hypocotyls of cotton seedlings with a suspension of *Xanthomonas malvacearum*. (*b*) Inoculated seedlings in the glasshouse. (*c*) Reactions to hypocotyl inoculation, showing a somewhat resistant seedling (right) in a generally susceptible line.

PLATE 4

(*a*) A cotton flower and green boll. (*b*) A ripe boll, ready for picking.

PLATE 5

Tie ridging (*a*) on the hill-sand soils of western Tanzania; (*b*) on the red ferrallitic
soils of Uganda (note the muddy water).

PLATE 6

Soil conservation measures. (*a*) Graded contour strips. (*b*) Bench terraces. (*c*) Part of the original farm layout showing strips 9 and 10 (see p. 253) in the middle of the photograph.

PLATE 7

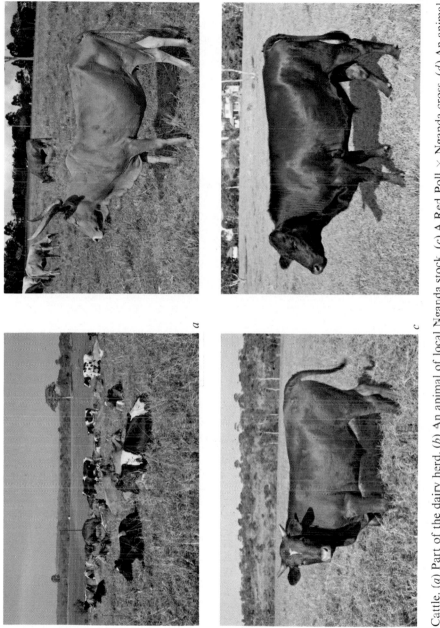

Cattle. (a) Part of the dairy herd. (b) An animal of local Nganda stock. (c) A Red Poll × Nganda cross. (d) An animal from a first backcross to Aberdeen Angus stock.

PLATE 8

Small holdings near Namulonge from the air.

Towards a working model

Use of existing data. The preceding discussion shows that the next step is to put the model to work to simulate growth and predict the yield of a crop. In its simplest form a model must meet the following specifications:

(*a*) initiate leaves, branches and flower buds;

(*b*) increase the dry weight of these organs;

(*c*) abscise leaves, flower buds and vegetative buds;

(*d*) develop flower buds
  initiate anthesis,
  increase dry weight of the boll,
  abscise, mummify or dehisce the boll;

(*e*) take account of differences among varieties in boll growth rate and fruiting structure development;

(*f*) take account of water supply to the crop;

(*g*) take account of temperature and solar radiation;

(*h*) take account of competition for water and light;

(*i*) have adequate internal controls to allocate carbohydrates to competing sinks and to senesce the crop.

The model in Figure 4.3 has been converted to a daily sequence of operations set out in Figure 4.7.

Temperature will control the rate of many processes in the model. Boll growth rate and the rate of respiration will be functions of temperature. Certain processes, for example the initiation of a node and the maturation of a boll, will be completed when the requisite number of day degrees or heat units have been accumulated. Thus the number of mainstem nodes will increase as a function of day degrees. Flower buds will increase in number as a function of nodes and plant population.

Simulation of the carbohydrate economy will involve computing photosynthesis and respiration. Photosynthesis will be a function of leaf area and radiation using the de Wit model. Respiration will be a function of temperature and crop dry weight. The partition of the carbohydrate available for dry weight increase will depend on satisfying the bolls' needs first, and dividing the balance emprically between leaf and non-leaf tissues. A day's production of leaf tissue will be subsequently abscised as a function of time.

The model will compute plant water status which will be able to switch off several processes when it drops to certain values. The photosynthetic rate will drop to half when minimum daily relative water content falls to 82 per cent, and will stop when it reaches 72 per cent. The production of nodes, and hence flower buds will stop when relative water content at any time during the day falls to 82 per cent.

(a)

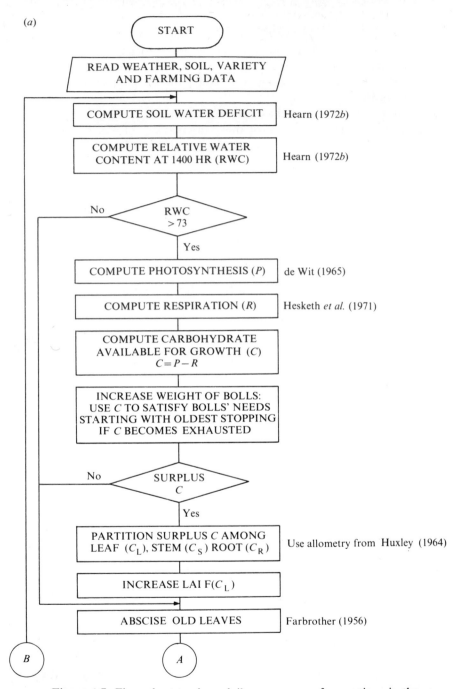

Figure 4.7. Flow chart to show daily sequences of operations in the
proposed working model.

(b)

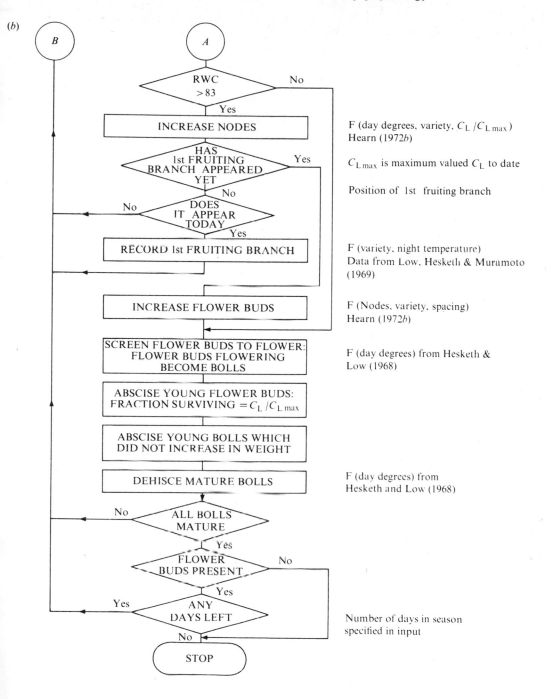

B

A

RWC > 83 — No

Yes

INCREASE NODES

F (day degrees, variety, $C_L / C_{L\,max}$)
Hearn (1972b)

$C_{L\,max}$ is maximum valued C_L to date

HAS 1st FRUITING BRANCH APPEARED YET — Yes

No

Position of 1st fruiting branch

DOES IT APPEAR TODAY — No

Yes

RECORD 1st FRUITING BRANCH

F (variety, night temperature)
Data from Low, Hesketh & Muramoto (1969)

INCREASE FLOWER BUDS

F (Nodes, variety, spacing)
Hearn (1972b)

SCREEN FLOWER BUDS TO FLOWER: FLOWER BUDS FLOWERING BECOME BOLLS

F (day degrees) from Hesketh & Low (1968)

ABSCISE YOUNG FLOWER BUDS: FRACTION SURVIVING $= C_L / C_{L\,max}$

ABSCISE YOUNG BOLLS WHICH DID NOT INCREASE IN WEIGHT

DEHISCE MATURE BOLLS

F (day degrees) from Hesketh and Low (1968)

ALL BOLLS MATURE — No

Yes

FLOWER BUDS PRESENT — No

Yes

ANY DAYS LEFT — Yes

No

Number of days in season specified in input

STOP

5-2

The crop will terminate when all the bolls are mature and there are no more flower buds. The crop will also terminate if time runs out, i.e. the model has run through the days for which meteorological data are supplied. The second situation represents the real case in an indeterminate crop such as cotton where it is still growing actively at the end of the season, either because the crop has set insufficient bolls to check vegetative growth or because it has regrown after the bolls mature.

The model has been developed independently of those of Stapledon (1970) and Baker (McKinion *et al.* 1975). All three employ the key concepts first used in the simple model for the distribution of carbohydrates developed at Namulonge (Hearn 1969); the models use crop development data to generate a sink formed of bolls and then estimate its capacity before allocating carbohydrate to it. The model in this chapter differs from the other two because (*a*) it has no pool of assimilates (current production is allocated immediately to increase the dry weight of plant parts); and (*b*) it is sensitive to changes in variety, plant population density and water supply.

There are enough data available for the proposed model to simulate the growth of a crop. Agreement between predicted and actual crop performance would not prove that the model correctly represents the real system. However, if the model can be verified by agreement in several contrasting sets of conditions, we could tentatively accept the model as valid and predict the performance of combinations of varietal attributes and cultural treatments in any season and at any site for which there is meteorological data. Combinations can be included that would take many years to test experimentally at great expense. Breeders and agronomists can then limit their selections and experiments in the field to those attributes and treatments that the model suggests are promising.

Such a policy would not, therefore, obviate the need for agronomic experiments, but it could, by providing a rigorously developed and comprehensive hypothesis, put agronomic experiments on a sounder theoretical basis. The model could assist with the economic appraisal of research proposals on a cost–benefit basis; by providing a means to integrate the expected research results into the total crop production system, the model could help to show whether the research, if successful, is likely to pay for itself in terms of increased national income.

Future developments. The model assumes that other factors are either not limiting, like pests and diseases, or limiting at a constant level as implied in some of the empirical constants. As suitable data become available, pest and disease damage can be included by decreasing the number of buds and bolls, and the amount of leaf tissue by appropriate amounts.

Likewise concepts to extend the model to include nitrogen have been developed, using data from elsewhere (e.g. Basinski *et al.* 1971), and are outlined below. It may be possible to use the data from investigations into the nitrogen economy of the crop at Namulonge (Chapter 3). The complete model would integrate the carbon, water and nitrogen economies of the crop. For each, the supply and demand are calculated and compared. Four categories of sinks – leaf, stem, fruit, and root – are recognized. It is postulated that the sink nearest to the source of the least well-supplied commodity will take priority for carbon and nitrogen. For example with inadequate nitrogen or water, the root would take priority. The model would maintain a nitrogen pool in the plant, to which it adds by taking up nitrogen as a function of soil supply and recent root growth. The pool is depleted by nitrogen used for growth. At the beginning of each day a fraction of the leaf nitrogen is remobilized into the pool. At the end of the day the remobilized nitrogen is restored to the leaves starting with the youngest, until either the pool is exhausted or the leaves are inadequately illuminated (below the compensation point). These concepts allow the more complete model to account for some of the features not covered by the simpler model. For example, the leaf sink can take priority over the fruit with a liberal water and nitrogen supply, allowing what has been termed non-stress shedding to occur. Also, leaves can decrease in nitrogen content as they age to simulate senescence and would abscise when their nitrogen content falls to a critical value. The model would allow a greater soil nitrogen supply to result in the production of larger leaves and bolls.

Gaps in our knowledge. A feature of a comprehensive model of the type described is that in the attempt to construct it, attention is drawn to gaps in our knowledge. Several such instances have already been mentioned, and may be summarized:

(*a*) factors that affect the distribution of assimilates among leaf, stem and root;

(*b*) factors that prevent the boll taking priority for assimilates, leading to non-stress shedding;

(*c*) nitrogen economy of the crop;

(*d*) senescence of leaves;

(*e*) the contribution to boll dry weight by the boll wall through photosynthesis under various field conditions.

Several of these topics are interrelated. Data are required in the form of functional relationships, which can then be incorporated into the model.

Conclusions and summary

The aim of this chapter has been to put Namulonge's contributions to crop physiology into perspective. A model has been presented in an attempt to unify diverse work and to transcend the narrow confines of academic disciplines.

The construction of the model outlined in Figure 4.5 has given a conceptual framework within which the physiological work could be reviewed. The two major lines of research at Namulonge into the physiology of growth were photosynthesis and the accumulation and distribution of dry matter in crop plants, and morphogenesis and the development of crop fruiting structure. In addition, some work was done on plant hormones. At times these lines have tended to diverge and form independent fields of study. This review has attempted to integrate the results attained. The three lines concern respectively sources of carbohydrates for growth, sinks for carbohydrates, and control of the relationships between sources. Clearly all contribute to a better understanding of crop performance and its variation. The model has shown how some environmental and genetically controlled factors could affect growth and thence yield. A recurring theme has been the dominant role of water. Not only is water dominant, but variation in its supply confers a uniqueness to each season. By moulding the pattern of carbohydrate supply and thus requiring a unique pattern of sink development in order to maximize yield, variation in water supply causes the near-insoluble problems imposed by interactions of environments with genotypes and agronomic treatments. The review suggests that the most important event in the development of a crop is not one described by classical phenology but occurs when the capacity of the sink formed by the bolls becomes equal to the carbohydrate supply. Amongst the gaps revealed in our knowledge are the internal nitrogen economy of the crop and non-stress shedding, which suggest important questions for future research. As these questions are answered we shall be nearer to the interrelated goals of crop science and agronomy. The former seeks to predict or explain yields; the latter to adapt crops to suit the environment or to modify the environment to suit crops.

5

Entomology

W. REED

Very few original plans for research projects survive to fruition, unaltered by accrued experience and changing circumstances. In a review of any long-term venture, the benefit of hindsight will provide a critic with ample evidence of errors and ill-judged diversions. The criteria for judging any project should be whether the successes outnumber the failures and the benefits outweigh the cost. It is too much to expect of any twenty-year programme, conducted by a series of entomologists, that a critic could truthfully declare that all had gone well and that the course could not have been re-plotted with advantage. It is with these points in mind that this review of the work of the Entomology Section at Namulonge has been written and should be read.

Formation of the section and its policy

The Entomology Section at Namulonge was founded when E. O. Pearson arrived in Uganda in 1949. His first recorded task was to formulate the policy for the Section which was recorded in the 1949–50 *Progress Report*. This is so pertinent to the present review that it is well worth reproducing here:

Much progress towards minimizing pest damage has been made by methods which depend on detailed knowledge of the ecology of the pests and their wild and cultivated hosts, and on the place of the latter in the farming system. This knowledge is fundamental to a proper attack on pests, from whatever angle. But control by evasion is distinctly limited by the place of cotton in African peasant farming, especially in the rain-fed areas, and we must look to other methods. The first of these is the search for resistance, already very successful in certain cases. The second is direct action by insecticides. Hitherto, this has been impracticable, but with the introduction of new and more potent insecticides, and with the expected development in patterns and methods of African farming, the outlook is more hopeful, and research on the technique of direct control is today an urgent need. . . .

African cotton pests fall roughly into two groups, those which take the crop itself after the main plant structure has been established, and those which attack the vegetative parts of the plant before the main framework is established and reduce fruitfulness by destroying or distorting the structure. The first group includes bollworms and stainers, which are of most importance in the savannah regions of Africa with a single rainy season. Species of *Lygus* and *Helopeltis* fall into the second group, and are pre-eminently pests of cotton in the moist equatorial zone. Jassids, though belonging to the second group, are a limiting factor in the savannah areas.

Since Namulonge is situated in the moist equatorial zone, the second group, and *Lygus* in particular, must obviously have prior attention. The study of *Lygus* is complicated, however, by the effects of bacterial blight and climatic factors. *Lygus*, bacterial blight and the Uganda climate all tend to modify the structure of the cotton plant in the same way. Hence the first problem facing the entomological, plant pathological, and crop husbandry sections is the disentangling of the effects of *Lygus*, bacterial blight and climate. Studies of the bionomics of *Lygus*, will run concurrently with the assessment of its role in determining crop behaviour. The data provided by these investigations will provide an estimate of its present importance as a pest, its probable future behaviour under changing agricultural conditions, and the prospects of control either directly or by the development of resistant varieties.

Work on wider issues will be of two kinds, studies of widespread problems which will enable guiding principles of general application to be laid down, and basic entomological investigations that can best be tackled by a central organization. In the immediate future, insecticidal work in co-operation with the Uganda Department of Agriculture, the East African Agriculture and Forestry Research Organization, and the Colonial Insecticides Committee will take precedence over basic entomological work, and it is hoped that much time will be saved in other territories by the establishment of guiding principles in the use of insecticides on rain-fed peasant crops of tropical Africa.

An important local problem for the entomological and pathological sections is the protection of experimental plots from disturbance by pest and disease attack.

Co-ordination of programmes in different territories, is of particular importance for the success of the Corporation's entomological work. The Namulonge programme will have its applications in all African cotton growing territories, British or non-British. On the other hand, it will be incomplete in itself, and will be complemented by entomological investigations carried on as far afield as Nigeria, Nyasaland and eastern Tanganyika.

An important project connected with the co-ordination of entomological work is the preparation of a Handbook on Cotton Pests. Considerable progress has been made and it is hoped that the work will be completed as soon as the staff position becomes easier.

By 1972, it should be easy to fault a policy formulated in 1949, for the rapid development of entomology and methods of pest control over the past twenty years should tend to make statements of research policy made in 1949 at least dated and perhaps even rather ridiculous. It is a fitting tribute to Pearson's foresight and judgement that in 1972 it would be difficult to formulate a better blueprint for future research, or even to change the priorities. It may also be an indictment of the past twenty years that the research needs of 1949 have still to be fulfilled in 1972. Perhaps Pearson should have recorded his policy, not in a progress report, but in large indelible print on the wall of the entomologist's office! Nevertheless the fact that the needs for research today are largely those of more than twenty years ago does not mean that no progress has been made at Namulonge. Much of value has been achieved, but a great deal is left to be done.

In this review it is convenient first to consider what Pearson described as 'direct action by insecticides', because the trials and tribulations of insecticide use at Namulonge have tended to govern the policy of the entomology section and to some extent the policy of other research at Namulonge.

Insecticide use

DDT at Namulonge

The entomology section at Namulonge was founded opportunely, for DDT was becoming widely available and was to become the most valuable weapon against cotton pests throughout the world during the next decade. DDT had been used so successfully in public health work during and after the war that large-scale production had lowered costs sufficiently for the pesticide to be of economic use against plant pests, even on field crops of relatively low value, such as cotton.

Early trials

DDT was first used at Namulonge in a trial by Pearson in 1950–51, in the first full season after the official opening; a 25 per cent increase in yield was recorded from its use, and an intensive programme in insecticide research was underway.

After the 1950 trials, when DDT was sprayed weekly for the control of lygus, Pearson was joined by two more entomologists, bringing the section up to its full intended professional complement. Q. A. Geering was an entomologist recruited by the Corporation and K. S. McKinlay was seconded from the crop insecticides unit of the East African Agriculture and Forestry Research Organization with the intention of developing the insecticide research programme at Namulonge. In the 1951–52 insecticide trials it was noted that:

Each application of the insecticides resulted in an immediate reduction of the *Lygus* population as a whole. The population quickly rose again however, the extent of the recovery being a measure of the immigration rate . . . A striking feature of the trials was that following sprayings, *Lygus* populations fell to almost the same extent on the untreated controls as on the treated plots. The plot sizes used were of the order of 1/40th acre and it is evident that the normal ambit of *Lygus* adults is sufficiently large for the population to adjust itself with great rapidity over all the plots in the field, the population on the untreated plots being drained off by dispersal to the treated plots, from which there is no compensatory counter-movement, the population having been killed. Thus, although the insecticides killed many *Lygus* the combined effects of the rapid distribution of survivers, and the sustained immigration of fresh adults impartially to all plots, resulted in only small differences of average populations between treated and untreated plots. As might be expected, these differences were larger the larger the plot size and resulted in contrasts in plant appearance and crop formation which were detectable by eye in plots of 1/17th acre and well marked in plots of 1/5th acre.

Thus at a very early stage at Namulonge the limitations of small plot trials, when dealing with a mobile pest such as lygus, were fully appreciated. A similar effect was recognized in the Sudan by Joyce (1956) who referred to it as an 'interplot effect' and later showed that very large plots of up to three hectares had to be separated by more than 150 metres, before the effect could be completely overcome (Joyce & Roberts, 1959). At Namulonge, however, McKinlay (1953) concluded that the use of large plots to overcome this interplot effect would seriously reduce the precision of an experimental comparison and would be too costly in time, materials and labour.

Single-row spraying technique

McKinlay noticed that where small units of cotton (two rows of twenty plants each scattered at random throughout a largely unsprayed field) were sprayed to 'run-off' level each week with 0.1 per cent DDT, they were quite free from signs of lygus damage, while untreated plants, even in adjacent rows showed signs of very heavy damage. He concluded that this protection must have been caused by a repellent effect of the insecticide, since DDT took up to twenty-four hours to kill a lygus adult. From these observations he developed his single-row spraying technique. This technique was described by McKinlay (1954) and consisted of spraying single rows of cotton to run-off weekly; the single-row plots being separated by seven or more rows of unsprayed cotton. It was postulated that the insecticide repelled the lygus adults and killed the nymphs and bollworms on the sprayed rows but had little effect on the overall population of lygus in the field. It was therefore considered that sprayed and unsprayed cotton could be compared without the complicating 'interplot effect'.

Both McKinlay and Geering left Namulonge in 1954, but the insecticide investigations were continued by T. H. Coaker, a Corporation entomologist who had joined the section in 1953. Evidently the repellent action of DDT against lygus was unsatisfactory, for the best result that had been obtained by 1954 was in a trial where the best treatment reduced the index of damage to leaves to 36 per cent of that on the controls. In 1954 Coaker added toxaphene to the DDT sprays to improve the repellence and succeeded in reducing the damage-index on the sprayed to 18 per cent of that on the unsprayed cotton.

The single-row spraying technique was continued by Coaker until 1958–59, his last season, when he returned to randomized block trials with ¼-acre plots (0.1 ha), within which one-tenth acre plots were sampled for yield. With Coaker's departure, insecticide trials at Namulonge were discontinued.

Early trials

Up to 1959, in seven years' trials, the sprayed cotton had outyielded the unsprayed by 11 per cent on average, but the differences were not consistent and in two years the sprayed yielded less than the unsprayed. In reviewing this series of trials, McKinlay & Geering (1957) and Coaker (1957a) noted that in most seasons the sprayed cotton produced a heavier early crop than the unsprayed but the unsprayed recovered or compensated for early loss and produced a heavier later crop. As a net result, in most years there was little difference between the yields from sprayed and unsprayed: only in years where late bollworms were of major importance, or when the rains ended early, could spraying be expected to be worthwhile. Since neither of these events could be predicted, the spraying of cotton at Namulonge could not be relied upon to give economic yield returns.

Second series of insecticide trials

By 1965, yields on the farm at Namulonge had dropped, and the eight-year average was 501 kg seed cotton per ha. This level of yields was embarrassingly low for the Central Research Station and, not unnaturally, the wisdom of continuing an extensive and intensive research programme on cotton, in an area where cotton could not be grown to a reasonable yield level, was questioned. The role of insect pests in depressing the farm yields was again queried and entomologists from the Uganda Department of Agriculture were invited to reinvestigate the potential use of insecticides on Namulonge cotton. This decision was taken because a new generation of more toxic insecticides was by then available, and progress had been made in improving plant growth by fertilizer use.

This second series of trials was carried out by Ingram in 1965 and 1966, McNutt in 1967, Nyiira in 1968 and continued by Reed from 1969 to 1972. These trials all used randomized block designs, including unsprayed controls, with plot sizes of 0.1 ha. The response to spraying in this seven year series was almost identical with that obtained in the first series, a mean yield increase of 11 per cent, but with two years within which the sprayed cotton yielded less than unsprayed cotton. Ingram (1969), when reporting the results from the 1965 and 1966 trials

Figure 5.1. The Ransomes Cropguard Super 105 sprayer mounted on a high-clearance Massey Ferguson 165 tractor. This sprayer was used with great success at Namulonge.

and reviewing the earlier work, concluded that routine spraying at Namulonge gave no benefit and in some seasons could be harmful. This conclusion would appear to be very reasonable when the results from the nine years' trials then available were considered.

Increased yields from insecticide use

By 1966, however, Arnold (1967) decided that a determined effort to increase the Namulonge farm cotton yields was needed and he consequently increased the level of inputs. Contrary to the evidence obtained from spraying trials, and against the advice of the entomologists, Arnold decided to include insecticides as a routine input on the farm cotton. The results from the use of better husbandry, improved varieties and regular use of fertilizers and insecticides were spectacular. In the six years from 1966 to 1972 the farm cotton averaged well over 1400 kg per ha, compared with the previous six-year average of less than 500 kg seed cotton per ha.

The contribution that the use of insecticides made to this startling yield increase cannot be directly measured, for much of the increase was undoubtedly associated with the much better level of farming. In four of the six years, however, a block of cotton was left unsprayed, although treated in the same way as the rest of the farm cotton. The yield differences between this unsprayed block and the sprayed farm

Table 5.1. *Yields of seed cotton in kg per ha at Namulonge for the seasons*
1950–51 *to* 1971–72

Season	Farm bulks			Spraying trials (small plots)		
	Unsprayed	Sprayed	Difference	Unsprayed	Sprayed	Difference
1950–51	648	.	.	.	.	.
1951–52	1097	.	.	878	939	+61
1952–53	770	.	.	.	.	.
1953–54	767	.	.	920	1048	+128
1954–55	405	.	.	191	237	+46
1955–56	1083	.	.	448	525	+77
1956–57	1133	.	.	1199	1531	+332
1957–58	713	.	.	1340	1336	−4
1958–59	460	.	.	847	842	−5
1959–60	632	.	.	.	.	.
1960–61	788	.	.	.	.	.
1961–62	232	.	.	.	.	.
1962–63	536	.	.	.	.	.
1963–64	365	.	.	.	.	.
1964–65	390	.	.	.	.	.
1965–66	604	.	.	962	934	28
1966–67	.	1266[a]	.	1462	1305	−157
1967–68	355	1164	+809	571	969	+398
1968–69	1144	1633	+489	1641	1879	+238
1969–70	.	1295	.	1322	1445	+123
1970–71	1119	1637	+518	1618	1816	+198
1971–72	931	1940	+1009	1954	2254	+300

[a] From the 1966–67 season the cotton received higher levels of inputs in addition
to spraying (see Chapter 9).

cotton were much greater than the differences between sprayed and
unsprayed plots in the spraying trials, in those years. All of these yield
comparisons are summarized in Table 5.1.

Valid comparison of sprayed and unsprayed cotton

There now appears to be little doubt that the use of insecticides on
cotton at Namulonge is well worthwhile. The results from the two series
of spraying trials, however, conflicted with this conclusion and the
inference is that the validity of yields from unsprayed control plots (at
Namulonge and elsewhere) is open to question (Reed 1972). The
comparison of yields from sprayed and unsprayed cotton must be taken
from plots that are large enough and sufficiently isolated to be unaffected
by each other. In addition, the economic value of spraying can only
be estimated if both the sprayed and unsprayed cotton are each grown
and treated at their level of optimum economic return; this may well
involve the comparison of closely sown, heavily fertilized, sprayed
cotton with widely sown, unfertilized, unsprayed cotton, sown at

different dates. Such comparisons may also involve the use of different cotton varieties.

With the benefit of hindsight, it is tempting to argue that the delay between first testing of DDT in 1950 and its first widescale use at Namulonge with such obvious benefit in 1966, should not have occurred. The critical decision was obviously that made in 1953 when McKinlay decided to use his single-row technique. It is probable that many entomologists, when following up the results and observations recorded from the 1951 trials (Pearson, Geering & McKinlay 1952), would have planned future trials differently, and would have compared a large sprayed area with a large unsprayed area. Such an approach would have forsaken that of formal statistics, but the difference between the yield of sprayed and unsprayed cotton at Namulonge is so large that formal statistical comparisons are unnecessary. In this case, the quest for a statistical measure of the effects of spraying appeared to hold back the development of crop protection at Namulonge for over ten years.

Effect of insect control on field experiments

From the time that crop protection with insecticides became routine on cotton at Namulonge, research on other aspects of cotton growing was affected. Increases in yield from the use of fertilizers, for example, were obtained from sprayed fertilizer trials, but not from unsprayed trials. This observation shows how crop management could well have been a factor in the lack of response to fertilizers recorded in earlier, unsprayed trials (see Chapter 3). When studying the effects upon the cotton plant of various treatments the work of the physiologist and agronomist is simplified if the complicating effects of pests can be eliminated by insecticide use. The multiplication of seed from the cotton breeders' selections is more rapid and reliable when losses caused by insect pests are minimized. It is perhaps appropriate to mention at this point, however, the danger of using insecticides when evaluating new selections. The use of insecticides on breeding material during the selection and testing stages could result in the selection of cottons that do well under sprayed conditions but badly under unsprayed conditions. Although the well-known work on jassid resistance appears to be the only real success story in selecting pest resistant cottons in Africa, it is likely that the present commercial varieties in Uganda have a reduced susceptibility to many pests; this becomes obvious when exotic cottons are grown next to BPA and SATU under unsprayed conditions at Namulonge. By selecting and testing under sprayed conditions, cottons may well be selected that are more susceptible to pests than our present commercial varieties. Selection under entirely sprayed conditions could therefore be disastrous in a country such as Uganda where

c. 2 mm

Figure 5.2. *Earias biplaga*, the spiny bollworm of cotton, which was adequately controlled by insecticides at Namulonge.

most of the cotton is still unsprayed. This danger was well appreciated, however, and all lines that appeared to be suitable for inclusion in a commercial seed issue were extensively tested in trials over the full geographical and ecological range in which the cotton was to be grown; these trials were at varying levels of management and several were not effectively sprayed. Additionally, testing of new, promising strains was latterly replicated under sprayed and unsprayed conditions.

Pests not controlled by DDT

DDT use at Namulonge was mainly directed against lygus and the American and spiny bollworms, but other pests had to be controlled at times. In the 1956 season, and in some subsequent seasons, severe attacks of red spider mite (*Tetranychus telarius* complex) were observed on sprayed cotton at Namulonge. Similar outbreaks of mites have been recorded on cotton elsewhere and it is generally accepted that such outbreaks result from the insecticide killing the predators of the mites, but not the mites themselves, thus leaving the mites free to multiply without check from their natural enemies. The use of insecticides to control the outbreaks was studied at Namulonge by Coaker (1957a) and elsewhere by Ingram (1962). In recent years the outbreaks of mites on cotton at Namulonge have been adequately controlled by spot spraying with dimethoate, and these mites do not appear to present such a serious threat to sprayed cotton as was once feared, providing the initial outbreaks are dealt with in time.

c. 2 mm

Figure 5.3. The false codling moth (*Cryptophlebia leucotreta* Meyrick) which is only 10 mm long, photographed in the laboratory. This pest was not adequately controlled by insecticides.

Stainers (*Dysdercus* spp.) are occasional pests at Namulonge and major pests in some areas, particularly in the drier areas of western Uganda. Carbaryl has been shown to give very good control of these pests and yield increases in irrigated trials at Mubuku, where this insecticide was used, were spectacular. Cotton sprayed with DDT gave just over 2000 kg seed cotton per ha, but more intensive spraying, including carbaryl, produced more than 4000 kg, with one small subplot yielding just under 9000 kg per ha (Reed 1971). This trial also showed the interdependence of insecticide and fertilizer use, for in previous trials using a limited spraying regime, there was no worthwhile return from fertilizer use. It therefore became apparent that responses to fertilizer were well marked when the major pests were controlled, but small or non-existent where no pest control was afforded.

The only pest at Namulonge that was not controlled satisfactorily with insecticides was the false codling moth (*Cryptophlebia leucotreta*). In a series of trials using a wide range of insecticides, the best control recorded was a 50 per cent reduction of damage using such toxic insecticides as monocrotophos and parathion. This degree of control is inadequate and the use of such toxic and expensive insecticides was impracticable on Uganda cotton. Although the search for an insecticide to control the false codling moth continued, it seemed likely that control measures that do not involve insecticides would prove more useful.

Insecticide use on cotton in Africa

It is evident that Pearson's hope that work at Namulonge would save time and trouble 'in other territories by the establishment of guiding principles in the use of insecticides on rain-fed peasant crops in tropical Africa' was not realized. The interpretation of results from the insecticide trials resulted in an impasse, which was not broken until the successful use of insecticide on cotton in other territories and throughout most of Uganda was already well established. The trend-setters in insecticide use on raingrown cotton in Africa have undoubtedly been J. P. Tunstall and G. A. Matthews, working in southern Africa with such success (Matthews & Tunstall 1968).

Compensation

Although the early work on insecticides at Namulonge appeared to be largely abortive, a great deal of useful knowledge was obtained in the course of the studies designed to investigate the apparent failure of insecticide use. Much attention was paid to the assessment of crop loss and the clear exposition of compensation for early loss of crop (McKinlay & Geering 1957) is of great value. The failure of insecticides to increase crop yields also diverted the attention of the entomologists to control methods that did not involve insecticides and this led to studies of the ecology of the pests, and their relationship with, and effects on, cotton.

Future of insecticides

Having at last accepted DDT as the main weapon against cotton pests at Namulonge, the major aim of further work was to find an acceptable replacement. The objections to continued DDT use are now well known, for its main advantages to the farmer, of cheapness, availability and persistence, are in themselves major disadvantages from the viewpoint of environmental pollution. The presence of minute traces of DDT in cotton-seed cake and oil are now easily detectable and are unacceptable in many markets. It seems likely that at some time in the not too distant future, the use of DDT on cotton will have to be discontinued.

Alternative insecticides

Many trials of possible alternative insecticides were carried out over the seasons 1969–70 to 1972–73, both at Serere and Namulonge. The replacement for DDT had to be safe, cheap and effective, but no insecticide was discovered that could be substituted for DDT, to the benefit of the cotton growers of Uganda. A few insecticides were tested which proved to be at least as effective as DDT, but all these were much more expensive and some more toxic. The best alternative to DDT was endosulfan, which controls the same pest spectrum as DDT but is more

Figure 5.4. The battery-powered sprayer which was successfully used to apply insecticides at ultra-low volume in experiments. This method of spraying is quick and needs little manual effort, but the insecticide formulations required are relatively expensive.

toxic to spiny bollworm and in addition, is effective against aphids. Endosulfan was used regularly at Namulonge in the control programme for cotton pests, replacing DDT when counts showed that aphids or spiny bollworm were reaching damaging populations. The much higher cost of endosulfan and its greater acute mammalian toxicity, however, will preclude its use by farmers until the use of DDT is prohibited.

Fixed or variable spray regimes?

Over most of Uganda the use of a fixed spray regime of four sprays of DDT at 1.1 kg active ingredient per ha in 90 litres of water, starting five weeks after germination and repeated at two-week intervals, was recommended. This minimum fixed regime was effective against lygus and early American bollworm attacks and was shown to give economic benefits to most farmers (Ingram 1965). Packs of DDT, each sufficient to spray an acre of cotton four times, were heavily subsidized by Government and sold cheaply to cotton farmers through Co-operative Unions; during the late 1960s and early 1970s, about 400 000 'acre packs' were distributed annually, but even this massive supply was sufficient to spray less than a third of Uganda's cotton with the minimum effective regime.

Figure 5.5. The simple and cheap Plantector sprayer which is used by most
farmers who spray their cotton with DDT in Uganda.

It was known that more effective control of pests, than that afforded by the four sprays of DDT, could be thoroughly worthwhile on well grown cotton. Davies (1970) showed that up to twelve sprays, using a range of effective insecticides, could be used with profit. Throughout Uganda, however, variations in pest attack, in species and severity, in different seasons, are such that a fixed regime cannot be ideal. An optimum fixed regime will be inadequate under heavy pest attacks and wasteful if pest attacks are light. The method developed and used so successfully by Matthews & Tunstall (1968), in which the cotton is scouted for pests, and sprayed according to the pests present, is obviously the most economic means of insecticide use. A similar method was latterly used for determining the insecticide use on the farm at Namulonge and between six and eight sprays, mostly of DDT, were used on the bulk of the cotton during the 1970–71 and 1971–72 seasons.

The extension of a spraying recommendation, based upon pest counts, to Uganda's cotton farmers was not, however, acceptable. Such a recommendation would have involved a huge exercise in farmer education which was beyond the resources of the already overstretched extension service, and would have required a major and costly modification to the existing scheme of subsidies for insecticides.

Studies on individual pests

While the use of insecticides can, temporarily at least, give control of most cotton pests, other methods of control usually involve individual species of pests and so require the study of each separately.

Very many insects and other arthropods are found on cotton at Namulonge. Many feed upon the plants, but only a few are numerous and destructive enough to merit individual attention as pests. In most years the species involved and their pattern of attack are similar. An infestation of lygus (*Taylorilygus vosseleri* Popp.) on young cotton is usually followed by American bollworm (*Heliothis armigera* Hubn.) and spiny bollworm (*Earias biplaga* Walker) attacks during flowering and fruiting stages, then false codling moths (*Cryptophlebia leucotreta*) and stainers (*Dysdercus* spp.) damage the maturing bolls (Figure 5.6). Attacks by aphids (*Aphis gossypii* Glov.) can be damaging during prolonged dry spells and red spider mites can defoliate the cotton if the initial outbreaks are not swiftly controlled. The early lygus attack is often augmented by other species of mirids and other heteroptera, but *Taylorilygus vosseleri* is by far the most dominant species in the complex of early sucking pests.

Studies of the ecology and control of each of the major pests have been carried out at Namulonge and it is convenient to consider progress in relation to individual pests, rather than as an historical progression.

Lygus

Taylorilygus vosseleri, formerly known as *Lygus vosseleri*, has been commonly referred to as lygus for so long and so widely that it can be conveniently referred to as such here.

Lygus feeds upon young tissues, particularly on flower and leaf buds. The damage caused to cotton has been well described in detail by Hancock (1935) and Taylor (1945) and the early work has been summarized by Pearson (1958). In brief, the symptoms of lygus attack are unmistakable tattering of the leaves, caused by the insect feeding on the leaf buds, and later the characteristic tall plant with few retained fruits. Many workers have considered lygus to be the most destructive pest of cotton throughout most of Uganda, but Bowden & Ingram (1958) thought that the role of bollworms in causing crop loss had been underestimated and that some of the symptoms of leaf damage attributed to lygus could be caused by other pests, such as young spiny bollworms and aphids. The view that the damage caused by bollworms in Uganda had been underestimated is fully accepted, and the perhaps excessive attention devoted to lygus studies resulted in a neglect of bollworm research. Nevertheless, it is difficult to agree with the theory that the symptoms formerly regarded as typical of lygus damage could be caused by biting pests, such as bollworms. The damage symptoms caused by lygus and other mirids appear to be unique and do not occur in areas where lygus bugs are absent or rare but bollworms and aphids are common.

Figure 5.6. The relationship between the incidence of insect pests of cotton and the agricultural calendar at Namulonge. A diagrammatic representation based on observations from 1969 to 1972.

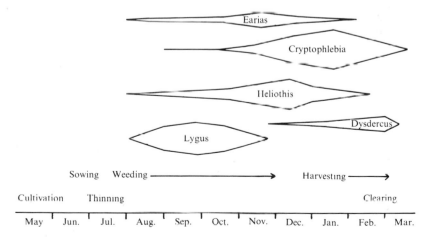

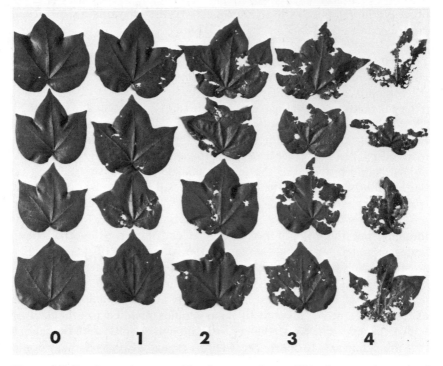

Figure 5.7. Leaf tattering caused by the cotton lygus. This photograph was used as a guide when recording the degree of leaf tattering in experiments designed to test the relative susceptibility of different cotton varieties to lygus attack.

A great deal was known and written about lygus as a cotton pest in Uganda well before Namulonge was established. The most complete record was by Taylor (1945), who acknowledged earlier work by Hargreaves, Gwynn and Stephenson. Consequently the entomology section at Namulonge inherited a problem upon which much had already been done, but which was complex enough to occupy much of the attention of Namulonge entomologists for the next twenty years.

Pearson's policy statement emphasized the need to disentangle the effects of lygus, bacterial blight and climate. Such complications became less apparent, however, for bacterial blight was largely eliminated by the introduction of the resistant Albar cottons, and long-term studies of the effect of different components of the climate, in the absence of pest damage, were made possible by the use of frequent insecticide applications to agronomic trials. Logan & Coaker (1960) showed that lygus could disseminate bacterial blight, but this factor is no longer of any importance.

Detailed studies of the mechanism of lygus feeding and the damage caused to the cotton plant were recorded by Dale & Coaker (1958) in a series of pot experiments with lygus reared in the laboratory, using a method developed by Geering (1953). A method for assessing lygus activity by recording an index of the leaf damage was developed by Coaker (1957*b*) from an earlier method first used by Gwynn (1936).

Much of the work on lygus at Namulonge prior to 1958 was confined to studies of crop loss, in conjunction with single-row insecticide trials, already described. The absolute assessment of crop loss caused by lygus is obviously impossible, for the effects of such damage are modified by the subsequent pest attack and climatic conditions. The effect of protecting cotton against lygus attack can be measured in any year at a particular site but this effect will not be typical of the effect at other sites and in other years; the assessment of crop loss by McKinlay & Geering (1957) and Coaker (1957*b*) clearly showed this variation at Namulonge, while Dale & Coaker (1961) found a similar situation in east and north Uganda.

Lygus-resistant cottons

Pearson's intention that the first priority of the entomology section should be the search for resistance to lygus was by no means a novel proposal; the search had already been under way for several years within the Uganda Department of Agriculture. Taylor (1945) reviewed the earlier work on lygus resistance by Stephenson, Gwynn and himself and concluded that although heritable differences in susceptibility had been clearly demonstrated, further progress would depend upon the identification of 'what we are breeding for', and he suggested two possibilities: 'hairiness' and 'toughness of tissues', either or both of which might confer resistance. He was of the opinion that the prospects of finding a useful resistant variety were not good and noted that rapid-growing, succulent plants were more prone to attack than the slower-growing, tougher plants. Taylor's studies on lygus susceptibility were continued by H. L. Manning at Kawanda, who also noted the correlation between plant height and lygus attack.

At Namulonge, genetic differences in lygus damage scores were recorded in a trial when 28 types of cotton, including *Gossypium barbadense* L. and *G. arboreum* L. were compared in the 1951–2 season, but subsequent to Pearson's departure in the following year, there was no further mention of lygus resistance or susceptibility in any *Progress Report* until the 1962–63 season when Stride (1964) resumed the work. Thus the first priority of Pearson's research plan did not survive his departure!

Stride's approach to the search for lygus resistance was to 'identify strains of cotton less susceptible to lygus attack than the present commercial varieties'. He initially tested 100 widely differing types and by the 1968–69 season had shown that plants obtained from crosses between UKA/2 MB.N and Barhop 11 were less susceptible to lygus damage than BP52 and that these differences were transmitted to their progeny.

Resistance mechanism

The simple selection of less susceptible types formed only a small part of Stride's lygus studies. He was convinced that the real breakthrough could be made by the identification of the mechanism of resistance or susceptibility. Stride considered that such a mechanism was likely to lie within the chemistry of the plants, that the cotton plant contained phago-stimulants and phago-inhibitors, the balance of which determined the susceptibility to lygus feeding. By fractionating extracts of cotton plants, and testing the fractions in lygus feeding trials, he found that a red pigmented extract inhibited feeding. Paradoxically, however, this inhibitor was most concentrated in the young terminal leaf buds, where lygus preferred to feed.

Stride's quest for inhibitors of lygus feeding was initiated despite Taylor's (1945) view that 'A variety of cotton which is resistant on account of a lack of attractiveness or palatability is out of the question'! Taylor also considered that cotton was particularly well suited as a host for lygus but in a series of experiments Stride (1968) clearly demonstrated that cotton was not a highly preferred host. Subsequently Stride (1969) investigated the use of a more attractive host, *Cissus adenocaulis* Steud. ex A. Rich, as a trap crop. In a preliminary trial, plots with *C. adenocaulis* replacing every tenth row of cotton suffered slightly more damage than the control plots. Presumably the alternative host attracted, and maintained, large populations of lygus which 'spilled over' onto the cotton. In subsequent trials, the *C. adenocaulis* rows were sprayed with DDT: this gave better protection against lygus than in the control plots, where every tenth row of cotton was sprayed. This method could be of advantage, for only one-tenth of the area is sprayed and the bollworm's natural enemies would be preserved, if they restricted themselves to the cotton. It is unlikely that this method would be accepted and used by farmers, however, for a tenth of their land would be occupied by a useless weed and some spraying would still be essential. In contrast, great success has been reported in the USA where, under irrigated conditions, strips of lucerne between cotton fields attract virtually all the *Lygus hesperus* Knight and reduce lygus damage on the cotton to acceptable levels (Stern, Mueller, Sevacherian & Way 1969). There, however, the alternative host is of value to the farmer and insecticide is not needed. A similar system cannot be used in most areas of Uganda for the lucerne would have to be irrigated if it were to remain succulent and attractive to lygus.

The only other potential trap crop would appear to be sorghum, for it is more attractive to lygus than cotton, is of value and grows well in most cotton areas in Uganda. Unfortunately lygus bugs are only attracted to sorghum during the short period between flowering and seed maturation and as soon as the seeds dry, the large populations of lygus which have fed on the sorghum will disperse onto the cotton. Successive sowings of sorghum could provide a continuous, attractive trap-crop, but the later sowings would be virtually destroyed by the sorghum pests, particularly shoot fly, stem borers and midge.

Further work was concentrated upon selection for reduced lygus susceptibility from within the most advanced BPA lines and commercial issues. If such selection were in any degree successful then it would be likely that the selected plants would be of immediate commercial use; whereas a strain selected purely for its lygus resistance from outside the cotton breeder's pool of commercial varieties, would be unlikely to possess the required quality and yield characters. This selection work showed some promise, for one selection from BPA had resistance comparable with Stride's best selection and was much more vigorous. Unfortunately, in the first yield trial under unsprayed conditions, in which a heavy attack of lygus was induced, the most susceptible selections outyielded the most resistant selections by more than 100 per cent (Reed 1974). The lygus work at Namulonge appeared to thrive upon a series of contradictions!

Again with hindsight, it is tempting to speculate that had Pearson's policy been followed consistently from 1950 onwards, continued selection for lygus resistance might have produced by 1972 a commercial seed issue with considerable lygus resistance. Furthermore, if such a programme had failed to produce any worthwhile resistance, we should by then at least have had enough data to decide whether the continued search for resistance was justified.

Cotton stainers

Since jassids (*Empoasca* spp.) had been controlled by breeding for hairy plants, the only other major sucking pests of cotton at Namulonge, apart from the lygus complex, were the stainers (*Dysdercus* spp.). Stainers were at first considered to be serious pests late in the season at Namulonge and continued to be serious pests in parts of western Uganda.

Geering (1956) developed a useful method for the controlled breeding of stainers in the laboratory so that their biology could be studied. Geering & Coaker (1960) examined the effect of various hostplant foods on the fertility and development of stainers. The role of stainers in infecting bolls with *Nematospora* is well known and Leakey & Perry (1966) demonstrated that other boll rots could also be introduced by cotton pests, including stainers.

From 1966, however, stainers ceased to be major pests at Namulonge

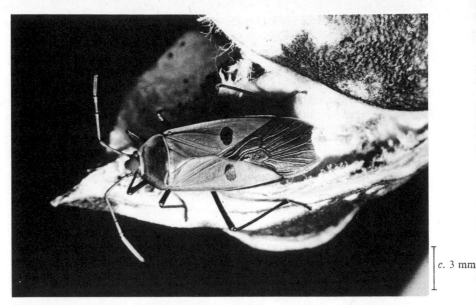

c. 3 mm

Figure 5.8. *Dysdercus superstitiosus*, a cotton stainer which was a damaging
pest of unsprayed cotton in many parts of Uganda.

and no further work on this pest was considered necessary, although
trials of insecticide use for stainer control were continued in western
Uganda. At Namulonge, stainers were breeding within the cotton and
reached damaging populations by the end of January. From the time
that the bulk of the Namulonge cotton was sprayed, however, the cotton
crop was set much earlier and virtually all of the crop was picked, if
labour was available, before the end of January. Thus although DDT
and endosulfan were not very effective against stainers, by protecting
the crop from lygus and bollworms they permitted the production of
an early crop, which was harvested before the stainer populations were
large enough to cause serious damage.

American bollworm

Heliothis armigera, the so-called American bollworm, which is a pest
of raingrown cotton throughout Africa, has been extensively studied
at Namulonge. By 1953, Geering and McKinlay considered that the
bollworms, particularly *H. armigera*, could reduce cotton yields to a
similar or greater extent than lygus because, although the cotton was
able to compensate for losses caused by lygus, it could not do so for
late bollworm attacks. Later, however, Coaker (1959a) could show no
correlation between yield and *H. armigera* larval populations and
concluded that the bollworms were not important factors in determining
yields at Namulonge.

Coaker studied the populations of *H. armigera* on cotton and other crop hosts throughout each year. He found that peak attacks on cotton generally occurred in November, but that the populations of this bollworm were sparse in comparison with those in some other countries in Africa. He concluded that because this bollworm was active at Namulonge throughout each year on a wide range of hosts, it was accompanied by parasites, predators and diseases which stabilized the bollworm populations at relatively low levels.

Heliothis *virus*

These factors of natural control, which Coaker reported, included at least 17 parasitoids and a nuclear polyhedral virus. In a series of field experiments with this disease, Coaker (1958*a*) showed that it was possible to use the virus in sprays with some success. This was the first attempt to control a *Heliothis* sp. with a virus on a field scale and the work is often referred to by the many workers who are now developing virus insecticides for the control of bollworms in America and elsewhere. The use of virus for bollworm control has been greatly restricted because of the success of the chemical insecticides. In Uganda, DDT still gives excellent and cheap control of American bollworm infestations, but in many other countries the *Heliothis* spp. are now extremely resistant to DDT and several other insecticides, so the use of virus as an insecticide is of great interest.

False codling moth

Apart from *H. armigera*, other bollworms were not regarded as serious pests of cotton at Namulonge during the early years, except the spiny bollworm, *Earias biplaga*. Pink bollworm, *pectinophora gossypiella* Saund. was recorded regularly, but never reached damaging populations, for it was held in check by uprooting and burning the old cotton.

The false codling moth, *Cryptophlebia (Argyroploce) leucotreta* was formerly regarded as a serious pest. First recorded in Uganda by Hargreaves in 1921, Taylor (see Tothill 1940) regarded this as the most important of the bollworms, for it was common in all cotton growing areas in Uganda. At the end of the 1934–35 season samples of nearly mature green bolls collected within a radius of forty miles of Kampala showed a mean infestation of 27 per cent.

At Namulonge in the 1950s, the false codling moth was recorded as a very minor pest. In the 1950–51 season this pest was recorded as light, in 1953–54 it was found in 'low numbers in bolls', but at a time when nearby maize was heavily infested, for 44 per cent of the cobs contained larvae one month before harvest (Geering, McKinlay & Coaker 1954). It was considered, therefore, that maize was more attractive than cotton and acted as a trap-crop. In 1954–55 Coaker found

Figure 5.9. A rotting boll with a hole made by a fully grown larva of the false codling moth when it left the boll after feeding inside. Most of the boll rotting at Namulonge was shown to be initiated by insect attack which allowed micro-organisms to invade the boll.

that false codling moth larvae constituted only 2 per cent of the total bollworm larvae recorded and from that season, *C. leucotreta* was not mentioned in *Progress Reports* again until 1967–68, when Stride noted that this pest was not controlled by the standard insecticide spray regime and considered it to be of sufficient importance to merit special study. By 1969, *C. leucotreta* was by far the most damaging pest of sprayed cotton; examination of green bolls late in the season showed that this pest was causing considerable loss, over 90 per cent of the late bolls being damaged.

Thus the false codling moth at Namulonge appears to have changed its status from a very minor pest to the most damaging. The increase in infestations is not clearly recorded, but there appears to have been a rapid increase since the widescale spraying of cotton was started in 1966. The false codling moth lays its eggs on the cotton bolls, the young larva quickly bores into the boll wall and from that time it is usually totally protected from contact insecticides until the mature larva emerges from the boll prior to pupation.

Insecticide failure

It was not known whether the failure to control *C. leucotreta* resulted from its immunity to the insecticides used, or from lack of contact with them. Attempts were therefore made to improve the spray cover on the bolls by using tail-boom sprayers, which had been found to give better coverage and better red bollworm control in southern Africa. At Namulonge, however, these sprayers gave little improvement in *C. leucotreta* control and reduced yields slightly by damaging the cotton (Reed 1970). In laboratory tests Miss M. Yeates of the Anti-Locust Research Centre showed that much higher concentrations of the insecticides tested were required to kill this pest than were required to kill *H. armigera* and *E. biplaga*.

Hence, the spraying programme latterly used at Namulonge failed to control the false codling moth adequately, but evidence that some control was achieved was gained from spraying trials where the examination of green bolls showed that the highest infestations were always on the unsprayed 'control' plots. More frequent spraying and the use of toxic insecticides such as monocrotophos, reduced populations on sprayed plots to 50 per cent of those on unsprayed controls.

If insecticide use at Namulonge were exerting some control of false codling moth, why did the infestations of this pest increase so dramatically from the time that spraying became routine on cotton at Namulonge? The answer may be that the pest was formerly controlled by biological factors that were subsequently destroyed by the spraying programme. Some support for this theory was given by counts that showed that the proportion of bolls containing live *C. leucotreta* larvae on the large unsprayed block was much lower than that recorded from sprayed cotton, at any one time (Reed 1973). This biological controlling factor was not identified; four insect parasitoid species were recorded from the larvae, but these were all so uncommon, even on unsprayed cotton until the end of the cotton season, that they were unlikely to have had a noticeable effect upon their host's populations.

Another reason for the increase of false codling moth on cotton at Namulonge may have been the changed cropping programme. Latterly, maize was grown almost entirely during the first rains and not during the cotton season, as was at one time practised on part of the farm. Since it was noticed in 1953 that maize could act as a trap-crop for this pest, this change could have resulted in the concentration of infestations on cotton, whereas formerly they were concentrated on maize. Later experiments in which maize was grown next to cotton, however, failed to show any reduction in attack on the cotton; the maize and cotton appeared to be equally attacked.

Alternative control measures

Because insecticides failed to control *C. leucotreta*, other methods of control were considered and attempted. Populations were noticed to be very great on late sown cotton adjacent to early sown; this factor tends to make the yield data from conventional sowing date trials of doubtful validity for, despite weekly spraying, the last sown cotton in Namulonge trials has been found to have over 90 per cent of its bolls damaged. In an attempt to prevent the spread of the pest from early to later-sown cotton, all cotton at Namulonge grown in the 1970–71 and 1971–72 seasons was sown in blocks, and each block contained cotton sown within a short period.

The incidence of the pest throughout the year has been studied; it was found mainly in maize cobs from April to August, then in cotton bolls from September, building up to a peak infestation in January–February. By harvesting the maize early in July and ensuring that little cotton was sown before mid-June, it was possible to establish a break, in July–August, when no maize cobs and no cotton bolls were available. It was hoped that this would reduce the pest population by limiting the hosts available at that time, for *C. leucotreta* was thought to be almost entirely dependent upon fruits (Pearson 1958). This 'close season' proved to be of limited value however, for the larvae were found to feed successfully in the stems and buds of the young cotton in August.

C. leucotreta *and boll rots*

The larvae enter the boll wall soon after hatching, then tunnel around inside the wall for several days before penetrating the inner septum and

c. 5 mm

Figure 5.10. Biological control in action. An American bollworm being killed by a larva of *Cardiochiles* sp. a common parasitoid of *Heliothis armigera* on unsprayed cotton in Uganda.

feeding upon the seeds and lint. Many young larvae die or disappear, for a large proportion of green bolls are found with tunnels in the wall, but with no larvae present. Although the bollworm only eats a very small part of the boll, the entry hole made by the first instar larva also provides an entry for boll-rotting organisms and many bolls attacked by this bollworm are totally destroyed by boll rot. The boll rotting does not deter the pest; it is only this bollworm that is regularly found feeding inside completely rotten bolls.

In a joint project with S. J. Brown, examination of green bolls during the years 1970 and 1971 revealed that over 95 per cent of rotting bolls showed insect damage symptoms; of these symptoms, well over half were attributed to *C. leucotreta* damage. It therefore seems likely that much of the Namulonge boll-rotting problem, which had attracted the attention of pathologists, is primarily a problem for the entomologist. Similar findings have been reported from the USA by Bagga & Laster (1968), who demonstrated the role of a number of different insects in the initiation of boll rotting.

Biological or natural control

The use of insecticides and resistant cotton varieties, together with cultural practices such as shredding and ploughing-in the cotton stalks, were the main pest control measures developed and used at Namulonge.

In addition, natural control of the pests was studied, and some attempt made to supplement this control.

The work by Coaker using *H. armigera* virus has already been mentioned; he also drew attention to the importance of parasites in bollworm control. Similarly all of the other pests studied at Namulonge have been found to have natural enemies. Two attempts were made to exploit natural control agents. *Trichogramma brasiliensis* Ashmead, an egg parasite, was imported from India, reared in the laboratory on *H. armigera* and *E. biplaga* eggs and released in the fields at Namulonge. *Phytoselus persimilis* Athias-Henriot, a predatory mite, imported from South Africa, was released at Namulonge and elsewhere in the hope that it would reduce red spider mite populations. Neither of these introductions was of noticeable benefit.

The value of beneficial insects, the parasites and predators, was largely sacrificed, for the insecticides used to kill the pests also killed their enemies. It would have been preferable to have used natural control rather than insecticides, but unfortunately the latter proved to be far more efficient in helping to produce cotton crops at Namulonge of excellent quality and quantity. If the lygus problem could be overcome by plant resistance, then the use of insecticide early in the season might not be needed and the enemies of the bollworms might be efficient enough to make insecticide use totally unnecessary. Such a situation would be ideal, but is a doubtful prospect even for the distant future. In the meantime the more limited aim should be encouraged of developing integrated control measures in which the use of biological control methods could be developed so that the use of insecticides can be progressively reduced.

The future of entomology at Namulonge

With the handover of Namulonge to the Uganda Government, it was intended that the station would assume responsibility for research into a wide range of annual crops of likely value to the country. This inevitably meant that the work of the entomology section would be expanded and diversified as extra professional staff became available. Although the section's work in the past was not restricted to cotton, Coaker's published work on maize pests and storage (Coaker 1956, 1958*b*, 1959*b*; Coaker & Passmore 1958) being good examples of the way in which the section has involved itself in farm activities outside the cotton work, the future must involve the section in much more intensive research on other crop pests. It is hoped that this will not reduce the rate of progress of the programme in cotton entomology, nor lessen its chance of success, for a well-planned, overall study of the pest control needs of Uganda's annual crops could well benefit the cotton research.

Pearson's original policy can still be regarded as a sound basis for the future of the research programme in cotton entomology; it embodies integrated pest control and allows for the necessary use of insecticides in the short term. It implies that the search for insect resistance and better insecticide use should receive equal attention.

Hindsight

Shortcomings in the programme of entomological research have already been highlighted, but the circumstances that promoted these short-comings should also be considered. At first, the section had a comple-ment of three professional entomologists, but by 1955 the number had dwindled to one and stayed at that level until 1972, apart from a period in 1960–61 when there was no entomologist at all. It is difficult to determine whether the limited achievements of the section led to a staff reduction, or the staff reduction led to limited achievement. Undoubtedly, one factor was that in the early years, the failure to show that pests caused losses of yield, persuaded the administration that entomology was an expensive luxury, for no problem apparently existed!

The failure to demonstrate worthwhile increases in yield from the use of insecticides at Namulonge was repeated at Kawanda and a few other trial sites, but this lack of response was fortunately restricted to the elephant grass zone. J. Bowden, W. R. Ingram and J. C. Davies, en-tomologists employed by the Uganda Government who worked at Kawanda and Serere, were responsible for research and development leading to recommendations for the control of cotton pests throughout Uganda. In a series of well-planned and executed trials, they found that insecticides could greatly increase yields in all cotton growing areas other than those in the elephant grass zone. They developed methods for spraying the many small fields that are a feature of agriculture in Uganda, and issued recommendations accordingly. The practical suc-cess of these recommendations has been confirmed by the enthusiastic way in which they have been received by increasing numbers of growers. Many farmers in the elephant grass zone insisted that they also should enjoy the benefits of subsidized insecticides and sprayers, and they were encouraged to do so following the re-appraisal of pest control at Namulonge.

From the time that the entomology section at Namulonge was staffed by only one entomologist, there was a lack of sustained effort on any one particular problem. One man could hardly cope both with the prob-lems of controlling pests locally and with the development of guiding principles for research elsewhere. Moreover the needs of research on cotton pests at Namulonge were so diverse that no entomologist could hope to study all the problems in depth. As a result, each entomologist naturally tended to specialize in the facets that either interested him most, or appeared to offer the greatest chance of success. Since each entomologist's interests and skills were different from those of his predecessors, the work tended to change emphasis and direction with little real continuity.

Obvious omissions already noted were the lack of continuity in the programmes related to lygus resistance and insecticide use. Another regrettable feature of the past was the lack of records of pest incidence in each year on cotton. Although a light trap was installed in 1954, it was soon discontinued and another was not installed until 1969. There are no records of bollworm or sucking pest incidence at Namulonge from 1958 to 1969. Hence it is not possible to check back to establish correlations of insect attack, climate and yields.

On the credit side, however, much was achieved despite the staff shortage. Namulonge was the setting within which Pearson's book *The insect pests of cotton in tropical Africa* was prepared. This book has been invaluable to all cotton entomologists, for it summarized clearly and concisely the huge volume of recorded work on cotton pests in Africa. The basic biological studies, particularly of lygus and American bollworm, are of great value to present-day workers, both within and outside Uganda. The emphasis upon the team approach to cotton problems, first suggested by Pearson and lately encouraged by Arnold, has paid great dividends, for no problem in cotton growing could be claimed as the exclusive preserve of any specialist; the work of the breeders and agronomists was complemented by the interest of the entomologists and pathologists, so that maximum benefits could be obtained. The results of this co-operation could be seen in the Namulonge cotton fields: at one period, visitors were tactfully diverted into the adequate, well-equipped laboratories; in the last few years, however, they were shown around the cotton fields with pride!

6

Plant pathology

S. J. BROWN

Introduction

The study of cotton diseases and their control in tropical Africa must take into account the conditions in which the crop is grown. In Uganda cotton is a relatively new crop. Attempts to establish it commercially began in 1903 and by 1931 production had reached 204 000 bales. Apart from seasonal fluctuations, production then remained steady until the late 1940s when expansion of the area in cotton led to a gradual increase in production until the mid-1960s. It might have been supposed that farmers would have gained considerable knowledge during the sixty years or so of cotton cultivation in Uganda, but examination of crop production figures for different areas of the country suggests that many farmers have been growing cotton for a relatively short period. The change in the main cotton producing areas was caused by several factors. Two of the most important were the opening of new land along the upper reaches of the Nile following the eradication of the Mbwa fly, *Simulum damnosum* (Bowden & Thomas 1970), and the expansion of coffee in many parts of Buganda at the expense of the cotton crop (Kibukamusoke 1962).

Cotton is not an easy crop to grow in Uganda; it requires much arduous hand labour throughout the season. Under these conditions any imperfection in the crop is quickly noticed. For instance, the occurrence of only an occasional plant infected with Verticillium wilt or the presence of *Ramularia areola* Atk. (*R. gossypii* Speg. Ciferri) on the lower leaves is often a cause of greater criticism than other factors causing far greater crop losses. Farmer appeal would seem to be an important factor in expanding cotton production, and greater control of diseases than might otherwise be justified, is needed to achieve it. As farmers become more experienced, they may well become more aware of the relative importance of diseases compared with other factors causing poor yields such as sowing too late or spacing too widely.

The methods available for controlling cotton diseases are severely restricted in Uganda by the conditions under which the farmer works.

Practically all cotton is grown in small plots that vary in size from a few square metres to about 1 ha. Larger blocks of cotton are rare. Cotton is often interplanted with food crops and yields are, generally, poor. Spray programmes, using fungicides or bacteriocides, would not only be uneconomic but would also be beyond the farmers' technical capabilities (Chapter 10). Control of cotton diseases must, therefore, be based mainly upon techniques in which the farmer is not a direct participant: genetic resistance and chemical seed treatments.

The study of cotton diseases at Namulonge developed primarily along lines aimed at reducing, and eventually eliminating, disease in the commercial crop. The work involved close co-operation between cotton breeders and pathologists. This chapter describes the plant pathological aspects of the work. Chapter 7 describes the plant breeding aspects.

A secondary, and more recent aspect of the work, arose from problems experienced with the more intensive farming practised at Namulonge during the late 1960s, and concerned the occurrence of seedling disease caused by the fungus *Sclerotium rolfsii* Sacc. Although the results of investigations on this disease were especially relevant to Namulonge, they might well find application more widely.

Bacterial blight

The disease

Bacterial blight is a collective name for the disease of cotton caused by *Xanthomonas malvacearum* (E. F. Smith) Dowson. The bacterium can infect all aerial parts of the plant giving rise to the various manifestations of the disease known as seedling blight, angular leaf spot, bacterial wilt, blackarm, vein blight, bract spot, and bacterial boll rot. Diseased seedlings can arise from contaminated seed, volunteer plants or infected crop debris from the previous season. The disease spreads through a crop from these primary foci of infection, not only by water splash (aided by the strong winds which frequently accompany tropical rainstorms), but also by insect transmission (Logan & Coaker 1960) and possibly by wind-blown dust.

Although it is generally accepted that resistant varieties constitute the ultimate solution to control, other methods have often had to be adopted until such varieties are available. The most successful and economic have been based on preventing the carry-over of infection from one season to another. Seed dressing, involving the use of dry dusts, has been very successful in reducing seedling blight which, in turn, leads to reduced secondary infections and increased yields. Such treatments will not, however, eliminate the small proportion of internal infection which, under favourable environmental conditions, can lead to severe epidemics. The close season introduced for the control of bollworms (Chapter 5) helps to eliminate infection of a new crop by volunteer plants. The ability of trash to infect a subsequent crop depends largely on rainfall between cotton seasons. With little or no rain, trash-borne infection can be severe but sufficient rainfall to cause only partial decomposition of crop debris will eliminate or reduce the potential infectivity to a level where it is relatively unimportant compared with seed-borne infection. Ploughing debris into the soil will also prevent trash infecting a subsequent crop (Arnold & Arnold 1961).

History in Uganda

The disease was first identified in Uganda in 1925 (Snowden 1926) by which time it had become firmly established and had probably already contributed to crop losses (Thomas 1970). Snowden suggested six lines of approach for investigations into controlling the disease. Two of them, chemical disinfection of the seed and breeding resistant varieties, were initiated after the severe epidemic in the 1929–30 season. Considerable success in reducing seedling disease was achieved with seed dressings but the work was abandoned after five years. The reasons for rejecting such a potentially effective method of control are not clear. Certainly bacterial blight was not very serious in the seasons from 1930 to 1934, when seed dressing, although reducing the disease considerably, did not lead to increased yields. This led Hansford & Hosking (1938) to the conclusion that regular treatment of the whole seed supply would be uneconomic, even if it proved economic in occasional seasons. There were also 'other practical and political considerations which rule out this method of dealing with the disease'. Jameson (1952), however, stated that the work on seed dressings was abandoned in favour of breeding resistant varieties because all the most effective treatments were based on dangerous mercury compounds which were unsuitable for general use.

Selection of resistant plants was continuing in parallel with the work on seed dressings. The line B31, selected by Nye in 1929, was subsequently shown by Knight in the Sudan to possess the resistance gene B_2 (see Chapter 7). In 1929, the level of resistance in commercial varieties was very low. Nye & Hosking (1930) reported that two Teso district variety trials, which were sited on poor soil, were complete failures after blackarm had reduced most of the plants to bare sticks. Other centres had much reduced yields associated with heavy bacterial blight development. Identifying resistant plants in such conditions presented no problems but when disease development was less severe and not uniformly distributed, techniques had to be devised to assist the plant breeders in recognizing genetically resistant plants from fortuitous escapes. Hansford (1933) showed that stem lesions were by far the major cause of crop loss in Uganda. Selection of resistant plants was continued with considerable success using various scores based on counts of mainstem lesions, and the seed was soaked before planting to give higher rates of infection. Later, stem lesion counts were combined with Knight & Clouston's (1939) leaf inoculation technique, which worked well at Serere but was not successful at Kawanda where the cool, cloudy weather retarded leaf lesion development.

The whole problem of the control of bacterial blight was reconsidered in 1948 after the realization that mainstem infection in the first two months of the life of the plant could be critical for yield. This suggested that attempts to reduce primary seedling infection would be worthwhile and a large programme of testing seed dressings was initiated at Serere in 1949. Results were extremely encouraging and led to the decision to begin seed dressing on a large scale in the 1950–51 season (Jameson & Thomas 1953).

Work at Namulonge

Research on bacterial blight commenced at Namulonge in 1951 after twenty-one years of investigations by the Department of Agriculture, during which the foundations of control methods had been laid. Considerable progress had been made in reducing losses by breeding moderately resistant varieties and by dressing the seed. Work at Namulonge was intended to give further understanding of some of the more fundamental aspects of disease development as well as being of practical value to Uganda. Accordingly the following problems were defined for investigation:

(1) Methods of infection of leaves, stems and bolls; factors influencing infection and the rate of tissue invasion; means of spread; factors influencing the extent of carry-over from one season to the next; and the effectiveness of control by seed disinfection.

(2) The physiology of parasitism by *X. malvacearum* and the nature of resistance; the development of sensitive tests of inherent resistance for use in plant breeding.

(3) The assessment of the relative importance of bacterial blight and the cotton lygus in causing crop loss in Uganda.

Inoculation techniques

The programme was ambitious by any standards and was intended to be relevant to all African commonwealth countries. However, local problems urgently needed attention and work was immediately started on devising satisfactory inoculation techniques for more accurate classification of plants in resistance-breeding programmes. In the first season in which different techniques, including seed, leaf, stem, and boll inoculations were compared it was concluded that 'resistance to one form of attack does not necessarily imply resistance to others, and it follows that at present no single criterion can be advocated, under

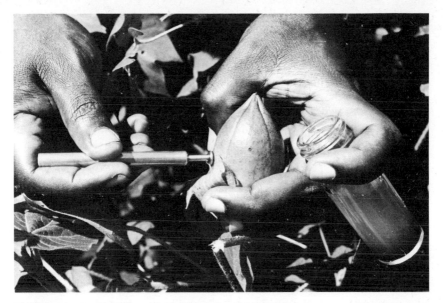

Figure 6.1. Inoculating a cotton boll with *Xanthomonas malvacearum*. A needle, which had been dipped into the bacterial suspension in the bottle, was inserted into the boll wall.

Uganda conditions, as a measure of effective resistance to the disease' (Wickens 1952). This hypothesis was reaffirmed on the basis of results from later seasons (e.g. Wickens & Logan 1956, 1958) when different rankings of varieties in resistance to different phases of the disease were the rule rather than the exception. It was concluded that selection methods for resistance in different countries should be based on the most damaging phase prevalent in the particular environment. In practice this was the only possible course open to breeders who relied upon field infection.

All phases of the disease occurred in Uganda and it was not certain which caused the most serious loss of crop. Selection of resistant plants at Namulonge was initially made on the basis of seed, stem, leaf and boll inoculations. Considerable progress was made, particularly with selections from Nigerian Allen from which the Albar stocks, and utimately BPA and SATU, were derived (Arnold *et al*. 1968). The seed inoculation and leaf spray inoculation techniques were soon discontinued because they were unreliable. Logan (1958) introduced a modification of the boll inoculation technique whereby a short needle, which had been dipped in a suspension of *X. malvacearum* was inserted into the surface of almost fully expanded bolls. This method was not only more reliable than the brush inoculation of younger bolls (Wickens 1953) but accurately identified genotypes with small differences in resistance.

Hypocotyl inoculations were also introduced and proved of great value in later studies of host–parasite relationships (Arnold & Brown 1968).

In the United States of America, Weindling (1948) had also advocated the use of a range of inoculation techniques in breeding programmes because of the apparent lack of correlations among expressions of resistance to the various phases of the disease. Knight & Clouston (1939), Bird (1950) and Lagière (1960), however, reported positive correlations. A satisfactory explanation for these anomalies was not put forward until effects of changes in environment on disease development were more fully appreciated (Arnold 1963; Arnold & Brown 1968). It was found that changes in environment, both external and internal, could influence disease development to such an extent that clear-cut relationships between resistance to one phase of the disease and another would not necessarily be apparent, even if resistance in all tissues had a common genetic basis. In such circumstances the only practical way of determining the genetic relationship of resistance to the various phases is to study the relative changes in resistance to one phase associated with genetic changes in another. Arnold (1963) found that twenty-three UKA strains, which had been selected for resistance using stem inoculations only, showed related increases in resistance when boll inoculated. There were also indications of an association between stem and leaf resistance. In several cases of the reported lack of association between resistance to one phase and another critical investigations of this type have not been made. Nevertheless, mechanisms that are under different genetic control, such as differences in the development of callus tissue in the bark (Arnold 1963), almost certainly affect the expression of symptoms and would tend to complicate measurements of the relationship between resistance to one phase and another.

Prior to 1967, leaf inoculations (Arnold & Brown 1968) were used only in experimental work. Following the discovery of more virulent types of the pathogen, however, leaf inoculation was used extensively in resistance breeding (Brown 1971). Plants showing satisfactory resistance to leaf inoculations were usually boll-inoculated at a later date giving the opportunity to screen for resistance under different environmental conditions. Stem inoculations were discontinued because of the difficulty of identifying differences in resistance at the more resistant end of the scale.

Epidemiology

The inconsistent and often inexplicable course of disease development following various inoculation techniques prompted further studies of the epidemiology of bacterial blight. Observations made on natural field infections and of disease development after seed and leaf inoculations (Wickens & Logan 1957) emphasized the complexity of the disease.

Observations on the development of boll rot indicated that vascular infection, either from infected branches or bracts could give rise to internal boll rot. It was concluded that this form of systemic infection gave rise to internally infected seed. Russell (1955) and Wickens (1956) had shown that *X. malvacearum* was capable of entering the vascular system and rapidly spreading throughout a cotton plant. With the benefit of hindsight it is easy to belittle the importance of such work, when the simultaneous development of blight lesions in leaves can often be seen above another stem or leaf infection. In some susceptible varieties the position of such leaf lesions can be accurately predicted according to the branching of the main leaf-veins. It should be remembered however that there were some workers, who, even in 1964 believed that *X. malvacearum* could only spread through a plant within parenchymatous tissue. The description of vascular movement of *X. malvacearum* thus contributed substantially to a better understanding of disease development within the plant.

Seed dressing

After the introduction of large-scale seed dressing in 1950, progressively greater quantities of seed were dressed each year until, in 1955, about 21 000 tonnes were dressed, enough for the whole country. The seed was dressed with a cuprous oxide formulation using large rotary-drum mixers sited at convenient centres throughout the producing area (Green 1960). Soon after the introduction of seed dressing, claims were made that dressing increased yields by 50–75 per cent in areas where blight was normally severe (Jameson & Thomas 1953). Their estimates of the value of crop lost through bacterial blight and of the benefits from seed dressing were given wide publicity (Colonial Insecticides, Fungicides and Herbicides Committee 1951; Pottie 1953; Holmes 1955; Padwick 1956, 1959), but were not substantiated by yields from dressed seed in subsequent seasons. The introduction of seed dressing in the Northern and Eastern Regions coincided with the release of S47 which was somewhat more resistant to bacterial blight than the variety grown previously. A later assessment, in which production in the Eastern Region for the eight years prior to the introduction of seed treatment was compared with the following eight years, showed an average increase associated with seed dressing of 27 per cent (Logan 1960). This increase was less than had originally been claimed for the new variety S47 alone (Jameson & Thomas 1951).

There is no doubt that seed dressing considerably reduced levels of primary infection. Table 6.1, taken from a study of Wickens (1957), shows data obtained over a period of six years from surveys by the Uganda Department of Agriculture of the percentage of plants with blight lesions on the mainstems at about eight weeks after sowing. By

Table 6.1. *Bacterial blight surveys*

Percentage of plants with primary mainstem infections (Wickens 1957).
Figures in italics refer to dressed seed

Seed distribution area	Year					
	1950	1951	1952	1953	1954	1955
West Nile	—	23	—	40	56	*9*
West Acholi	—	35	—	48	50	*4*
East Acholi	—	26	—	45	57	*5*
Lango	32	28	25	48	52	*4*
North Teso	37	*8*	*8*	*5*	7	*5*
Segregated areas[a]	*15*	*6*	*2*	*3*	*4*	*3*
South Teso	*21*	10	*5*	*5*	7	*2*
North Mbale	42	40	31	*15*	*14*	*9*
South Mbale	—	47	7	*8*	*10*	*15*
Busoga E–W	—	—	16	38	60	*10*
Busoga S–N	—	—	31	47	47	*8*
Bugerere	—	—	—	—	64	7
Kyagwe	—	—	—	*6*	*19*	*9*
Bulemezi	—	—	28	—	*11*	*5*
Entebbe	—	—	*4*	47	*8*	*4*
Masaka	—	—	21	17	3	*3*
Singo	—	—	—	49	65	*6*
Mubende	—	—	—	46	60	*6*
Bunyoro	—	—	—	59	63	7
Number of seed-dressing stations in operation	1	1	3	4	6	12

[a] Segregated cotton production areas used in the seed multiplication scheme (Chapter 8).

1954 there was, in spite of the earlier work, considerable doubt as to the economic justification of a large-scale seed-dressing programme. Evidence deduced from following the course of disease development showed that even with reduced levels of primary infection, secondary infection often built up, albeit more slowly, to the same level as that in crops grown from untreated seed. It was felt that a search for more efficient seed-dressing compounds combined, if necessary, with different methods of treating seed would be worthwhile. Seed-dressing trials often give inconsistent results. They are usually exposed to natural spread of the disease, which varies from place to place and season to season. Plot sizes must be sufficiently large to allow the effect of the disease on yield to be measured, and they must be far enough apart to prevent cross infection. Other factors, such as the length of the growing season (determined by rainfall), may influence the results; a crop severely infected in the early stages of growth may show substantial yield losses when the growing season is short but may almost entirely compensate for early crop loss when the growing season is extended. The generally more consistent results obtained at Serere were to a large extent attributable to environmental factors favouring blight development, but could also be related to a shorter growing season. Seed-dressing trials were usually sown later than other cotton in order to take advantage of more favourable weather conditions for blight development during the early stages of crop growth. The cessation of the rains is more clearly defined at Serere than at Namulonge and would, in most years, allow less compensatory growth.

In spite of a high proportion of trials in which disease development was insufficient to obtain reliable estimates of the effect of seed dressing, certain trends eventually emerged. Foremost was the demonstration of the efficiency of cuprous oxide treatments in reducing losses under conditions favouring blight development. The magnitude of the relative yield increases from seed dressing left no doubt that seed dressing on a commercial scale, which cost the equivalent of only 3 kg seed cotton per ha, was economically justified (Wickens 1958). Further evidence on the effect of seed dressing was obtained by Manning & Kibukamusoke (1958) who estimated, from fourteen years of data, that seed dressing had increased yields in Uganda by 8.9 per cent. Most of this increase had occurred in the S cotton growing area (northern and eastern Uganda), where bacterial blight was usually more prevalent and more severe.

Organo-mercurial compounds were consistently more efficient than those based on copper. In some instances where considerable secondary infection built up from diseased seedlings in plots treated with cuprous oxide, plots treated with mercury-based compounds remained disease free throughout the season. Serious consideration was therefore given to changing the whole seed-dressing programme to mercury based compounds. Although the advantages were clear, the disadvantages were less well defined; there was little evidence on which to assess the risks associated with the release of seed dressed with a toxic compound in relation to the benefits which could arise from increased yields. Capital expenditure would be required for modifying or replacing the existing seed-dressing equipment in order to protect the operators but little could be done to control the handling of treated seed once it was issued to the farmers.

In a large-scale trial sown in the Western Region, disease development was much less severe in an area sown with mercury-treated seed than in a nearby area sown with seed dressed with the current formulation of cuprous oxide. The mercury treatment was applied at a commercial dressing station using the standard rotating drum. Analysis of the mercury content of the air around the dressing machinery showed that far greater modifications to the existing equipment would be required, than had hitherto been considered, in order to reduce the risk of operator contamination (Wickens & Logan 1960a). Soon afterwards, it was decided not to use mercury-based seed dressings, primarily because of the lack of control over seed not sown by farmers, but also because of the heavy capital expenditure required for modifying existing seed-dressing equipment. This decision can now only be applauded. With a greater consciousness of environmental pollution, any risks involving human and animal contamination with mercury, either direct from the seed, or from cattle cake and oil products derived from treated seed, would be unacceptable.

Different methods of applying seed dressings showed no advantages over the standard method of mixing seed with a dry dust. The process of acid delinting, whereby the seed coat fuzz is burnt off the seed with sulphuric acid, is extremely effective in destroying external infection, but was rejected at the outset because of the high costs involved in drying the seed after treatment. Machine-delinted seed, which it was hoped would become more evenly covered with a dry dust, gave a greater incidence of seedling disease than fully fuzzed seed. It was concluded that the process of machine delinting distributed pockets of infestation among previously uninfested seed. Liquid treatments, involving soaking the seed in solutions or suspensions, although shown to be very effective in the Sudan, where sun drying of the seed is possible (Tarr 1953, 1956), were also rejected because of the high costs of drying in Uganda (Wickens 1957). The slurry treatment, in which small volumes (of as little as 1 per cent by weight of the seed) of concentrated solution or suspension are applied, avoided the drying problems but did not produce greater disease control than dry dust treatments (Wickens 1958 and unpublished data).

Detailed observations of disease development in several seed-dressing trials confirmed earlier observations (Hansford 1933, 1934) that stem lesions (or the blackarm phase) contributed most to reduced yields. Logan (1960) calculated a disease index for the severity of stem lesions at the cotyledonary node or the node immediately above. The linear relationship between the derived disease-index and the mean number of bolls per plot, when plants were twenty weeks old, accounted for 52 per cent of the total variability but, at final pick, accounted for only 37 per cent, suggesting that factors other than mainstem lesions were affecting yield. Nevertheless estimates of the effect of seed dressing on yield were obtained by extrapolation and it was estimated that seed dressing with cuprous oxide would increase yield by 8 per cent. In another experiment the effect of severity of stem lesions on the yield of individual plants was obtained. The most severe lesions reduced total yields by 60 per cent and the smallest lesions by 25 per cent. The percentage loss dropped with successive picks indicating that diseased plants had partly compensated for earlier loss of crop. The figures agreed very closely with those obtained by Hansford (1933, 1934). Such estimates of loss of crop, however, can be extremely misleading because they take no account of the disease on the crop as a whole.

In Tanzania it was shown that the relationship between lesion grade and the number of bolls produced per plant differed markedly between plots grown from treated and those grown from untreated seed. In plots grown from untreated seed, increased boll production by healthy or slightly affected plants compensated to a large extent for plants killed by the disease and for the lower boll production of severely infected plants. In one case, the average numbers of bolls per healthy plant in dressed and undressed plots were 7.0 and 3.5 respectively and estimates of yield obtained from boll counts agreed very closely with measured yields, which indicated an increase in yield associated with seed dressing of 33 ± 16 lb seed cotton per acre (37 ± 18 kg per ha) or about 10 per cent. In further trials, the difficulties associated with spread of the disease from one plot to another were avoided. Four varieties differing in their levels of resistance to bacterial blight were used in a series of variety trials grown from dressed or undressed seed and the variety × site interaction was used to estimate yield differences associated with seed dressing. These trials indicated that the benefit from seed dressing could amount to as much as 343 ± 39.3 lb of seed cotton per acre (384.5 ± 44 kg per ha), or about 37 per cent, using cuprous oxide on seed of a susceptible variety in a season favourable for seedling infection. Even with a variety possessing some resistance, yield increases ranging from 7.8 to 20.2 per cent were indicated (Arnold 1965).

Although experiments of this type were not repeated in Uganda, there was no reason for supposing that the conclusions should not be applicable to large parts of the cotton growing areas of Uganda. Summing all the accumulated evidence, therefore, it was concluded that seed dressing could prevent severe losses of crop caused by bacterial blight and that on average, large-scale seed dressing was likely to be highly economic. Accordingly, assessments of the effects of seed dressing on yield were discontinued.

As the two resistant varieties, BPA and SATU, spread through the country, bacterial blight ceased to be a significant factor in crop loss. Seed dressing was continued, however, in order to keep the incidence of the disease at the lowest possible level and thereby reduce the chances of new, more virulent forms of the pathogen from spreading through the crop. There was also the possibility that seed dressing with cuprous oxide exerted some control on seedling diseases other than bacterial blight, although it was shown not to be effective in reducing seedling losses caused by *Sclerotium rolfsii*. Continued evaluation of newly developed, non-toxic seed dressings is therefore desirable.

Alternaria disease

A disease of cotton caused by or associated with *Alternaria* spp. was recognized in the 1930s in the cotton growing areas of Uganda close to the shores of Lake Victoria. Disease development is associated with extended periods of cool, wet weather which largely determine the stage at which the crop is damaged. The disease can occur in various forms according to weather conditions and varietal reaction. Leaf and cotyledonary infections on resistant varieties, or when the weather is less favourable for disease development, appear first as small, dry, light-brown lesions, usually conspicuous because of a dark-purple margin. The purple margin persists as the lesions extend and may become 2 mm wide. Final lesion-size at Namulonge appeared to be associated more with weather conditions than with host resistance. However, fewer lesions were found on resistant varieties. The central dead tissue of a lesion may show marked concentric zonation and may turn white, following secondary invasion by other fungi. Severe leaf infections can sometimes occur in which large areas of the lamina become chlorotic and wilt, leading to extensive premature leaf shedding. Boll infections are characterized initially by a small black lesion surrounded by a wide purple margin. As the lesion increases in size, the central blackened area dries out and produces a distinctive sunken lesion with a rough surface, while the purple margin often becomes narrower. Lesions often coalesce to form a large area of dead tissue which may extend through the boll wall and cause internal boll rotting. Lesions bounding or straddling carpel sutures cause them to split prematurely, allowing other species of fungi and bacteria to enter the immature boll, causing various degrees of boll rotting. As with leaf infections, boll lesions often become white from secondary invading fungi. Alternaria infections are also associated with increased levels of boll mummification in susceptible varieties.

There is some confusion over the naming of the species of *Alternaria* that causes disease in cotton. Hopkins (1931) originally attributed the disease to *Alternaria gossypina* (Thum) comb. nov., distinguishing this fungus from *A. macrospora* Zimmerm. on the basis of spore size. Since then various names, including *A. gossypii*, *A. cheiranthi* and *A. macrospora* have been ascribed to it. Classification of the *Alternaria* genus is difficult because of the lack of stable characteristics (Smith 1960). The conidia, more correctly called thallospores, vary considerably in size, shape and degree of septation even in a single culture and variations found on different culture media are even larger. In most reports, the disease has been attributed either to an *Alternaria* species with conidia that have a short beak or stem and are similar in size and shape to those of *A. tenuis* Nees, or to another form, with conidia that have an extended beak and are in other ways similar to those of *A. gossypina*. Isolation and examination of diseased leaf and boll tissues at Namulonge indicated that the long-beaked form of the fungus was a primary pathogen and the short-beaked form a secondary pathogen, associated with boll rotting after insect damage. More work is, however, required to clarify the situation.

There is no doubt that large differences in resistance exist among varieties. UKAN and UKA selections, bred in Tanzania, where the disease is of only sporadic importance, developed very severe boll rotting when compared with Namulonge locally adapted selections (Arnold, Walker & Tollervey 1966). Detailed examination of the UKAN stocks revealed strong associations between loss of yield through boll rotting and scores of the severity of Alternaria infections. The reactions of groups of progenies derived from individual strains suggested that there were heritable differences in resistance. The results also indicated how selection for yield in an environment favouring the disease would include a large component of selection for resistance to Alternaria boll rot (Brown 1966).

It seems probable that Alternaria disease was prevalent in certain areas of Uganda in the 1930s when it was first recognized. Little more was heard of the disease until 1966 when the UKAN and UKA selections were badly attacked and BP52 stocks appeared the most resistant. It therefore seems likely that the disease had been present during the intervening period and resistance had been attained in the BP52 stocks by selecting for improved yields (Arnold 1970b). It is fortunate that Bukalasa, Kawanda and Namulonge, where successive selections of BP52 were made, are all situated in areas conducive to disease development in most, if not all, cotton seasons.

Verticillium wilt

History in Uganda

Wilt diseases of cotton in Uganda were first reported at Bukalasa in 1932 (Nye 1933). In the following year, increases in the incidence of wilted plants both at Bukalasa and in nearby farmers' plots prompted work on selection for resistance. Marked differences in resistance among local selections were shown in trials grown on wilt-infested land. Thereafter a large programme of selection for resistance was carried out for several years by growing plants in wilt-infested plots. Attempts to increase the level of infestation in soils by ploughing in diseased plants met with some success. Uniform levels of infection were never achieved, but continual selection of apparently resistant plants over a number of years eventually gave rise to populations showing substantial increases in resistance. Two of the more important selections were

BP52 and B181. BP52 gave rise to the commercial variety for southern and western Uganda and, in later years, formed the basis of the NC seed issues from Namulonge (Chapter 8). B181 was used as a standard resistant variety for the transfer of resistance to a number of other stocks in Uganda and was used, with considerable success, in the development of resistant varieties in the USA (Bird 1974). Crosses with BP50, a susceptible selection made at Bukalasa, gave rise to the D and E stocks which, as DE715/6M, eventually replaced S47 in the north and east of Uganda in 1958. The relatively greater resistance of the DE715/6M stocks, compared with S47, was later confirmed by Wickens & Logan (1957).

The cause of the wilt disease was initially attributed to a species of *Fusarium* but in 1938 *Verticillium dahlia* Kleb was established as the prime cause (Hansford 1938). The species of *Fusarium* often isolated from wilted plants, were considered to arise from secondary infections. Subsequent identifications, based on the classification of Talboys (1960) showed that the fungus causing Verticillium wilt in Uganda has greater affinities with *V. dahlia* than with *Verticillium albo-atrum* Reinke and Berth. which causes a wilt of cotton in the USA, but there may be no justification for distinguishing between these two species (Isaac 1967).

Although the rapid spread of the disease caused considerable alarm at first (Nye 1936), yield losses were always considered to be small. Infection rates in farmers' plots rarely exceeded 10 per cent and were usually less than 2 per cent.

The Namulonge work

From 1950 onwards, the maintenance and, where possible, the increase of resistance to Verticillium wilt was considered an insurance against the further spread and development of the disease and the possibility of adaptation by the pathogen to the existing levels of resistance in the commercial crop. Extensive use was made of a stem inoculation technique, developed by Wickens (1951), whereby the fungus was inserted into a vertical slit made in the base of the stems of plants about six weeks old. Claims were made that reaction to artificial stem inoculation would accurately indicate the relative behaviour of strains under natural infection (Wickens 1951). Differences in resistance among varieties and selections in plant breeding material were demonstrated, but attempts at screening for resistance in the Albar stocks in the early 1950s were abandoned when no apparent progress was made. Little more was heard of the disease until 1968 when wilted plants were observed in experiments at Bukalasa and in two district trials. Although there was no indication of varietal differences in disease incidence at either site, it was decided that the resistance of all potential seed issues should be checked before release from Namulonge. In view of the minor importance of the disease in Uganda, it was decided not to use field inoculations which might unnecessarily infest land with the wilt organism. Stem inoculation in the glasshouse, using a technique based on those described by Erwin, Moje & Malca (1965) and Bugbee & Presley (1967), suggested that BPA 67 had slightly greater resistance than the old variety BP52 NC63 (Brown 1968) but that there was little variation for resistance in either variety. It was possible, however, that the technique of inoculating stems might have bypassed resistance mechanisms associated with the roots. Attempts were therefore made to find a root inoculation technique suitable for use on seedlings raised in the glasshouse. Various methods, aimed at producing infection through undisturbed roots, were tried. All failed to give consistent infection rates, and many produced no infection. Consistent seedling infection was induced, however, by washing out plants from the soil, dipping the roots in a micro-conidial suspension of the pathogen and replanting. This method did not give 100 per cent infection, but little variation was revealed either among varieties or among plants within one variety.

Later the opportunity was taken of growing plants on a plot of land that had become infested with *Verticillium dahlia* as a result of growing okra (*Hibiscus sabdariffa* L.) for a number of years. As expected, infection rates were much less than with root inoculation of seedlings but good differentiation among varieties was achieved. The high resistance of BPA and SATU was confirmed, but it was also shown that two reselections from Albar stocks, closely related to BPA, were susceptible. Differences between the most susceptible and resistant selections were striking, indicating that Verticillium wilt could be a factor limiting crop production in Uganda if susceptible varieties were grown. There were strong indications that resistance mechanisms associated with disease development after stem inoculation were less effective in the material studied than those associated with root inoculation. There were also indications that stem and root resistance were not necessarily associated. Although much more work is needed to clarify these aspects, the field inoculation technique was firmly established as a valuable tool in the resistance breeding programme.

Fusarium wilt

A cotton wilt caused by *Fusarium oxysporum* Schlecht. f. sp. *vasinfectum* (Atk.) Snyder and Hansen was confirmed in Uganda in 1957 (Wickens & Logan 1960*b*), when it was found in two localities in the Eastern Province. There were, however, several previous occasions when a *Fusarium* sp. similar, if not identical with the pathogen, had been isolated from diseased plants, but had failed to cause characteristic wilt symptoms in inoculation experiments (Hansford 1929). The symptoms of both wilt diseases are similar. Difficulties in infecting plants artificially under experimental conditions and, as we now know, an association between soil type and pathogen species, apparently led to considerable confusion.

Following confirmation of the occurrence of Fusarium wilt, surveys were made, in conjunction with the Uganda Department of Agriculture, in the 1958–59 and 1959–60 cotton seasons. The surveys were designed to assess the extent and severity of infections in the major cotton growing areas. It was found that, with the exception of one site at Bulopa in Busoga District, the disease appeared to be confined to areas within Bukedi District. In laboratory examinations the fungus was isolated only from plants infected with the root knot eelworm, *Meloidogyne incognita* var. *acrita*, Chitwood (Wickens & Logan 1960*b*). Later experiments by Perry (1960, 1963) showed conclusively that the presence of a species of root knot eelworm increased the severity of Fusarium wilt in a manner similar to that reported in the USA. The reasons for the association of fungus and nematode in this instance are not clear. Histological studies suggested that the fungus was not attracted to nematode entry points nor did it colonize the egg mass or gall tissue in preference to other apparently healthy roots. This was in contrast to infection mechanisms involving the interaction of root knot with *Phytophthora parasitica* (Dast) var. *nicotiana* (Breda de Haan) Tucker on tobacco (Powell & Nusbaum 1960). Furthermore, the cotton nematode larvae caused very little physical damage or necrosis to the root tissue. Field trials with nematocides confirmed that reduction in the population of root knot nematodes in soils infested with *Fusarium* markedly reduced the incidence of wilt, but practical and economic factors ruled out the general use of chemical control.

Evidence collected by numerous observers during the wilt surveys indicated that Fusarium wilt was closely associated with the light, sandy soils, common in areas around the eastern shores of Lake Kyoga. Verticillium wilt was more often associated with the heavier soils in central Uganda and occasionally with other soil types that contained unusually large proportions of organic material. A broadly similar association of disease occurrence and soil type occurs in Tanzania, where Fusarium wilt is particularly prevalent in the sandy lake-shore regions.

The surveys also revealed that Fusarium wilt had not spread to all the areas of sandy soil in which root knot nematodes were active, suggesting that greater losses might be expected in future. However, results from the second year of the survey (1958–60) indicated a lower disease incidence in all areas, and from 1960–61 onwards no serious outbreak of Fusarium wilt was reported. It appears that the replacement of the susceptible variety S47 by the resistant varieties BP52 in Busoga District and DE 715/M elsewhere, started in the 1959–60 cotton season and completed the following year, may have contributed either wholly, or in combination with weather conditions, to the spectacular reduction in disease incidence. The relatively high levels of resistance of BP52 and DE 715/M were not, however, the result of conscious selection. Both varieties had been developed from lines containing good resistance to Verticillium wilt and neither had been tested for Fusarium wilt resistance prior to their release for commercial use (Chapter 7).

Although it is believed that Fusarium wilt could only become endemic in relatively small areas around the eastern shores of Lake Kyoga, the potential for increased cotton production in these areas means that it is important to check for resistance both to Fusarium and Verticillium wilt before releasing seed into commercial production.

Following the outbreak in 1957, reliable techniques for root inoculation with Fusarium wilt, were rapidly developed at Namulonge (Wickens & Logan 1960b). Although they were not used extensively in Uganda, they were later adopted in Tanzania as part of a major programme for the development of wilt resistant varieties (Perry 1962; Wickens 1964).

Seedling diseases

Seedling diseases of cotton were generally considered to be of minor importance in Uganda. Damage by *Rhizoctonia (Corticium) solani* Kuhn, causing damping-off in seedlings up to the first true-leaf stage and 'sore-shin' on older plants, was reported from time to time. Very occasionally dead seedlings were found with a white fungal mycelium around the hypocotyl, at or just below the soil surface. Laboratory isolations from such plants revealed the presence of a species of *Rhizoctonia* as well as a *Fusarium* sp., supporting the local belief that seedling deaths could be caused by a disease complex involving both fungi. However, attempts at reinoculating seedlings produced no conclusive results.

In the late 1960s, it appeared that the frequency of seedling deaths associated with white mycelium (later shown to be *Sclerotium rolfsii*) was increasing on the Namulonge farm. Moreover the disease appeared along rows of seedlings rather than on isolated plants and it was feared that stand losses in small-plot experiments could be excessive.

Severe losses of stand in a crop of French beans in 1969 stimulated investigations into both the potential importance of seedling diseases caused by *Sclerotium rolfsii* and possible measures for their control (Brown 1970). Local opinion was that losses of cotton seedlings were greatest after a French bean crop. However, the spectacular incidence of the disease on beans in the second rains of 1969 was associated with

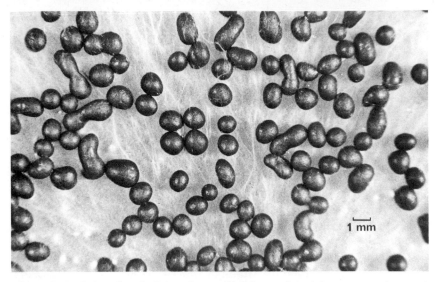

Figure 6.2. Sclerotia of *Sclerotium rolfsii* formed in laboratory culture. A regular supply of sclerotia was essential for experimental work in the field and glasshouse.

profuse growth of *S. rolfsii* on debris from the preceding mixed elephant grass and short grass ley. Furthermore the fungus was found only on debris which was just below or at the soil surface, suggesting that deep burial of crop debris might reduce disease incidence. In all subsequent outbreaks of seedling damage caused by *S. rolfsii* at Namulonge, including those on cotton, French beans, soya beans and groundnuts, there was always an accompanying development of mycelium on debris from the previous crop.

In an experiment designed to study the development of *S. rolfsii* on crop debris, French beans sown on land nine and twenty-nine days after breaking-up an elephant grass ley yielded at the rate of 313 and 1400 kg per ha, respectively. The lighter yield was associated with very large stand losses in seedlings and young plants. Practically all dead plants were infected with *S. rolfsii*. A trapping technique, in which pieces of dead stems of elephant grass were placed at random in the soil, showed that mycelial growth of *S. rolfsii* was far greater on all plots that had been broken-up on the later date. Comparisons of bare fallow plots with plots sown to beans suggested that the beans may have caused slightly greater growth of the fungus, but the differences were relatively small in comparison with those associated with dates of breaking-up the elephant grass ley. Furthermore, results from plots containing natural infestation and those from plots supplemented with sclerotia grown in the laboratory indicated that the fungus was a common and evenly distributed component of the soil microflora.

Other experiments were aimed at finding suitable chemical control methods. A glasshouse technique for screening compounds was developed and field tests with a few compounds gave very encouraging results (Brown 1971; Brown & Beteise-Hbenye 1973), although only one effective non-mercurial compound was commercially available at the time. All compounds were tested as dry seed dressings in spite of the fact that some were formulated as wettable powders and some were not entirely suitable for use on fuzzy seed. Ironically one compound which tended to separate from seed after application consistently gave good disease control.

Far more work is required to assess the economic importance of *S. rolfsii*. All experimentation has so far supported the theory, built up from casual observation, that the fungus is primarily a saprophyte. Sometimes it becomes a large and dominant component of the microflora on crop debris in the early stages of decomposition. If seedlings of cotton, French beans, soya beans and groundnuts come into contact with actively growing mycelium of *S. rolfsii*, they can be invaded and killed. The disease has not been observed on sorghum or maize seedlings at Namulonge.

The introduction of more intensive farming methods at Namulonge may have been responsible for the increase in the incidence of the disease. Crops such as maize and sorghum produced far greater quantities of stem and leaf debris than formerly, and in some instances crop debris could be ploughed in only a few days before sowing the next crop. Thus maximum mycelial growth coincided with seedling establishment and consequent seedling losses. It seems possible that modification of crop rotations or cultivation practices would be sufficient to control the disease but if this were impracticable or uneconomic, chemical control methods are available. All evidence indicates that *S. rolfsii* will only colonize debris contained in about the top four inches of soil, where most, if not all, seedlings are infected. However, the deep incorporation of debris from crops, such as maize or sorghum, is fraught with practical difficulties and it is almost impossible to incorporate elephant grass rhizomes deep enough to prevent the growth of the fungus.

The inability of *S. rolfsii* to grow on deeply buried trash could be associated with its rather specific requirements of relatively high oxygen and low carbon dioxide concentrations (Griffin & Nair 1968). Any treatment, such as timely irrigation, which alters the balance of oxygen and carbon dioxide in the soil may be beneficial in certain instances.

The boll rots

Two types of boll rotting have always been distinguished in Uganda: one caused by primary pathogens capable of infecting intact bolls, and the other by secondary micro-organisms which can infect only damaged bolls. The differentiation of primary and secondary pathogens is, however, imprecise. For example, *Glomerella cingulata* (Stonem.) Spauld and von Schrenk (*Colletotrichum gossypii* Southw.) can behave as a very weak primary pathogen but does far more damage when it develops after insect feeding (Leakey & Perry 1966). *Nematospora gossypii* Ashby and Nowell is transmitted by cotton stainers (*Dysdercus* spp.) (Frazer 1944). The fungus infects lint hairs, which are stained yellowish brown and are weakened. In the hotter, drier areas of Uganda, lint staining, in some seasons, can be extensive. Unlike most secondary boll pathogens, *N. gossypii* has evolved as a successful pathogen in association with the cotton stainer insects. However, neither *G. cingulata* nor *N. gossypii* could contribute to crop losses if boll-feeding insects were absent, and for practical purposes can be regarded as secondary invaders. The division of boll-rotting organisms into primary and secondary pathogens is therefore useful.

Of the primary pathogens *X. malvacearum* had caused considerable yield losses and extensive lint staining prior to the introduction of BPA and SATU, and Alternaria disease had occasionally contributed to yield losses, particularly in the more humid Lake Shore regions. Later issues of BP52 were resistant to *Alternaria* and with the introduction of BPA and SATU boll rotting initiated or directly caused by primary pathogens was virtually eliminated. However, there were still extensive, and sometimes severe losses caused by a wide range of secondary pathogens. Accurate estimates of yield losses from boll rotting in commercial production are not available. Detailed observations at Namulonge in the 1969–70 and 1971–72 seasons indicated a minimum loss of about 6 per cent in cotton that had received at least eight insecticidal sprays, and up to 24 per cent in an unsprayed trial. The assessment of boll damage, on which these estimates were based, included the direct effects of insect feeding, which would have been apparent even if subsequent rotting had not occurred, but did not take into account shed bolls. Studies of the association between boll rotting and insect damage suggested that sucking and biting insects provided practically all the entry points for secondary pathogens; other factors, such as damage caused by spraying and weeding operations, were unimportant (Brown 1970, 1971; Brown & Beteise-Hbenye 1973).

Because complete control of insect damage to cotton bolls seemed unlikely, the possibilities of developing varieties with greater resistance to boll rotting after insect attack were examined. It was not anticipated, however, that genotypes would be found that would resist rotting after extensive damage by insects. Rather the search was aimed at identifying genotypes with less severe rotting after relatively minor insect damage. At first it appeared that differences in the incidence of boll rot could be detected in material derived from wide range crosses (Chapter 8), but examination of a greater range of material in the following season did not reveal any worthwhile differences. The material examined varied widely in other characters which could have indirectly affected the incidence of boll rotting. Differences in boll size, numbers of bolls per plant and attractiveness to insect pests could all affect the extent of boll rotting. Moreover, intra-seasonal fluctuations in insect populations could markedly influence the incidence of rotting in varieties that matured during different periods.

Other evidence obtained in the 1970–71 and 1971–72 cotton seasons strongly indicated that Albar breeding stocks contained very little, if any, variation for resistance to boll rotting caused by secondary pathogens. In neither season could differences be detected among commercial varieties including BP52 NC63, SATU 65 and a range of BPA seed issues, in trials which received routine applications of insecticide (sprayed) and those which received none (unsprayed). As expected, there were large differences in yield between the sprayed and unsprayed trials. In 1970–71, boll rot incidence was slightly greater in the sprayed trial, but there was more than twice the number of bolls in the sprayed trial at the time of picking when rotting assessments were made. In contrast, the incidence of rotting in the unsprayed trial in the 1971–72 season was more than four times that in the sprayed trial. All the varieties behaved similarly even though BP52 NC63 was originally developed under conditions of no insect control and SATU 65 was developed in a different ecological zone. Suggestions that resistance to boll rot and to boll-feeding insects in the later BPA seed issues had deteriorated as a result of being selected under increasingly intensive regimes of insect control appeared, therefore, to be unfounded.

The contents of the developing boll form an ideal substrate for the growth of many micro-organisms and there appeared to be little hope of increasing resistance to boll rotting easily or rapidly, even in bolls subjected to relatively minor damage. Most infected bolls rot so severely that laboratory examinations cannot identify any particular fungal or bacterial cause. In less severely infected bolls, species of *Nematospora*, *Alternaria* and *Fusarium* were the most common fungi found at Namulonge.

Conclusions

The work in pathology at Namulonge spanned a period during which a major disease, bacterial blight, was effectively eliminated from the commercial cotton crop, but neither the pathologists nor the cotton breeders can claim sole responsibility for this achievement. Rather it was the result of the close teamwork that characterized the Namulonge approach. Not all the cotton pathology work was so successful, however, nor did it lend itself so readily to the team approach. The original programme concerned only bacterial blight. It was formulated against a background of twenty years' observation of the disease, but on relatively little experimentation. Only some of the items in the programme were tackled immediately. Others, such as the study of the physiology of parasitism, could never have been investigated in depth with the manpower and facilities available.

Practical problems concerning the development of improved inoculation techniques and control of the disease by seed treatment were given priority. In retrospect the merit of this policy can clearly be seen: seed dressing was established as a routine procedure of economic benefit and the inoculation techniques were used extensively both in resistance breeding programmes and in more fundamental studies. Following the successful work on immediate practical problems, however, increasing amounts of time and effort were devoted to some of the more fundamental aspects of variation in disease expression (Chapter 7).

Although bacterial blight was the most damaging disease of cotton when the programme was formulated, other diseases were subsequently investigated, not because they had become established as major causes of crop loss on a national scale, but because it was feared they might become so. Work on these diseases was inevitably intermittent, commencing after a locally severe outbreak and stopping either when other work was considered relatively more important, or when interest was lost. The depth of the work often corresponded with the particular interests of the individual, which did not always result in long-term continuity of approach. As we have seen, after some initial work on Verticillium wilt in the early 1950s no further work was done until 1968, when two unusually severe local outbreaks caused renewed interest in the disease. Even so, it was an accidental infestation of a plot of land with *Verticillium dahliae* that finally gave the opportunity to assess resistance levels in breeding stocks. It will be interesting to see if, for similar reasons, there is a resurgence of interest in Fusarium wilt, which was studied intensively from 1958 to 1962.

With bacterial blight under control, greater attention could well be given to other diseases. Conclusions about Alternaria disease, for example, were based almost entirely on field observations, and the procedures for routine screening of breeding material that were developed were based on natural infection. There is a need for studies on the epidemiology of the disease, as well as on the life cycle of the pathogen. Similar remarks apply to the leaf blight caused by *Ramularia areola* which is locally important, particularly in the northern and eastern areas of Uganda and to which recent cotton selections appear to vary in resistance.

Whatever disease may loom into prominence, however, there is no doubt that resistant varieties provide the only immediate answer to disease control under Uganda conditions. Nonetheless, work in plant pathology should continue to adapt to the needs of changing agricultural systems, in which the scope for disease control may become wider and the relative importance of different diseases may change.

7

Resistance breeding

M. H. ARNOLD, N. L. INNES and S. J. BROWN

Introduction

One of the biggest contributions that the plant breeder has made to crop improvement has been through the provision of disease resistant varieties. Although successful projects have been recorded in nearly all crops, procedures for resistance breeding have, in most cases, followed the same general pattern: first, the development of a suitable screening technique; second, a search for resistant material (usually extending outside the species); and third, the transfer of resistance to advanced breeding stocks, often involving a programme of back-crossing.

During the course of such projects large numbers of major genes for resistance have been described. Indeed, failure to detect Mendelian segregation has sometimes been considered to be a disadvantage. In our work in East Africa, however, we came to regard the transfer of major-gene resistance as the exception, and to look upon resistance breeding as no different, in general terms, from breeding for other attributes of economic importance, such as increased yields or better quality.

There are three main aspects to the changes in approach that are involved. The first is to distinguish between immunity and resistance: with immunity, no disease symptoms are expressed; with resistance, disease symptoms are expressed but the damage done is restricted. The second is to regard the resistant state as normal and the fully susceptible state as rare: if a variety suffers from a disease, it is often described as susceptible, but if it is not entirely destroyed by the disease it is more useful to think of it as possessing some resistance. The third aspect is to search for small differences in resistance, so that they can be exploited by appropriate breeding methods.

If this approach is adopted, the problem then becomes one of breeding with continuous variation; of partitioning variation into its genotypic and environmental components; of studying the interaction of genotypes with environment; and of selecting and recombining to effect progressive changes in the population mean. These procedures are common to most plant breeding programmes. The factor that distinguishes resistance breeding is that variation is encountered in two organisms instead of one. Genetic variation in the host must be studied in relation both to genetic variation in the parasite and to the effects of environment on the association between the two. It is variation in the host–parasite relationship, therefore, that is the key to resistance breeding.

Resistance to bacterial blight

We developed these concepts from work on resistance to bacterial blight of cotton (see Chapter 6 for description of the disease). They applied in part, however, to breeding for resistance to other diseases as well as to certain insect pests of cotton, and there seems to be no reason why they should not have wide application in resistance breeding in other crops. To consider the principles in more detail, however, we must first describe how they were applied to breeding in cotton for resistance to bacterial blight.

Resistance genes

Heritable resistance to bacterial blight was first demonstrated by Knight & Clouston (1939) in the Sudan. Subsequently Knight screened more than 1000 different wild and cultivated accessions of diploid and tetraploid species of *Gossypium*, and published his results in a series of papers in the *Journal of Genetics*. He identified ten major genes, to which he ascribed the symbols B_1 to B_{10} (Knight 1957, 1963). Of these genes (described in Table 7.1) eight were dominant or partially so. An unnamed gene of low potence transferred by Knight from *G. herbaceum* L. to *G. bardadense* was given the symbol B_{11} (Innes 1966). The B_6 gene from *G. arboreum*, originally described by Knight as a modifier with no effect by itself, but which intensified resistance to near immunity when combined with B_2, was found to be a recessive gene which conferred intermediate resistance in a homozygous state (Saunders & Innes 1963). With the exception of B_8, a recessive gene in *G. anomalum* Waw. & Pey., all the *B* genes were transferred to *G. barbadense* stocks used in the Sudan breeding programme. In addition, in the Sudan, most of Knight's *B* genes have been, or are now being, transferred to upland cotton (*G. hirsutum* L.). No *B* gene by itself conferred immunity, although several gave strong resistance, and digenic and trigenic combinations were synthesized to produce near immunity and to provide a widely-based gene pool (Innes 1974*b*). Other genes for resistance, given the symbols B_9 and B_{10}, were found in the upland variety Allen (Lagière 1960). (A key to the names of cotton varieties and code letters for selections mentioned in the text is given in Appendix 8.1.) These genes are not homologous with Knight's B_9

Table 7.1. *List of major genes or polygene complexes conferring resistance to bacterial blight as described in the literature*

Gene symbol	Description and source	References
B_1	Weak, dominant gene obtained from Uganda B31 (*Gossypium hirsutum*)	Knight & Clouston (1939)
B_2	Strong, dominant gene from Uganda B31; also recorded in Albar from west Africa, UKBR from Tanzania, varieties in the USA. (All *G. hirsutum*)	Knight & Clouston (1939) Innes (1965*b*) Innes (1969*a*) Brinkerhoff (1970)
B_3	Partially dominant gene from Schroeder 1306 (an off-type *G. hirsutum* var. *punctatum*)	Knight (1944)
B_4	Partially dominant gene from Multani strain NT 12/30 (*G. arboreum*)	Knight (1948*b*)
B_5	Partially dominant gene from Grenadine White Pollen (a perennial *G. barbadense*)	Knight (1950)
B_6	Recessive gene from Multani strain NT 12/30 (*G. arboreum*); and possibly from Tanzania, UKBR 61/12 (*G. hirsutum*)	Knight (1953*a*); Saunders & Innes (1963) Innes (1969*a*)
B_7	Gene from Stoneville 20 and other stocks from the USA (*G. hirsutum*). (Dominance of this gene is dependent upon the genetic background)	Knight (1953*b*) Green & Brinkerhoff (1956); Innes & Brown (1969)
B_8	Recessive gene from *G. anomalum*, an uncultivated diploid species from Africa	Knight (1954)
B_{9K}	Strong, dominant gene from Wagad 8, an Indian commercial variety (*G. herbaceum*)	Knight (1963); Innes (1965*a*)
B_{9L}	Strong, dominant gene from Allen 51–296 from west Africa (*G. hirsutum*)	Lagière (1960); Innes (1965*a*)
B_{10K}	Weak, partially dominant gene from Kufra Oasis in Libya (*G. hirsutum* var. *punctatum*)	Knight (1957); Innes (1965*a*)
B_{10L}	Weak gene from same source as B_{9L}	Lagière (1960); Innes (1965*a*)
B_{11}	Weak gene from same source as B_{9K}	Innes (1966)
B_{In}	Dominant gene from an unknown variety from the USA (*G. hirsutum*)	Green & Brinkerhoff (1956); Brinkerhoff (1963)
B_N	Dominant gene from Northern Star, a variety from the USA (*G. hirsutum*)	Green & Brinkerhoff (1956)
B_S	Dominant gene from Stormproof 1, a variety from the USA (*G. hirsutum*)	Green & Brinkerhoff (1956)
B_{Sm}	Polygene complex found in Stoneville 2 B and Empire, varieties from the USA (*G. hirsutum*)	Bird & Hadley (1958)
B_{Dm}	Polygene complex found in Deltapine, a variety from the USA (*G. hirsutum*)	Bird & Hadley (1958)

and B_{10}, and to avoid confusion were redesignated B_{9L}, B_{10L} and B_{9K}, B_{10K} respectively by Innes (1965a).

In the Sudan environment, it was shown that the key to resistance in the important Albar stocks was the major gene B_2 fortified by genes of small effect (Innes 1963; 1965b). In the USA, Green & Brinkerhoff (1956) identified three major genes in upland varieties; these were given the symbols B_{In}, B_N and B_S (Brinkerhoff 1963). Polygenic complexes for resistance in the upland varieties Stoneville 2B and Deltapine were found by Bird & Hadley (1958) and were given the symbols B_{Sm} and B_{Dm} respectively. Table 7.1 summarizes the genes for resistance which have been described in the literature.

Discontinuous and continuous variation

Because Knight was unable to detect any segregation for resistance in *G. barbadense* stocks in the Sudan, he was faced with the need to transfer resistance from *G. hirsutum* and from the wild species of *Gossypium* to the locally-adapted, long-staple varieties of *G. barbadense*. At that time, consumers of Sudan cotton would not accept any change in lint quality, and Knight thought that the quickest way of producing acceptable resistant varieties would be through a programme of backcrossing to the local varieties. Although he recognized the importance of minor genes in intensifying the resistance conferred by major genes, his breeding methods were such that minor genes would have been lost in successive backcrosses to the susceptible parent, particularly when they showed no dominance for resistance. Indeed, the presence of minor genes was to some extent regarded as a complicating factor which tended to obscure major-gene segregation.

Knight followed the segregation of major genes in backcross generations by scoring disease severity on a scale of 0 to 12 after leaf inoculation (Knight & Clouston 1939; Knight 1946), and dividing the segregating populations at the positions of minimum frequency in the distributions.

When Knight's techniques were applied to cotton in Uganda, however, it soon became clear that his results were not repeatable. The leaf-spray technique was used for screening material during the 1940s in the cotton breeding programme at Kawanda Research Station. The results were difficult to interpret and major-gene segregation could not be demonstrated.

After Knight had visited Kawanda to see the work for himself, it was generally accepted that environmental conditions were such that the expression of symptoms was irregular and that natural spread of the disease made it impossible to distinguish between lesions arising from the inoculation and those arising naturally (Sir Joseph Hutchinson, personal communication). This contrasted with the situation in the

Sudan where consistent environmental conditions and the absence of rainfall during the incubation period gave very consistent results (Hutchinson 1959).

It was these conclusions that gave the impetus to the work on inoculation techniques at Namulonge (Chapter 6). Even with the more precise techniques that were developed, however, results were not entirely consistent from one test to another, and crosses between local material and varieties carrying Knight's major genes seldom showed definitive segregation patterns in F_2 populations.

Application of the new inoculation techniques to breeding material in Tanzania led to several important discoveries. First, it was shown that resistance to bacterial blight had evolved in cotton selected for increased yields in an environment where the disease was prevalent. Second, using replicated progeny rows to partition variation for resistance into its genotypic and environmental components, high heritabilities (in the broad sense) were demonstrated and, after several generations of intensive selection, a series of resistant lines was produced (Arnold 1963; 1970b). Third, when representatives of these resistant lines were evaluated in the Sudan, major-gene resistance was detected and the resistance was shown to be little different from, if not identical with, that conferred by the gene combination $B_2 B_6$ (Innes 1969a).

Here, therefore, we had the first demonstration of a situation that had already been inferred: that what could be treated as major-gene segregation in one set of environmental conditions could more usefully be regarded as a problem of continuous variation in another. But, perhaps more important for the plant breeder, it became clear that valuable resistance genes were present and waiting to be exploited in the locally-adapted populations of cotton.

Variation in the pathogen

Up to this time no specific study had been made in East Africa of variation in virulence of the pathogen. Several workers had observed differences among isolates of *Xanthomonas malvacearum*, both in the characteristics of the cultures on artificial media and in apparent virulence when used for inoculation. All of those concerned with resistance breeding had worked against a background of knowledge of the possibility of a breakdown in resistance, such as that known to have occurred with black stem rust of wheat (caused by *Puccinia graminis* Pers. f. *tritici*). For this reason, a mixture of isolates was always used for inoculation so that a broad-based resistance was more likely to be selected, which was less likely to be broken down by more virulent forms of the pathogen. In Tanzania and in Uganda infected material was collected from a wide area of the commercial crop, new isolations made for each series of inoculations and never less than six different isolates mixed to form the inoculum. In the Sudan, infected trash was collected from a wide area of the Gezira scheme. As more staff became available, however, variation in the pathogen was studied in greater detail.

In Tanzania, Cross (1963, 1964) compared isolates of *X. malvacearum* obtained from the more resistant (UKA) and the less resistant (UK) varieties and concluded that more virulent forms of the pathogen had arisen. Hayward (1964) showed that two types of the bacterium occurred which could be distinguished both by biochemical tests and the patterns of lysis obtained with a series of bacteriophages. Marked differences in virulence were also shown to occur among isolates of the bacterium made in Uganda (Brown 1964).

Although it became essential, therefore, to gain some understanding of the magnitude of such differences in relation to the magnitude of genetic variation for resistance, it was known from experience already gained that such studies would be difficult to interpret unless due account was also taken of differences associated with environment. The work on the host–parasite relationship that was started at Namulonge in 1962 was therefore designed to study the interrelationships of these three factors: variation in resistance of the host, variation in virulence of the pathogen and variation in components of the environment (Arnold & Brown 1968). These studies constituted an important support programme for the continued release of resistant commercial varieties and we must briefly discuss the results.

The host–parasite relationship

It was shown that virulence in *X. malvacearum* was not related to phage-type and that variation was such that it was more useful to regard virulence in the pathogen, like resistance in the host, as showing continuous variation. It was suggested that the host–parasite relationship could be regarded as a dynamic system in which disease expression was a function of the interactions of environmental factors and two polygenic systems, that of the host and that of the parasite. Although more virulent forms of the pathogen occurred they did not completely break down the resistance of the more resistant genotypes and the variability was such that further advances in resistance could be made, by appropriate breeding methods.

The situation revealed by these results was, however, complex. The levels of resistance of the various host populations, as well as their relative levels, varied with the pathogen culture used and with the environmental conditions. Nonetheless a considerable degree of order could be seen in the data when each host–parasite combination was regressed on the mean result over all combinations (Figure 7.1). This analysis was essentially that proposed by Yates & Cochran (1938) and developed by Finlay & Wilkinson (1964) for analysing genotype–environment interactions (Chapter 8). In this case, however, it was used to express the interaction between environmental conditions and the host–parasite relationship. The fact that patterns could be detected in these complicated results was further illustrated by relating variation in disease expression to easily measured components of the environment, such as temperature measured in °C h per day for the fifteen-day period from inoculation to scoring. Simple quadratic equations accounted for between 49 and 88 per cent of the observed variation, and the derived parabolas indicated optimum temperatures that were consistent with field experience of disease expression.

In the same series of experiments it was shown that variation for resistance expressed in one environment was not equally expressed in another. Populations that showed considerable variation for resistance in East Africa showed virtually none under Sudan conditions, an

Figure 7.1. Linear regressions ($y = bx + c$) of single host–parasite combinations on the mean of all six studied using NCC 38 (from Arnold & Brown 1968).

	IL 47/10	UK58	A(57)12
b	1.51±0.21	0.77±0.18	0.72±0.10
c	−54.8±11.8	31.1±10.4	24.7±5.9

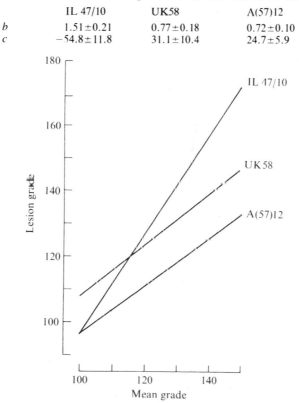

observation that went a long way towards explaining the anomalies previously found between Knight's results and those of workers in Uganda. Moreover the results also showed how selection for resistance under conditions that maximized intrapopulation variance could result in the production of genotypes that expressed their resistance under conditions in which the unselected, parental population had shown no worthwhile resistance. It was subsequently concluded that such genetic assimilation (Waddington 1961) could have been an important factor in the evolution, not only of disease resistance but also of other desirable characteristics of crop plants (Arnold 1970*b*).

Although these results went some way towards explaining the observed complexities there was still a great deal to be understood about the genetics of resistance. We developed the hypothesis that resistance to bacterial blight was essentially polygenic and that some of the polygene complexes could segregate as genes of large effect. The situation we envisaged was similar to that described by Thoday (1961) and Spickett & Thoday (1966) for the inheritance of sternopleural chaetae in *Drosophila*. It remained, however, to discover enough about the distribution of genes and the types of gene action involved so that we could devise breeding systems that would ensure the maintenance of resistance in commercial seed issues.

The genetics of host resistance

The statistical treatment of the grades or measurements of lesions that develop after inoculation has been discussed by Arnold (1963), Arnold & Brown (1968) and Innes (1974*b*). Data from stem, boll and hypocotyl inoculation show heterogeneity of variances which can usually, but not always, be eliminated by transforming to a logarithmic scale. Leaf lesion grades do not normally require transformation. Even after transformation, however, results often contained apparent anomalies which were difficult to interpret.

Figure 7.2 illustrates some examples of results from studies on the genetics of resistance (M. H. Arnold and S. J. Brown, unpublished). The diagrams show percentage frequency distributions of F_1, F_2 and the first backcross progenies from a cross between Bar 12/16 (homozygous for $B_2 B_6$) and a susceptible inbred line selected from IL 47/10. Results from three separate experiments are shown. The experiment shown in Figure 7.2*a* was boll-inoculated in the field and the other two were hypocotyl-inoculated in the greenhouse on two different occasions. Figure 7.2*a* shows frequency distributions typical of polygenic inheritance. The F_1 mean lies near the value of the mid-parent, suggesting mainly additive inheritance and the distributions give no indication of major-gene segregation. In Figure 7.2*c* the position of the F_1 mean relative to the mid-parental value suggests a high degree of dominance for resistance, while the bimodal distributions of the F_2 and of the first backcross to the susceptible parent clearly suggest major-gene segregation. The frequency distributions in Figure 7.2*b* are in several respects intermediate between those of Figures 7.2*a* and 7.2*c*. In general, greater variability was expressed in the third experiment than in the other two.

In these three experiments, samples of the same seed lots were used and the plants were all inoculated with the same bacterial isolate. Differences among the results can therefore be attributed, apart from sampling errors, to differences in environment both internal, embracing the differences between boll tissue and hypocotyl tissue, and external, embracing effects of field and greenhouse conditions.

Although the parental populations were highly inbred, their phenotypic variances differed from one test to another and in Figure 7.2*c*, the F_1 population shows a wider distribution than expected from a cross

Figure 7.2. Percentage frequency distributions of lesion grades, after inoculation with *Xanthomonas malvacearum*, in parental populations of Bar 12/16 (resistant) and IL 47/10 (susceptible); together with F_1, F_2 and first backcross progenies derived from them. (*a*) Results of boll inoculations; (*b*) and (*c*) results of hypocotyl inoculations from two further experiments using samples of the same seed. In each case, the broken line represents the mid-parental value.

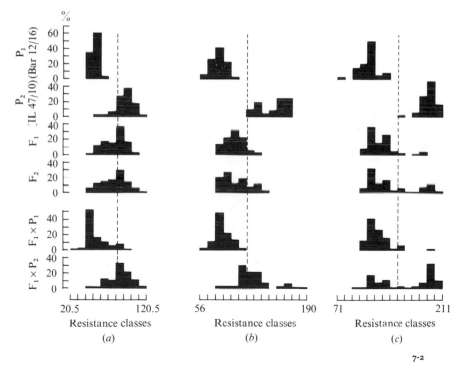

Table 7.2. *Demonstration of variation for resistance in a 'susceptible' variety*

Estimates of genotypic variances (g_{ii}) and broad-sense heritabilities (g_i^2) for resistance of three cotton varieties to two cultures of the pathogen. The estimates were obtained from the differences between seed plants and clonal plants in their reactions to boll inoculation. The numbers of bolls (n) mean lesion grades ($\bar{x}$) and intrapopulation variances (s^2) are also shown

| Culture | Variety | Seed plants | | | Clonal plants | | | g_{ii} | g_i^2 |
		n	$\bar{x}$	s^2	n	$\bar{x}$	s^2		
NCC 38	IL 47/10	55	83.0	106	73	87.3	35	71	0.67
	BP52 NC63	42	81.2	239	44	86.2	169	70	0.29
	A(57)12	68	68.7	258	91	72.6	103	155	0.40
NCC 40	IL 47/10	77	79.9	183	95	67.6	91	92	0.50
	BP52 NC63	38	72.9	361	38	80.5	220	141	0.61
	A(57)12	68	53.6	183	106	45.7	2	181	0.99

between two homozygous parents. The question that arises is whether part of the parental variation was genotypic and, fortunately, some independent evidence can be adduced to provide an answer. Innes (1968) discussed the apparent genetic break-down of a line of *G. barbadense* (Bar 14/41) that was initially described as homozygous for $B_2 B_6$. Taking into account the distribution of reaction grades that developed from this line from one generation to the next, his failure to recover typical $B_2 B_6$ resistance upon repeated selection and self-fertilization as well as Arnold's (1963) demonstration of the evolution of polygenic resistance, Innes inferred that one or other of the genes, probably B_6, was a polygene complex that segregated under certain circumstances. If this interpretation is correct it would not be unlikely that some segregation would have occurred in the population of Bar 12/16 maintained at Namulonge.

As far as IL 47/10 is concerned, S. J. Brown (unpublished) demonstrated in more than one experiment that this variety, which was used as a susceptible control, was segregating for resistance. In an attempt to eliminate genotypic variation in a population of plants, he developed a technique for reproducing cotton plants vegetatively, so as to produce from a single plant a clone of about 120 plants. Boll lesions that developed after inoculation of clones were then compared with those that developed from a random sample of the same number of plants grown from seed. The results, which are summarized in Table 7.2, clearly showed that in IL 47/10, previously thought to be entirely susceptible, genotypic variance could be detected comparable to that in material known to be segregating for resistance, such as the other entries in the Table.

Most of Knight's original work on major genes was done with *G. barbadense* and, under Sudan conditions, clear patterns of segregation were obtained. Even in the Sudan, however, such clear-cut patterns are not obtained when the genes are segregating in a background of *G. hirsutum*. In Knight's original grading system of 0–12, grade 11 was later omitted and reference to grade 11 in the results presented in this chapter corresponds to his grade 12. In grades 0–3 lesions are minute, reddish in colour and never wet. At the other end of the scale, grades 8–11 refer to lesions that are at least 2 mm across, black, angular and wet at first. Although grade 11 lesions are found after inoculating susceptible lines of *G. barbadense*, grades found in *G. hirsutum* lines rarely exceed 10.

Figure 7.3. Percentage frequency distributions of leaf lesion grades, after inoculation with *Xanthomonas malvacearum*, in parental populations of *Gossypium barbadense* and *G. hirsutum* as well as the F_2 populations derived from them. The broken lines indicate the mid-parental values. (Diagrams constructed from data presented by Innes 1968.)

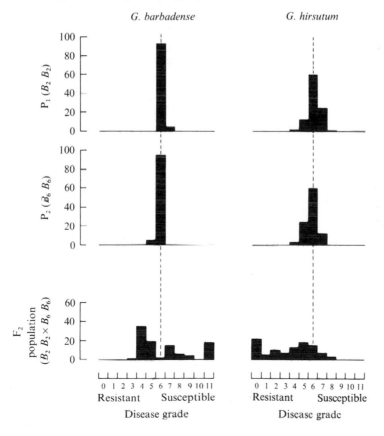

Figure 7.3 shows examples of results obtained in the Sudan from crosses between lines homozygous for B_2 and others for B_6, either on a background of *G. barbadense* or of *G. hirsutum*. Results for the *G. barbadense* cross give F_2 segregation in close agreement with the expected proportions of nine highly resistant, four intermediate and three susceptible. With the same cross made between lines of *G. hirsutum*, however, the parental populations show wider distributions and segregation in the F_2 generation is far less clear-cut.

As more and more data were accumulated, it became apparent that results in the Sudan were not all as distinct as had been presumed from Knight's original publications. Furthermore, major-gene segregation could sometimes be detected under Uganda conditions, where earlier tests had indicated only continuous variation. In addition to the example given in Figure 7.3, discontinuous segregation patterns were detected in F_2 populations derived from a number of different crosses (Innes & Brown 1969).

Although genes of large effect (or closely-linked blocks of polygenes) clearly occur, therefore, Mendelian segregation can be detected only under particular conditions. These conditions include the absence of additional genes of minor effect and the presence of environmental conditions necessary for adequate penetrance of the host genes. Consequently, instead of attempting to control these complicating factors in order to monitor the occurrence of major genes in the breeding programme, it seems more sensible to breed under conditions which allow the full exploitation of genes of small effect. In order to do this, some knowledge of the types of polygenic action involved must be gained, and an approach to genetical analysis adopted that includes the estimation of biometrical parameters.

Biometrical analysis

Studies were made under field conditions in the Sudan and in Uganda using diallel sets of crosses of upland varieties, some of which were homozygous for known *B* genes (Innes & Brown 1969; Innes, Brown & Walker 1974). Although some of the segregation patterns of F_2 populations suggested the presence of a gene of large effect, in most of the F_2s the distribution of lesion grades indicated either polygenic inheritance or major-gene segregation against a polygenic background. Results further emphasized the importance of genotype–environment interactions, and although most of the non-additive genetic variance for resistance could be accounted for by dominance, genotypes carrying known *B* genes did not maintain the same dominance relationships in the two countries as can be seen from the W_r/V_r graphs shown in Figure 7.4*a*.

Hybrid populations involving the upland variety 101–102B, which was shown to be almost immune to the disease, demonstrated that high levels of resistance, built up by the selection of minor genes, could be effectively transferred to more susceptible varieties. The evidence from these experiments suggested that a stable polygene complex had been synthesized in 101–102B which had been produced in the USA from a cross between a line carrying polygenic resistance and one homozygous for Knight's $B_2 B_3$.

It was also shown that genetic variances for resistance varied with the isolate of the pathogen used. For example, dominance variance was important for the isolate NCC 38 but not for NCC 40 (Figure 7.4b). However, further information about the genetic relationships of different isolates was obtained from other experiments.

Figure 7.4. Results of diallel analyses (F_1 data) (from Innes, Brown & Walker 1974). (a) Leaf disease grades: the regression of array covariance (W_r) on array variance (V_r) at Namulonge and Wad Medani. 1, Acala 4.42; 2, Bar 7/1; 3, Bar 24/5; 4, 101–102B; 5, Acala 1517BR; 6, Bar 12/16. (b) Boll lesion data: the regression of array covariance (W_r) on array variance (V_r) for culture NCC 38 and culture NCC 40. Varieties as in (a) above.

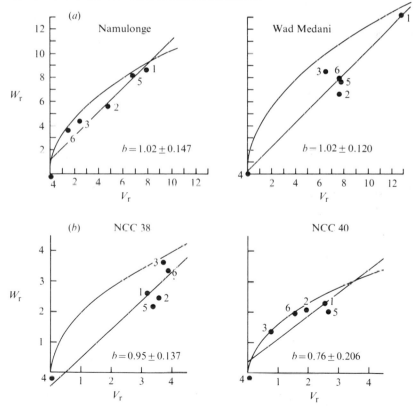

Resistance to different cultures of the pathogen

Although mixed isolates were always used for inoculation in the main breeding material, it was important to know the extent to which selection for resistance to one culture of the pathogen might be associated with resistance to others. Starting with the three host populations and the two pathogen cultures that had featured strongly in studies on variation in the host–parasite relationship, S. J. Brown selected separately for resistance to the two cultures, namely NCC 38 and 40. The most resistant phenotypes in each host population were

Figure 7.5. Effects of selection for resistance to one culture of the pathogen on levels of resistance to another. (*a*) Populations selected for resistance to NCC 38. (*b*) Populations selected for resistance to NCC 40. ●, IL 47/10; ×, UK 58; ○, A(57)12.

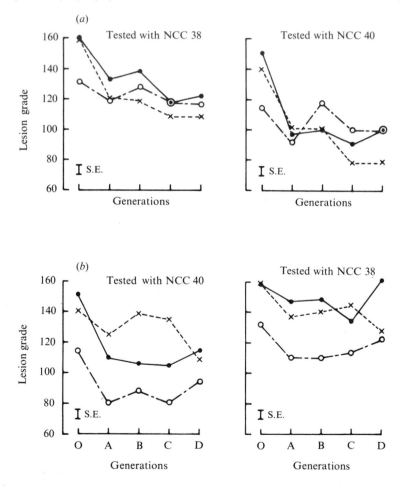

Table 7.3. *Correlated changes in resistance to different isolates of the pathogen*

Differences in hypocotyl resistance between selected and unselected populations of IL 47/10 and UK 58 (upper half of table) and IL 47/10 and A(57)12 (lower half of table). Both populations were separately selected for resistance to NCC 38 (R38) and NCC 40 (R40) and then evaluated for resistance to these two cultures as well as to a range of other cultures

		Test cultures (NCC numbers)										
		38	40	109	112	37	32	89	36	31	34	Mean
IL 47/10	R38	−64	−39	−16	−30	−23	−32	−68	−19	−18	−12	−32
	R40	−4	−21	+8	+9	−4	+6	−21	+19	−2	−3	−2
UK 58	R38	−38	−52	−19	−17	−29	−16	−59	−13	+4	−18	−26
	R40	−22	−30	−8	+7	−19	−9	−34	+3	−2	−2	−12
S.E.						± 8.3						±5.0
IL 47/10	R38	−43	−57	−19	−30	−10	−17	−75	−24	−21	−29	−32
	R40	+3	−34	+14	+2	+27	+23	−23	+3	−17	−12	−1
A(57)12	R38	−6	−10	−13	−1	0	−13	−4	+3	+1	−8	−5
	R40	−3	−21	+4	−1	−8	−7	−17	−3	−12	1	−7
S.E.						±10.9						±5.4

chosen and self-fertilized for four successive generations. Residual seed was then used to evaluate all the selected populations simultaneously with the original populations. Progenies selected for resistance either to NCC 38 or to NCC 40 were evaluated separately for resistance to the same two cultures. Results are summarized in Figure 7.5.

These results illustrated a number of important points. They confirmed the inference that the so-called susceptible variety IL 47/10 possessed genetic variation for resistance which could be exploited by selection. They showed that selection for increased resistance to NCC 38 was accompanied by increased resistance to NCC 40, but not consistently vice versa. Indeed, selection for resistance to NCC 38 gave a bigger increase in resistance to NCC 40 than selection for resistance to NCC 40 itself, while selection for resistance to NCC 40 gave no effective improvement in resistance to NCC 38 in some cases. Apparent irregularities in the results, such as the failure of selection consistently to improve resistance from one generation to the next, might well be associated with environmental conditions giving differential gene penetrance from one occasion to another. As we have seen, changes in environmental conditions will affect not only phenotypic variation in the population under selection, but also the relative levels of resistance in the final evaluation.

In further experiments, other seed lots from plants selected in the same way gave rise to populations of seedlings that were evaluated for resistance to a range of other isolates of the pathogen. Some of the results are summarized in Table 7.3. These results show a complex situation in which change in the population mean caused by selection for resistance to one culture of the pathogen not only shows different changes when evaluated for resistance to other cultures of the pathogen, but these changes are not consistent from one host population to another. All of these observations would imply that resistance is under the control of a relatively large number of genes.

Breeding methods

Conclusions on methods of breeding for resistance to bacterial blight reached as a result of observations on variation in the host–parasite relationship (Arnold & Brown 1968) were largely endorsed by subsequent work. Broadly, our methods consisted of subjecting breeding material to inoculation in as many different ways and under as many different environmental conditions, as could reasonably be accommodated in a practical breeding programme, while at the same time creating opportunities for recombination of resistance genes. Furthermore, the inoculum used included as broad a spectrum as possible of pathogen virulence. It is significant that this method of breeding does little more than rationalize the system adopted by breeders whose resources, particularly in terms of skilled manpower, are limited.

We do not question that it is possible, under certain conditions, to classify *X. malvacearum* into races characterized by their patterns of reaction on a set of differential host varieties such as those described by Hunter, Brinkerhoff & Bird (1968). We simply question the value of devoting scarce resources to this approach when it is difficult to see how a system can be devised that would have universal applicability and how, in any case, such knowledge can contribute to a more effective breeding programme under local conditions. Indeed, it may well be that the desire for precise results with internationally recognized races of the pathogen has served, in some cases, only to increase the chances of a break-down in host resistance, because the genetic basis of that resistance has been too narrow.

We think that there is a case for regarding variation in the host–parasite relationship in a manner similar to that in which the plant breeder regards any other character showing continuous variation. It is not essential to know how many genes are involved; only to know something about gene action. Adequate knowledge of this can be gained from the nature of frequency distributions in segregating generations and from relatively simple biometrical analyses.

Vertical and horizontal resistance

While it may well be that the distinction made between horizontal and vertical resistance (Van der Plank 1968) has served a useful purpose (Roane 1973) we do not regard resistance to bacterial blight as conforming to one or the other. On the contrary, it may be held to exhibit elements of both. Robinson (1971) defines vertical resistance in terms of the demonstration of statistically significant interactions between *pathodemes* (our host populations) and *pathotypes* (our cultures). In our tests, such an interaction could usually be detected, but was invariably accompanied by differences in mean levels of host resistance to a range of cultures, which would appear to correspond to horizontal resistance.

We have already seen (Figure 7.1) how regression analysis may be applied to give an overall picture of changes in the host–parasite relationship with changes in environment. Where information is available on the resistance of a range of genotypes to a range of cultures

Figure 7.6. Linear regressions ($y = bx + c$) of mean lesion grades of individual host populations on the mean of all host populations, when inoculated with a range of different cultures of the pathogen. (*a*) Results of boll inoculations. (*b*) Results of hypocotyl inoculations from a different experiment. (Regressions calculated from data presented by Arnold & Brown 1968.)

	b	c		b	c
IL 47/10	0.64±0.275	43.6±7.13	IL 47/10	1.10±0.200	14.8±14.2
Bar 7/8	1.56±0.165	−24.6±4.27	Bar 7/8	1.42±0.198	−33.1±14.1
Bar 12/9	1.52±0.161	25.4±4.18	UK 58	1.18±0.200	−16.1±14.2
A(61)28	0.82±0.169	−7.5±4.39	UKA G2.1	0.50±0.185	39.1±13.1
Bar 7/10	0.45±0.072	14.4±1.86	A(61)28	0.80±0.153	−5.2±10.9

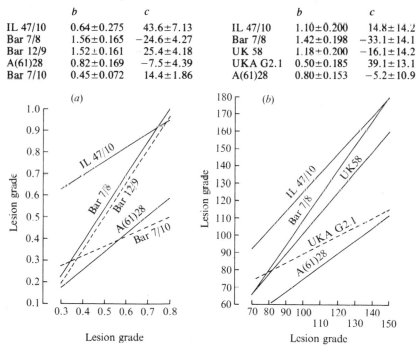

of the pathogen, regression analysis may be used to examine the nature of differences among genotypes. Figure 7.6 shows regression analyses of some results that were tabulated in an earlier paper (Arnold & Brown 1968). In the regression analyses, differences in slopes of the regression lines represent the interaction of resistance in the host populations with virulence in the pathogen (equivalent to effects caused by vertical resistance), while differences in the levels of the lines represent differences in resistance manifested to a similar extent over all cultures of the pathogen (equivalent to horizontal resistance). For example, in Figure 7.6a it can be seen that relative to the two locally-adapted populations, IL 47/10 (largely susceptible) and A(61)28 (strongly resistant), the two moderately resistant strains produced in the Sudan, namely Bar 12/9 ($B_2 B_2$) and Bar 7/8 ($B_2 B_2 B_3 B_3$), show greater resistance at the lower end of the scale than they do at the upper, even though they were derived from material of Uganda origin (Appendix 8.1). However, the presence of B_6 in a Sudan-bred strain confers strong resistance over the whole range, as can be seen from the slope and level of the regression line for Bar 7/10 ($B_2 B_2 B_3 B_3 B_6 B_6$). The results shown in Figure 7.6b obtained under different conditions, using a different inoculation technique, also show that Bar 7/8 responds differently from the locally-selected populations.

There is, however, an important difference between the behaviour of our material and that of material familiar to many workers who are faced with the problem of a single major gene conferring near-immunity to a disease, and hence giving rise to the problem described by Van der Plank as the vertifoliar effect. Clearly in such cases it is impossible to detect underlying polygenic segregation. The occurrence of this phenomenon in a range of crops may have given rise to the impression, not evident in our results, that there is necessarily something fundamentally different between resistance conferred by major genes and that conferred by polygenes.

Resistance to other cotton diseases

Some of the principles used in breeding for resistance to bacterial blight were successfully applied to other resistance breeding programmes in cotton. For example, when Fusarium wilt became a serious problem in Tanzania, preliminary comparisons of introduced resistant varieties with contemporaneous commercial stocks, suggested that a more detailed examination of variation for resistance within the locally-adapted material, should be an essential preliminary to any resistance breeding programme (Arnold & Arnold 1959). After suitable inoculation techniques had been developed, heritable resistance to Fusarium wilt was shown to occur in some, but not all, locally-adapted lines (Wickens

1964). Repeated selection in the variable material gave rise to highly resistant varieties which, when released for production in those areas where the disease was prevalent, made an important contribution to maintaining production. There is no doubt that these resistant varieties were produced far more quickly than could have been achieved by crossing with introduced resistant varieties that were neither adapted to the local environment nor of acceptable lint quality.

The work on breeding for resistance to Fusarium wilt in Tanzania was in many ways comparable with the earlier work on breeding for resistance to Verticillium wilt in Uganda (Chapter 6) and there is considerable circumstantial evidence to suggest that resistance to the two diseases is associated. Derivatives of the more important lines selected for resistance to Verticillium wilt were subsequently shown to be resistant to Fusarium wilt (Wickens & Logan 1960*b*) while a variety (S47) derived from material susceptible to Verticillium wilt was shown in both Uganda and Tanzania to be susceptible to Fusarium wilt as well. Later, material selected for resistance to Fusarium wilt in Tanzania was shown to be resistant to Verticillium wilt in Uganda (Brown & Beteise-Hbenye 1973). There are important implications in these observations which require further study. If resistance to both diseases should prove to be genetically associated, then there would be considerable scope for co-operative work between Uganda and Tanzania in breeding varieties resistant to the two diseases. Fusarium wilt is of great importance in Tanzania but unimportant in Uganda, while Verticillium wilt is of sporadic importance in Uganda, but unimportant in Tanzania.

Another example of the development of resistant lines by selection in predominantly susceptible material is afforded by work in the Sudan on resistance to the leaf-curl virus, transmitted by the whitefly, *Bemisia tabaci* Genn. Resistant lines were obtained by repeated selection in a Sakel line (NT 2) that had previously been described as fully susceptible. According to Knight (1948*a*) this unpublished work by S. H. Evelyn clearly illustrated the synthesis of polygenic resistance. Unfortunately, although later selections from the material were almost immune to the disease, they were never used commercially because of their unacceptable lint quality.

Both the fact that stable resistance to Alternaria disease has evolved in Uganda stocks with no conscious selection for it, and the nature of variation measured in genotypes exposed to natural attack by the disease suggest that resistance is under polygenic control (Chapter 6). Tanzania selections, although derived from similar basic material, are largely susceptible to the disease, reflecting the fact that *Alternaria* is of only sporadic importance in that country and, in consequence, there has been little unconscious selection for resistance.

Resistance to insect pests of cotton

While there have been several attempts to breed cotton varieties resistant to insect attack in Uganda (Chapter 5), and locally bred varieties have been shown to be relatively unattractive to lygus bugs (Reed 1974), it is with resistance to the leaf-sucking jassid (*Empoasca* spp.) that the greatest success has been achieved. Indeed it is only the availability of jassid-resistant varieties that has made the cultivation of cotton economically feasible in the drier regions of eastern and southern Africa. Resistance is related to the length and density of hairs on the plant surfaces (Parnell, King & Ruston 1949; Knight 1952), but length of hair is possibly the more important, because density without length affords little protection (Saunders 1965b).

In the Sudan, an intensive programme aimed at studying the genetics of hairiness while transferring genes for hairiness from a range of wild and cultivated species of *Gossypium* was initiated by Knight (1952) and continued by Saunders (1961, 1963, 1965a–c). Although their work led to the identification of six major genes controlling hairiness, Saunders emphasized the important part played by minor genes in modifying the expression of the character. He stressed the need to accumulate genes of small effect and acknowledged that the problem of breeding for hairiness was essentially one of handling quantitative genetic variation.

In East Africa, breeding for resistance to jassids followed a more empirical approach. In Tanzania, field assessments of hair-length and density were augmented by evaluating selected lines in trials designed to encourage attack by the insect (Arnold 1970b). Although during the early stages of the programme hairiness appeared to be associated with poor lint quality, this association was eventually broken and varieties with progressively higher levels of jassid resistance were released for commercial production (Peat & Brown 1961). These varieties played a vital part in expanding cotton production in Tanzania (Chapter 10).

In Uganda, jassid was in general not as serious as in Tanzania. Nevertheless hairiness was a criterion of selection from the earliest days of breeding programmes, and varieties released for production possessed adequate levels of resistance (Innes & Busuulwa 1973; Reed 1974). In some countries, however, one of the problems encountered in breeding for jassid resistance has been the associated increase in susceptibility to whitefly and to aphids which, in some environments, thrive on hairy cottons. Fortunately neither is a serious pest in East Africa, but in the Sudan and in certain parts of India, breeding for hairiness in cotton so increases its susceptibility to these pests that the problems created are greater than those solved, particularly because both whiteflies and aphids are more difficult than jassids to control with insecticides.

Conclusions

There is no doubt that breeding resistant varieties offers the only readily effective way of controlling pests and diseases of crops in many developing countries. Moreover, the increasing costs of chemical control give greater urgency to the production of resistant varieties, not only in the developing countries but in the developed countries as well. Instead of relying on the transference of resistance genes from introduced varieties, our work has repeatedly demonstrated the importance of examining locally-adapted material for relatively small amounts of variation under polygenic control. Possibilities for rapid improvement are most likely to occur with populations that have been exposed to long periods of natural attack by the pest or disease but, as the example of Fusarium wilt in Tanzania showed, even without this condition, a search for resistance may prove fruitful. Clearly, the scope for this approach is greater with outbreeding or partly outbreeding crops than it is with inbreeders. Nevertheless, the disadvantages of transferring resistance by wide crossing are such that, even with complete inbreeders, it would seem to be worth examining thoroughly the extent of variability for resistance among the full range of locally-adapted genotypes, as well as in the segregating populations that can be derived from them. In practice, this approach would be conducted in parallel with one that involved crossing with known sources of resistance, until it became clear which was the more likely to succeed. As we have seen, however, it might well be that breeding by exploiting the small amount of variability in the locally-adapted stocks, would give the more rapid progress towards a successful commercial variety.

8

Plant breeding

M. H. ARNOLD AND N. L. INNES

Introduction

Plant breeding has often been described as the management of evolution. Natural variation and selection, the essential ingredients of evolution, are replaced in plant breeding by systems of management designed to control and exploit genetic variation, so as to produce improved commercial varieties in the shortest possible time. Such systems of management must take into account the reproductive characteristics of the species as well as the agronomic background of the crop.

Cotton is primarily self-pollinating, but cross-pollination occurs to an extent that is mainly dependent upon the activity of honey bees. At Namulonge, the extent of outcrossing varies greatly from plant to plant (Hutchinson & Lawes 1953) but average values such as those reported by Arnold, Innes & Gridley (1971), ranging from 7.3 to 10.9 per cent, are probably typical. The reproductive parts of the flower are relatively large so that emasculation and pollination can be done fairly easily by hand. Self-pollination can be ensured by using one of several methods of sealing the flowers to prevent opening. Several genes causing male sterility have been described (Weaver & Ashley 1971) but no satisfactory system of inducing sterility and restoring fertility has yet been discovered. There is a limit to the amount of crossing that can be done by hand and the large-scale production of hybrid seed has only been attempted when a plentiful supply of cheap labour is available, such as in parts of India (Innes 1971). Nevertheless, it is possible to some extent with cotton to make use of plant breeding methods that have been devised both for self-pollinating and cross-pollinating crops.

A general system for cotton breeding, which is applicable to a varying extent to other crops, is illustrated in Figure 8.3. The organization of a breeding programme may be likened to that of a factory, in which raw material is taken through a series of processes before emerging as an end-product. In this case the raw material is genetic variation and the end-products improved varieties. The diagram indicates some of the interrelationships among the various processes

Figure 8.1. Self-pollination. The flower-bud was tied with cotton yarn during the afternoon or evening before the flower was due to open. The photograph shows the flower on the day following tying when it would otherwise have been fully open.

involved and gives a visual impression of the balance that is essential in the system as a whole. The larger conical structure in the centre of the diagram represents the machinery for routine selection. It is this machinery possibly more than any other aspect of the system that determines success or failure, and its scope and design have generated a great amount of controversy. Other parts of the system, such as the production of genetic variation, or the type of population finally prepared for release as a commercial seed issue, present problems of greater intrinsic interest to the geneticist, but it is the painstaking, monotonous, repetitive routine of selection and evaluation that requires the greatest effort from the successful plant breeder, and it is this that we shall consider first.

Selection and evaluation

Sceptics of the value of the subjective approach to selection have coined such phrases as 'horse-back plant breeding' or 'eye-ball genes' to describe the methods of the man who uses his judgement and experience to select or discard material in the field. At the other extreme, those who have stressed the importance of a statistical basis for selection have been accused of doing their selection on a calculating machine and of considering it unlucky to go into the field to look at the growing plants. While extreme views of this type make for vigorous and healthy argument, practising plant breeders seldom adhere rigidly to one approach or the other, but attempt to combine the best of both philosophies.

Many cotton breeders in Africa have attempted to increase the yielding potential of an established variety by reselection, and the extent of their success has depended both on the methods they have used and on the persistence of worthwhile genetic variation. Often, the breeder has relied on visual assessments in the early generations, but the desirability of measuring genetic variation led to the use of biometrical techniques in Uganda. The advantages and disadvantages of the two approaches can be illustrated by reference to J. E. Peat's programme of reselection from Mwanza local stocks in Tanzania and to H. L. Manning's BP52 programme in Uganda. (A key to the names of cotton varieties and code letters for selections mentioned in the text is given in Appendix 8.1.)

Pedigree selection in Tanzania

The selection methods developed in Tanzania involved two main phases. The first was concerned entirely with a subjective assessment of worth. Single plants (first generation) and progenies (second generation) were examined from time to time throughout the season, mainly

(a)

(b)

(e)

Figure 8.2. Controlled cross-pollination. On the day before flowers were pollinated, the corolla and part of the calyx were cut from the female flower (a) and the unripe anthers removed (b). The stigma was then washed (c) and a piece of drinking straw placed over it (d). The male flower was tied as if for self-pollination (see Figure 8.1) and, on the following day, the ripe pollen was transferred to the female stigma (e). The piece of drinking straw was then replaced to prevent uncontrolled cross-pollination.

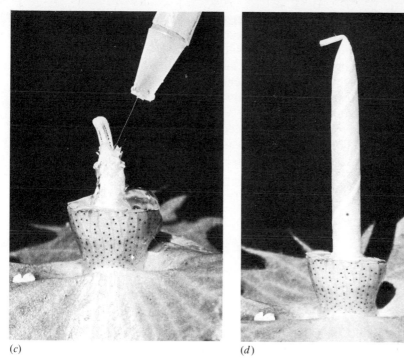

(c)   (d)

for their degree of hairiness as a measure of jassid resistance (Parnell *et al.* 1949). Towards the end of the season, well into the period of boll opening, a final assessment was made of cropping performance. This subjective assessment was influenced by such things as the number and size of the bolls, the extent of bad boll opening and other defects that caused the plant or the plot to have an unfavourable appearance. (Defects falling into this general category were not identified accurately until later in the programme, when it was shown that damage caused by bacterial blight was an important factor in the assessment.) In the first generation of selection, several thousand single plants were tagged initially and, of these, from 300 to 600 would be selected for growing as progenies in small plots in the following season. Some self-pollination was done (selfing) but no attempt was made to self all the material and many of the selections probably stemmed from open-pollinated plants.

In the second phase, selections made during the first phase were tested in replicated strain trials. Usually, strain trials were duplicated, one sown on well-manured land and the other on poor, exhausted soil so that differential responses to soil fertility could be detected. Yield and quality assessments in the strain trials were used as a basis for selecting those to be carried forward for more extensive testing in district trials. It was the performance of selections in terms of yield and quality, measured over a number of seasons in replicated district trials, that finally determined those to be included in a seed issue. (The constitution of seed issues is discussed on p. 241.)

Seed issues (designated UK for Ukiriguru) were made only when it appeared that worthwhile improvement had been effected. Seed issues were themselves tested in district trials where they were compared over several seasons both with each other and with the original commercial variety (MZ 561) which they replaced. Averaged over all available trials, the yield advances were impressive and a summary of the data gave the following results (Peat & Brown 1961):

Seed issue	Percentage increase over MZ 561	Number of seasons	Number of trials
UK 46	14	10	116
UK 48	18	9	97
UK 51	23	7	82
UK 55	31	4	50

The BP52 programme in Uganda

The approach adopted by Manning in the BP52 programme at Namulonge was quite different. Instead of examining a relatively large amount of material in the early stages of selection by somewhat crude and subjective methods, he argued that it was better to measure yield and its components accurately in a small number of progenies grown in replicated plots. The data could then be used to estimate genetic

Figure 8.3. Diagrammatic representation of a cotton breeding programme.

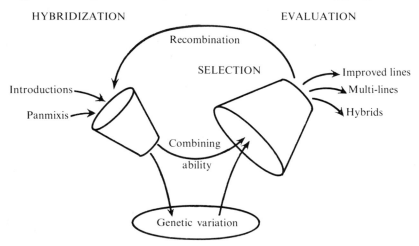

variation so that a selection procedure could be used that would lead to maximum genetic advance within the material studied.

Field management of the material was developed from the replicated progeny row system proposed by Hutchinson & Panse (1937). Selections were self-fertilized and reselected in each successive generation. This was accomplished by splitting the selfed seed from each single plant into two portions, one of which was used to sow a single row for selfing and the other to sow replicated plots. These plots, which were open-pollinated, gave rise to bulks of seed used for testing selected progenies in the following season (Figure 8.4).

Selection index. When the programme was started, quality in BP52 was already at a satisfactory level (Manning 1963), so that the aim was to increase yields without affecting quality. Experience had shown that measurement of yield in small plots was subject to considerable experimental error and it was held that greater accuracy and more rapid progress could be achieved by using a selection index that took into account the individual components of yield. In practice, four traits were measured: lint per plant (X_W), bolls per plant (X_1), seeds per boll (X_2) and lint per seed (X_3). Given acceptable quality, the first of these (lint per plant) represented economic worth (W): the others were components of it.

Figure 8.4. Breeding system used with a selection index.

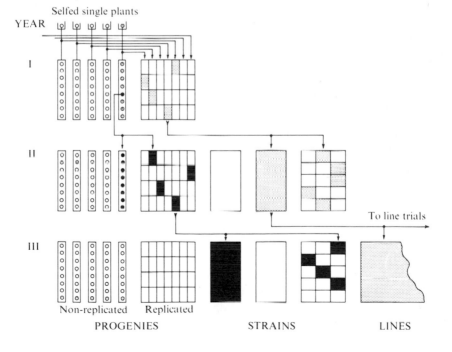

The assumption underlying the calculations in this selection procedure is that measured phenotypic traits can be adequately described in terms of genotypic (G) and environmental (E) effects so that phenotypic net worth (W), for example, can be written as:

$$W = G_W + E_W. \tag{1}$$

Because only G_W can be modified by selection, the problem is to use the information from measurements of all traits in such a way that G_W is most accurately determined, and the greatest genetic advance can be made by selection. It is necessary, therefore, to calculate a suitable selection index, and an index of the form:

$$I = b_1 X_1 + b_2 X_2 \ldots + b_n X_n \tag{2}$$

is used, in which the weighting coefficients ($b_1 \ldots b_n$) are calculated so as to maximize the relationship between I and G_W. This is done by taking into account the phenotypic variances and covariances of the traits as well as their genotypic covariances with net worth. Optimum values of b are obtained from the simultaneous solution of the derived set of equations (Smith 1936).

In the particular index devised by Manning, estimates of variances and covariances were partitioned, according to the pedigree tree, into components associated with the mean differences among groups of progenies derived from different lines, those among groups derived from different strains and differences among progenies all derived from the same strain. Only the 'progenies-in-strains' component was used in computing the b coefficients for the selection index so that contributions to variation arising from differences among homozygous grandparental plants were discounted. This is illustrated in Figure 8.5 from which it is clear that the extent to which the progenies-in-strains estimate of variance exceeds the error variance (the genotypic estimate) is a direct reflection of the heterozygosity of the grand-parental plant from which the progenies stem. The ratio of genotypic to phenotypic variances gave estimates of heritability (in the broad sense) which were used to determine those traits which were suitable for inclusion in an index. The trait bolls per plant (X_1) was subsequently excluded from the index because, assessed in this manner, its heritability was consistently low.

Manning's methods of assessing the improvement effected by selection contrasted markedly with Peat's simple annual comparison in district trials of his seed issues with the original commercial variety. There were three main aspects to Manning's assessment: calculation of the theoretical genetic advance at the progeny stage; comparison of progeny bulks (strains) with his modal bulk standard in replicated trials; and testing of the best selected lines in district trials.

The theoretical genetic advance. The theoretical mean genetic superiority of progenies selected on the basis of the index was calculated as

$$\frac{k\Sigma(b_i g'_{iw})}{\sigma_I} \tag{3}$$

expressed as a percentage of the mean yield ($\bar{X}_w$). Here, k is the selection differential, $(\bar{I}_s - \bar{I})/\sigma_I$ i.e. the amount by which the mean index of the selected progenies ($\bar{I}_s$) exceeds the mean of all progenies ($\bar{I}$), expressed in standard deviation units (σ_I); b_i are the coefficients derived for the selection index and g'_{iw} are the genotypic covariances of each trait with worth.

The theoretical genetic advance was calculated annually and, over the first seven generations of the programme, showed considerable fluctuation. The advance was estimated as 10.9, 12.3, 7.0, 8.2 and 4.0 per cent in the first, third, fourth, sixth and seventh generations respectively, but no advance could be detected in the second and fifth generations. Manning (1956*b*) summed these calculated advances and claimed that the cumulative expected advance after seven generations of selection amounted to 42.4 per cent, an average of 6.0 per cent per

Figure 8.5. Diagram illustrating the type of pedigree tree used for partitioning variances and covariances in the analysis of replicated progeny rows. Nomenclature for progenies, strains and lines in the BP52 (C) material is also illustrated.

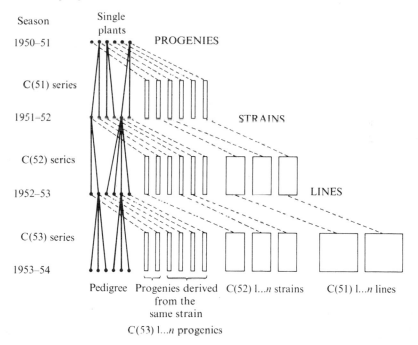

annum. Further advances were inferred from data obtained in the ensuing five years and the cumulative advance was recalculated as 50.5 per cent after twelve generations (Manning 1963).

Comparison with the modal bulk standard. Bulked seed from the selected progenies was compared each season with successive modal bulks at from one to ten sites to give estimates of performance over a wider range of environments. The aim of modal selection was to stabilize yield and quality in the original BP52 material and so produce a standard control for estimating advances made by selection (Manning 1956b). The procedure has been fully described by Walker (1964). Essentially it consists of taking a sample of from 200 to 500 plants from a population grown as an isolated, open-pollinated plot. Plants that deviate by more than one standard deviation from the mean are then discarded successively for the three traits: lint per seed, seed weight and lint length. Seed of the remaining plants is mixed together to form the modal bulk and the procedure is repeated in successive generations.

When certain modal bulks were compared with commercial BP52 stocks yield differences were evident, from which Manning inferred that yield had not been stabilized by modal selection but that progressive, though unexplained, yield increases had occurred from generation to generation. The magnitude of the increases was estimated from a regression analysis, and yield increases of the BP52 C series of selections were presented as the sum of the amounts by which they exceeded the corresponding modal bulk, plus the inferred yield advance of the modal bulk over the original stock (BP52/OMB). Over six seasons these figures indicated a total yield advance of about 36 per cent over the original stocks (Manning 1956b). Recalculated after ten generations, the advance was estimated as 40.9 per cent for control yield levels of 100 lb lint per acre (121 kg per ha) and 31.8 per cent for levels of 300 lb lint per acre (363 kg per ha) (Manning 1963).

In testing the best selections in district trials, Manning was limited by the design of the trials which, at that time, were laid down as 5×5 Latin Squares. This gave very little scope for testing the full range of material and testing of new selections was restricted to only one, two or three entries. Moreover, it was not always possible to include commercial BP52 stocks (Local) as control. To overcome this difficulty, Manning related the yields of his selections to the calculated yield of the original bulk (BP52/OMB). Deviations from recommended sowing dates, and differences in response at different yield levels, were allowed for using a multiple regression analysis, and the average yield advance realized after ten generations was estimated as 23.2 per cent (Manning 1963). This estimate, however, related to the best single selection in any one season, not to the succession of multi-lines released as commercial seed issues.

Commercial evaluation of BP52 NC seed issues. More direct assessments of the contribution of the NC seed issues to increased production must involve comparisons of the type made by Peat in Tanzania, in which new issues are compared with the old variety in replicated trials over the whole producing area. The fact that it cannot be proved that the original commercial variety remains completely unchanged over all the years of testing does not alter the fact that, in the absence of an alternative, the grower must use open-pollinated, unselected seed of the original variety from year to year. In the case of the BP52 programme, therefore, direct comparisons between the NC seed issues and Local stocks, maintained from year to year as isolated, open-pollinated bulks, give estimates of the benefits that are likely to be realized by the farmer in commercial production. To permit such comparisons, while also testing a worthwhile number of new selections, all district variety trials in the south and west of the country, were modified during the years 1962 and 1963 so that sixteen entries could be included in trials of 4×4 balanced lattice design.

Comparisons of NC seed issues with Local made over thirteen seasons in district trials are summarized in Table 8.1. Pooled estimates of standard errors have not been calculated because of the heterogeneity of variances among trials in one season and also because of marked genotype-environment interactions. The data are summarized from tables of results of district trials published annually in *Progress Reports*, to which reference can be made for estimates of the accuracy of comparisons among varieties in individual trials.

Of the succession of seed issues shown in Table 8.1, NC57 was composed mainly of lines derived from the C(54) series of progenies, representing the seventh generation of selection. The programme was continued until the fifteenth generation, which gave rise to the C(62) progenies, some of which were included as lines in NC65. Normal procedures were seriously impeded as a result of extremely poor seed production in the 1961–62 season and the programme was terminated. In the absence of C(63) lines derived from the selection index procedure, NC66 (the last of the NC seed issues) was constituted from equal proportions of three previous seed issues: NC61, NC64 and NC65.

We may now compare the yield increases indicated in Table 8.1 with those calculated by Manning. It is at once apparent that, with the exception of the 1961–62 season, the average yield increases of the NC seed issues over Local fall well short of those estimated by Manning. Indeed, in several seasons, the average yield of the NC seed issues is seen to be less than that of Local. In seeking an explanation of these

Table 8.1. *Mean lint yields of BP52 NC seed issues compared with former BP52 seed issues in district trials (each of five replications) over thirteen seasons*

| | | | | | | Increase of NC over Local | | | | |
| | | | Annual | | | Running average | | | | |
Season	Local[a] (kg per ha)	NC	(kg)	(%)	No. of trials	(kg)	(%)	No. of trials	NC seed issue
1958–59	271	276	+5	+1.8	21	+5	+1.8	21	NC57
1959–60	331	325	−6	−1.8	30	0	0	51	NC58
1960–61	316	353	+37	+12.2	25	+12	+3.9	76	NC58
1961–62	117	158	+41	+35.0	17	+19	+7.3	93	NC60
1962–63	282	319	+37	+13.1	22	+23	+8.7	115	NC62
1963–64	241	271	+30	+12.4	15	+24	+9.2	130	NC61 and 63
1964–65	313	332	+19	+6.1	8	+24	+9.0	138	NC61 and 63
1965–66	333	345	+12	+3.6	8	+22	+8.0	146	NC61 and 63
1966–67	415	390	−25	−6.0	25	+17	+5.8	171	NC61 and 65
1967–68	346	331	−15	−4.3	19	+13	+4.4	190	NC63
1968–69	324	298	−26	−8.0	28	+10	+3.3	218	NC63
1969–70	419	409	−10	−2.4	15	+8	+1.9	231	NC63
1970–71	423	484	+61	+14.4	7	+12	+2.8	238	NC63

[a] The term 'Local' is used here to refer to BP52 stocks in commercial production prior to the 1958–59 season. Seed for trials in the 1958–59 and 1959–60 seasons was of the variety BP52 K51 that had been maintained at Kawanda Research Station as open-pollinated, isolated bulks. Seed used in trials in the 1960–61 season was obtained by sampling from commercial ginneries in zones not growing NC seed issues. This stock was maintained at Namulonge as isolated, open-pollinated bulks and used in all subsequent trials.

discrepancies it is necessary first to recognize the magnitude of the influence of environmental conditions on the relative performance of two varieties, and, second, to understand the basis of Manning's calculations.

Taking the results shown in Table 8.1, we see that the average difference in yield between the NC seed issues and Local ranged from an increase of 35 per cent, measured over seventeen trials in the 1961–62 season, to a decrease of 8 per cent, measured over twenty-eight trials in the 1968–69 season. In the first instance, the yield level was 158 kg seed cotton per ha, in the second 298 kg per ha. When the data are studied in detail, however, it is not possible to establish a consistent relationship between yield level and the difference in yield between NC and Local. Nor are the yield differences between these two stocks entirely consistent from trial to trial within one season, showing that in addition to the seasonal effect, differences in environment from locality to locality also give rise to different relative yields.

In Manning's presentation of the results, inferences on the yield

Table 8.2. *Lint yields of modal bulks at Namulonge*

Season	Number of replica- tions	BP52 (Local) (kg per ha)	BP52 (MB) (kg per ha)	Difference (kg)	Difference (%)	MB generation
1962–63	10	290	332	+42±17.4	+14.5	14
1963–64	5	204	172	−32±17.3	−15.7	15
1965–66	5	274	278	+4±36.3	+1.5	17
1967–68	5	383	398	+15±25.0	+3.9	19
1968–69	10	443	376	−67±19.9	−15.1	20
1970–71	16	521	587	+66±11.7	+12.7	22
Average		352	357	+5	+1.4	

advance achieved were not always based on direct comparisons between Local and the new selections, but on comparisons between the
new selections and his modal bulk standard. Yields of Local or original
bulk (OMB) were calculated using a linear regression equation that
related the performance of successive modal bulks to the original
material. However, the regression equation itself was derived from
comparisons with Local and particular modal bulks in different seasons,
so that allowance could not be made for the different relative performance of the two stocks under different environmental conditions. Such
interactions of genotypes with environment provide the key to the
discrepancies between Manning's conclusions and those now given.

A detailed re-assessment of the performance of the modal bulk
material (Arnold 1972) indicated that the conclusion of a progressive yield
increase is untenable. When the earlier data are combined with the more
recent results, there is a strong inference that field selection of the 200
to 500 plants that provided the material for modal selection, resulted
in rapid selection for earliness. It now appears that in the particular
seasons in which comparisons were made, environmental conditions
favoured the quicker-maturing types to an increasing extent, giving the
illusion of a progressive yield increase with successive modal bulks.
The conclusion of a progressive increase was not tested, however, by
comparing a range of modal bulks simultaneously with Local over a
range of environments. Had this been done, or had the tests been made
under different seasonal conditions, the inferred relationship between
the modal standard and original BP stocks would almost certainly have
broken down as can be seen from the data in Table 8.2, taken as an
example of the more extensive data on this point presented by Arnold.

That such marked changes in the relative performance of two stocks
can occur, even at one locality, emphasizes the dangers of calculating
yields by interpolation, especially when the derived yield is to form part

of the base line against which yield improvement is to be measured. The fact that in many of the comparisons reported by Manning, Local yields were derived and not measured undoubtedly gave rise to an important source of error. The selection programme in BP52 therefore poses a range of important questions for the plant breeder, which include the efficacy of the selection index procedure as a practical breeding system, as well as the extremely complicated but vitally important problem of how to test and select in the face of genotype–environment interactions of the magnitude that have been shown to occur.

Limitations of the selection index procedure

It is now clear that the basic assumption underlying the calculation of a selection index (expressed in equation 1) represents an inadequate description of the biological situation actually encountered, where the interaction of genotypes with environment is of fundamental importance. It would therefore be necessary to construct a new index, based on a relationship of the form:

$$W = G_\mathrm{w} + E_\mathrm{w} + (GE)_\mathrm{w} \qquad (4)$$

and even this might well prove to be an oversimplification.

In any case, the introduction of the GE term into the equation takes the mathematics into a greater order of complexity and this, together with the impossibility of measuring GE effectively at the progeny stage, means that a practical breeding system for cotton could not be built around an index of this type. Moreover, the importance of genotype–environment interactions limits the usefulness of heritability estimates that are derived from a single experiment. What needs to be estimated, before meaningful assessments of genetic variation can be made, is the magnitude of genotypic differences in relation to the magnitude of the genotype–environment interactions that are encountered in the producing area. Equally a genetic advance calculated on the results of a single experiment is meaningful only in the context of that one particular environment. To sum such calculated advances over a period of years is clearly meaningless under Uganda conditions.

The selection index procedure had been adopted partly because of the difficulties of defining criteria for visual selection in BP52 and partly to investigate the extent to which biometrical techniques could be used to exploit relatively small amounts of genetic variation. It was held (Manning 1956b) that small, progressive improvements in an already well-adapted variety might produce greater returns than could be achieved from more spectacular strides in less adapted but more variable material. Under different environmental conditions this prediction might well have been fulfilled, but with the fluctuating patterns of rainfall experienced in Uganda other breeding methods were clearly needed.

Revised breeding programme

With these considerations in mind, the strategy of the breeding programme was revised from 1961–62 season onwards in order to broaden the genetic base of the material, to increase the chances of recombination, and to modify the system of selection and testing so that performance could be estimated over a wider range of environmental conditions.

Uganda has traditionally produced two varieties of cotton: the high quality, long-staple 'BP' (Bukalasa pedigree) types and the somewhat shorter 'S' (Serere) types. Although begun at Bukalasa, the BP programme was subsequently transferred to Kawanda before becoming the responsibility of Corporation staff at Namulonge from 1946 onwards. The breeding programme for S types at Serere became a Corporation responsibility in 1948.

The revised breeding programme made provision for developing material at three different levels of adaptation and variability. First, was the locally-adapted material (Group I) which had been developed at Namulonge and Serere from Albar 51 and from the related UPA and LA crosses. Selections from this material were tested alongside selections from the UKA stocks (Arnold 1963), imported from Ukiriguru. All four stocks were studied intensively from 1961 to 1965 in the hope of finding immediate, blight-resistant replacements for BP52 and S47. Second, more variable material (Group II) was produced by crossing, in various combinations, stocks that were either resistant to bacterial blight or were adapted to East African conditions. These included BP52, Albar, UPA, UKA, Reba and Bar 12/16. Third, a long-term reservoir of potentially useful variability (Group III) was established by starting a wide-range crossing programme, which initially included as parents Albar, UPA, UKA, Reba, Acala, Punjab, Barhop, Barwilds and Pima S_1 (Low 1964; Arnold, Smithson & Tollervey 1964) but to which other parental stocks were progressively added, including stocks derived from interspecific hybrids.

At both stations, a range of different types was selected and, as the programme developed, potential S types, selected at Namulonge were transferred to Serere for evaluation and reselection, while potential BP types, selected at Serere, were sent to Namulonge for similar treatment. At Serere, Group I material was maintained by modal bulking and by mass selection for the density of seed coat fuzz (Low 1964). After extensive testing in district trials, Albar 4MB was multiplied as the first SATU seed issue from Serere in the 1962–63 season. SATU subsequently became the standard variety for the north of the country (Figure 8.17).

At Namulonge, the replicated-progeny row system was at first retained for testing first generation selections in Group I material.

Although an index was calculated it was used as only one criterion of selection. In the 1961–62 season, for example, the selection A(61)6 was excluded by the index. It was nevertheless retained in the programme because it appeared to show the best combination of yield, quality and blight resistance. In the event, it was this selection that was subsequently released as BPA 66, the first blight-resistant replacement for BP52, which was eventually grown in the whole of the south and west of the country (Arnold *et al.* 1968).

Modified selection system. Recognizing the limitations of the selection index method, the problems that gave rise to its introduction were re-examined to discover what alternative solutions might be found. One reason for introducing the index was the unreliability of yield estimates measured in the very small plots used in replicated progeny rows. One of the factors contributing to relatively large experimental errors was the irregular stands often experienced, to the extent that it was not unusual for some plots to be deficient of 40 per cent or more of the nominal plant population. In these circumstances it was argued that the components of yield, rather than yield itself, could be more meaningfully measured. From 1962 onwards, however, greater attention was given to field techniques that would increase the reliability of yield measurements at the progeny stage.

Figure 8.6. Modified breeding system. For further explanation see text.

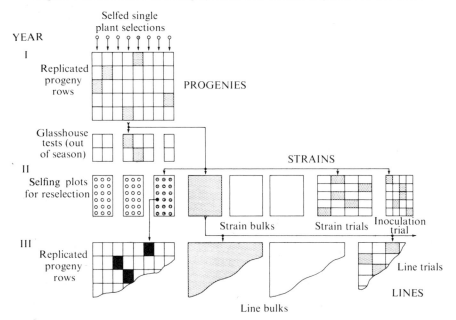

Several factors were involved. First, the production of seed from single plants was increased by incorporating greater insect control and new agronomic findings into the management of the breeding plots. Second, plot sizes at the replicated progeny row stage were increased to twenty stands, whenever possible. Third, the scale of the whole operation was increased so that from two to three times the number of selections could be included in replicated progeny rows. Fourth, when conditions for germination were marginal, plots were watered to assist germination, but not to an extent that could be regarded as materially altering the seasonal weather conditions. One of the results of these changes was that residual (error) variances were reduced to the extent that practically no benefit, in terms of relative efficiency, was recorded from lattice square designs. Consequently, it became possible, without loss of precision, to simplify procedures and use randomized complete block designs.

These changes coincided with a change of emphasis in the programme from the BP52 stocks to stocks represented by Groups I and II (p. 211) and it soon became apparent that, in this more variable material, heritable yield differences could be measured in replicated progeny rows. Nevertheless the problem arose of how to incorporate a routine selection procedure for resistance to bacterial blight. Figure 8.6 illustrates the selection system that was introduced in an attempt to overcome these problems. The system incorporated out-of-season testing for blight resistance in the glasshouse and the assessment of yield and quality factors in replicated progeny rows, but enabled testing over a wider range of environmental conditions by restricting reselection to alternate years. This meant that a selection could be tested as a progeny in one season and as a strain in the next, before a decision to retain or discard it was made. As had been the practice in the later years of the BP52 programme, strains were tested at more than one site whenever possible, giving an additional range of environmental conditions. With a limit on the number of replicated rows that can be handled in any one season, another advantage of selecting in alternate generations is that a wider range of material can be studied, one section being tested in replicated rows in one season and the other in the next.

The system also incorporated alternate generations of selfing and open pollination. Inbreeding is necessary in a programme of selection for resistance to bacterial blight because of the variable expression of dominance, under Uganda conditions, of the factors controlling resistance (Chapter 7). At the same time, outcrossing among selected lines is desirable as a means of increasing recombination. In an attempt to strike a balance between these two requirements, no provision was made for selfing at the progeny (first generation) stage and seed for the next generation of selection was derived from open-pollinated plants

8

in the replicated rows. A further advantage of the system is that all selfed seed from a single plant becomes available for the replicated rows, instead of only part of it, as was the case in the former system, in which both replicated and non-replicated rows were grown. A possible disadvantage of the system is that control varieties, that differ markedly from the breeding material, cannot be included at the replicated-row stage because of the undesirable outcrossing that might occur. In practice, this is not a major disadvantage because adequate controls at the progeny stage can be provided by including closely related material of known performance.

This selection system, and modifications of it, formed the basis for screening material in the Namulonge breeding programme for yield, quality and disease resistance from 1963 onwards. It was, for example, applied to a reselection programme in A(61)6 from the time of its successful establishment as the commercial variety BPA. This reselection programme will now be described because it illustrates the appli-

Figure 8.7. BPA breeding programme. SP, single plant; ST, strain trial; SB, strain bulk; RPR, replicated progeny rows; FB, farm bulk; LT, line trial; LB, line bulk; —, selfed; ---, open-pollinated.

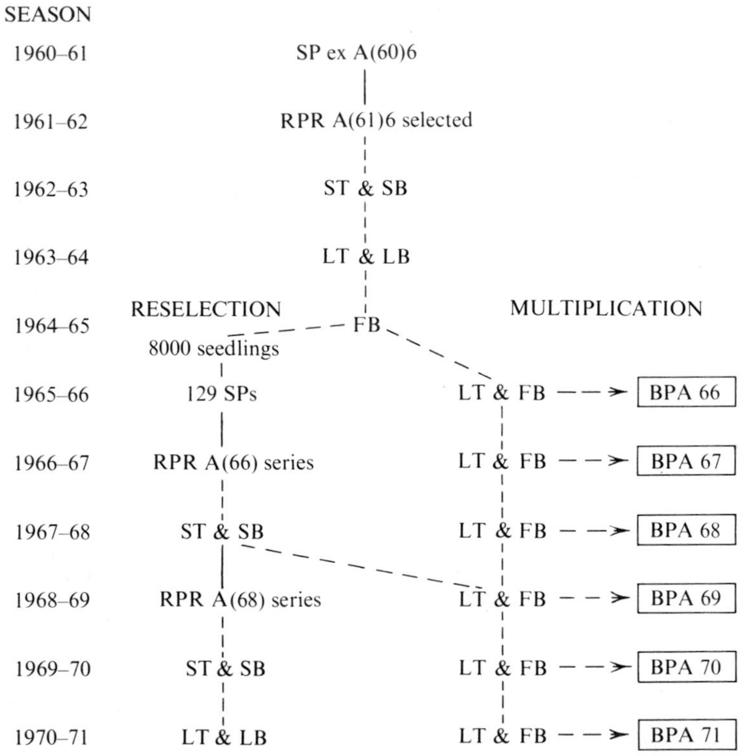

SEASON

1960–61		SP ex A(60)6	
1961–62		RPR A(61)6 selected	
1962–63		ST & SB	
1963–64		LT & LB	
1964–65	RESELECTION ----- FB		MULTIPLICATION
	8000 seedlings		
1965–66	129 SPs		LT & FB ——➤ [BPA 66]
1966–67	RPR A(66) series		LT & FB ——➤ [BPA 67]
1967–68	ST & SB		LT & FB ——➤ [BPA 68]
1968–69	RPR A(68) series		LT & FB ——➤ [BPA 69]
1969–70	ST & SB		LT & FB ——➤ [BPA 70]
1970–71	LT & LB		LT & FB ——➤ [BPA 71]

cation of the modified selection method and also provides another example, analogous to the BP52 improvement programme, of an attempt to improve an established variety by exploiting residual genetic variation.

BPA improvement programme. The establishment of A(61)6 as the new commercial variety BPA was not achieved without some misgivings. It was much more resistant to bacterial blight than BP52, had comparable resistance to jassid, Verticillium wilt and Alternaria boll rot, had proved to be heavier yielding in district trials and had a comparable ginning percentage. In lint quality, it met the commercial requirement for an increase in staple length over BP52 of from one to $\frac{2}{32}$ in. (except when growing conditions were very unfavourable), was comparable in bundle strength but was both coarser and more mature. The net result of these differences in fibre characteristics was a yarn that was comparable in appearance but slightly weaker, on average, than comparable yarns spun from BP52 (Arnold *et al.* 1968). There was consequently a need to follow up the first release of BPA as soon as possible with further seed issues giving improved yarn strengths. Moreover, A(61)6 itself was shown to be segregating for blight resistance. The initial aim of the BPA reselection programme, therefore, was to improve yarn strength and increase the uniformity of blight resistance in the variety, while maintaining yield and ginning percentage at the level already attained.

Figure 8.7 illustrates the selection programme in BPA from 1966 to 1972. Table 8.3 shows the composition of BPA multi-lines. The starting material was a bulk of A(61)6 that had been open-pollinated since selection as a progeny in the 1961–62 season. During each ensuing season there were opportunities for out-pollination with other, closely related Albar selections. Some 8000 seedlings were screened in the glasshouse for resistance to bacterial blight in 1965, and about 2000 resistant plants transplanted to the field and self-fertilized. Approximately 1000 of these plants were selected on the basis of the amount of selfed seed produced and the lint was tested for the three quality characters – micronaire value, fibre bundle strength and effective length. Correlation diagrams were plotted for pairs of these characters on a random sample of 200 of the selected plants. These diagrams illustrated the tendency for certain characters to be negatively associated and enabled rational selection limits to be set (Arnold & Church 1967). Evaluation of 129 progenies grown in replicated progeny rows in the ensuing season indicated that many of the progenies showed worthwhile improvements in yarn strength over BPA. Results for the three progeny row trials are summarized in Table 8.4.

As the programme of testing and reselection proceeded, however,

8-2

Table 8.3. *Composition of BPA multi-lines*

Per cent (w/w) seed of each component in initial mixture

	A(61)6	A(66)22	A(66)29	A(66)36	A(66)102	A(66)131	A(66)134	BPA 68	BPA 69	M1	M2	BPA 70
BPA 66	100	—	—	—	—	—	—	—	—	—	—	—
BPA 67	100	—	—	—	—	—	—	—	—	—	—	—
BPA 68	100	—	—	—	—	—	—	—	—	—	—	—
BPA 69	—	8.3	—	8.3	8.3	25.0	—	50.0	—	—	—	—
M1	—	20.0	—	20.0	20.0	20.0	—	20.0	—	—	—	—
M2	—	16.7	16.7	16.7	16.7	16.7	16.7	—	—	—	—	—
BPA 70	—	—	—	14.3	14.3	28.6	—	—	14.3	14.3	14.3	—
BPA 71	—	—	—	—	8.3	8.3	—	—	—	25.0	25.0	33.3

Table 8.4. *Performance of 129 single-plant progenies selected from BPA 66* (*1966–67 season*)

		Lint yield in kg per ha			Yarn strength at 40s		
Progenies	Range	Mean of progenies	Mean of BPA controls	Range	Mean of progenies	Mean of BPA controls
A(66) 1–43	461–653	539±4.2	493±11.2	2558–3044	2774±9.2	2599±18.9
A(66) 50–92	526–689	586±3.9	572±10.5	2456–2857	2680±5.6	2642±15.2
A(66) 99–141	429–656	519±4.1	484±11.0	2497–2949	2746±4.3	2696±11.1
129 progenies		548±3.6	516±9.5		2733±6.7	2646±15.3

Table 8.5. *Yield and components of quality of A(66)37 compared with those of the BPA material from which it was selected: means of data from fourteen district trials*

	A(66)37	Means of 2 BPA controls
Lint yield as % BPA	89	100
Ginning percentage	33.3	33.2
Effective length, 32nds inch	45.0	43.5
Maturity ratio	0.84	0.82
Micronaire value	3.6	3.6
Standard fibre weight, 10^{-8} g per cm	184	187
Fibre bundle strength[a]	22.2	21.6
Yarn strength at 40s[b]	2680	2371
Yarn appearance	3D	4C–D

[a] g per tex at ⅛ in. test length.
[b] This and other terms in this table are explained in detail by Lord & Underwood (1958).

it became clear that it was going to be difficult to make a substantial improvement in yarn strength while maintaining yield and ginning percentage at the level of BPA 66. As a general rule, selections with better than average yarn strength tended to have worse than average yield or ginning percentage as can be seen from the example shown in Table 8.5. Further attempts to find recombinant types within the material did not meet with notable success, but the indications were that, on balance, a modest improvement in yarn strength was achieved and transposed into commercial production through the seed issues BPA 69, 70 and 71. It seems probable that the genetic potential for quality in this material reached the limit of the environment to express it and although differences in favour of the new issues showed up in some tests, they did not consistently do so.

Similarly, differences in yielding capacity between the BPA reselec-

Figure 8.8. BPA cotton at Namulonge. Picking was delayed in this plot to demonstrate the yielding capacity of the variety.

tions and the original BPA stocks derived from A(61)6, were not very great. Averaged over twenty-two trials in the 1970–71 season, for example, BPA 70 yielded between two and three per cent more than BPA 68.

The aim of the programme was to follow up the first release of BPA with further seed issues showing improved yarn strength and greater resistance to bacterial blight. The urgency of the first requirement disappeared as BPA rapidly acquired an excellent reputation on world markets as a cotton well suited for processing with modern, high-speed machinery, and for blending with man-made fibres. The substantial differentials in price it commanded over BP52 had not been foreseen at the time of its release, but were of great importance in helping to maintain the price paid to the growers. As far as resistance to bacterial blight was concerned, the disease became progressively less important as each new wave of seed spread through the producing area.

Thus the initial aims of the programme of reselection in BPA were realized, but it was clear that the limit of worthwhile progress by reselection had been reached. In the more variable, but related material that gave rise to the SATU variety of Serere, reselection revealed some extremely promising lines. The breeding programmes that gave rise to SATU showed important differences from the prolonged period of pedigree selection that produced the initial BPA line, A(61)6 (Figure 8.9). In order to understand the SATU programme, however, we must first outline the history of Albar.

SATU reselection programme. The detailed history of Albar and related Allen stocks in Africa has been documented by Innes & Jones (1972). Briefly, Albar stocks in Uganda originated from selections from Nigerian Allen made at Namulonge in 1951 on the criterion of resistance to bacterial blight. There followed a period of line breeding in the material, designed both to stabilize resistance and to examine the material for yield and quality. Progeny rows were screened for resistance and for other field characters and tested for yield and quality in replicated strain trials in the following season. From the 1956–57 season until 1961–62, reselection was based on a replicated-progeny row and selection index system run in parallel with the BP52 programme. However, in order to retain a broad genetic base in the material certain selections, differing in their characteristics, were mixed together to give

Figure 8.9. Pedigree tree of Namulonge selections from A637.

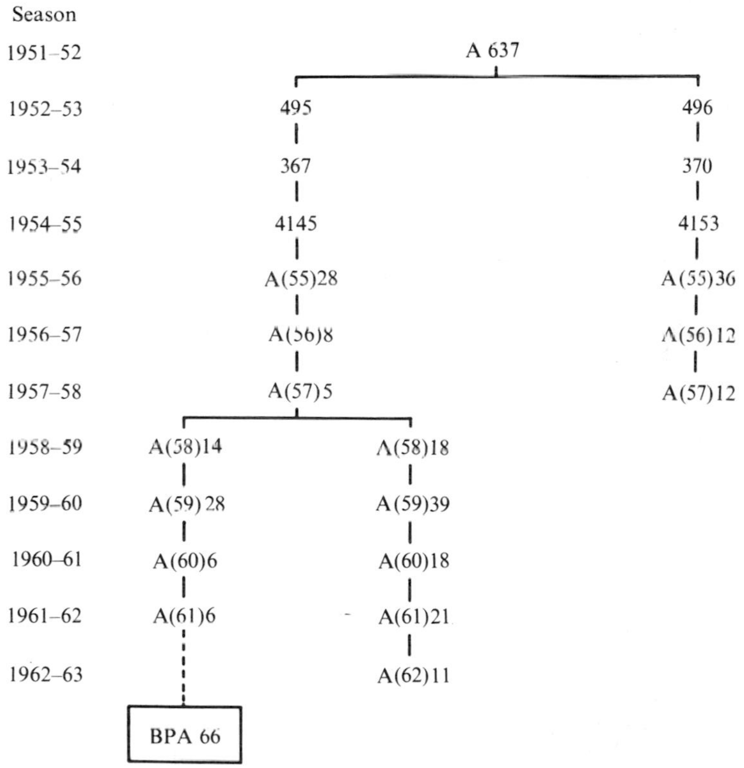

Season

1951–52		A 637	
1952–53	495		496
1953–54	367		370
1954–55	4145		4153
1955–56	A(55)28		A(55)36
1956–57	A(56)8		A(56)12
1957–58	A(57)5		A(57)12
1958–59	A(58)14	A(58)18	
1959–60	A(59)28	A(59)39	
1960–61	A(60)6	A(60)18	
1961–62	A(61)6	A(61)21	
1962–63		A(62)11	

BPA 66

a base population that was maintained initially both at Namulonge and at Serere by modal bulking (Figure 8.10). After the successful establishment of Albar 4MB and 5MB as the new variety SATU, work on this material was discontinued at Namulonge and expanded at Serere. SATU had greater resistance to bacterial blight and was higher yielding than its predecessor S47 under most conditions, but there was some criticism of its yarn strength, which fell short of the yarn strength of Albar stocks produced in Malawi. In consequence, as in the case of BPA for the south and west of the country, there was a need to follow up the early issues of SATU with further seed issues in the north having greater yarn strength.

A programme of reselection was started at Serere in 1965 (Costelloe & Riggs 1967) when the material had been grown as open-pollinated bulks for seven years, and further selections were made from populations of SATU seed issues. Selections from this programme showed wide variation. Not only did it prove possible to select heavier-yielding SATU types with considerably improved yarn strength (Jones 1973) but it was also possible to select from this material grown at Namulonge recombinant types that showed promise of giving the level of quality required in BPA seed issues, with substantially better yields (Innes 1974*a*). The programme (Figure 8.10) shows how a series of improved lines were brought together and allowed to interpollinate for several generations before attempting further selection. It points clearly to the possibilities for developing more sophisticated systems of recurrent selection for cotton and illustrates the wealth of variation that can be released from a single basic stock when appropriate breeding methods are used.

Selections from hybrid derivatives. Methods used for selecting in the segregating material produced in the crossing programme (Groups II and III) have varied somewhat with the resources and manpower available. A great deal of roguing was done in the early generations on the basis of eye judgement for such characters as susceptibility to jassid or to Alternaria boll rot. As much material as possible was inoculated with *X. malvacearum* using one of the various spray techniques for leaves or needle-prick techniques for stems and bolls. It became clear that, even with intensive selection, many of the resultant lines were not well adapted to local conditions. At Namulonge a large proportion of the selections were discarded at the F_4, F_5 and F_6 stages when it became clear that, even after several generations of reselection, the material was still highly susceptible to *Alternaria*. Other lines were discarded both at Namulonge and Serere because they were unproductive, too weak in the stem, were susceptible to bacterial blight or, when tested in replicated trials, were shown to have inferior lint quality, yield or ginning percentage.

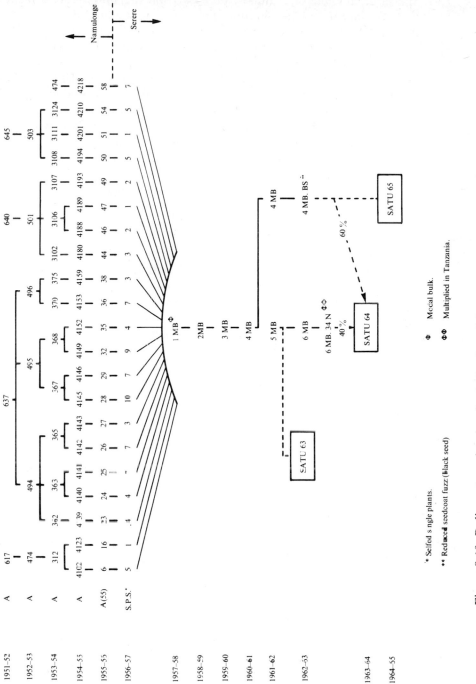

Figure 8.10. Fedigree tree showing the derivation of SATU seed issues (from Innes & Jones 1972).

Of the large amount of material, derived from the first series of crosses in Groups II and III, nothing that completely matched up to the requirements of a commercial variety had been found by the 1971–72 season. CA lines were obtained, however, from crosses between BP52 and Albar that were similar to BPA in lint quality, were heavier yielding and gave improved ginning percentages (Innes 1973*b*). Unfortunately their seed was more fuzzy than seed of BPA and produced more seed coat neps. Interesting recombinant types were produced from other crosses and used as parents in further crossing. The two most promising stocks from Serere were the PB types derived from a cross between Pima S_1 and Bar 12/16 (Jones & Fielding 1971), and the Albacala types derived from crosses between Albar 51 and Acala 4–42 (Low 1966). Useful stocks that emerged from the Namulonge programme were Briga types derived from a cross between IL47/10 and Bar 12/16, and the AH lines derived from a cross between Albar 51 and Barhop.

Evaluation of lines and varieties

The difficulties of evaluating breeding material, both with respect to the influence of resultant seed issues on commercial production and to assessing the possibilities for worthwhile improvement by reselection or other breeding methods, have already been touched upon in considering the results of the BP52 selection programme. These difficulties can be illustrated more fully by reference to results of extensive tests in district trials of BPA material. The problems exemplify those that face the plant breeder for any crop that is to be produced in the broad equatorial belt of Uganda, where fluctuations in weather conditions from season to season and from place to place, mean that the plant is subjected to widely different growing conditions, particularly with regard to the frequency and duration of periods of water stress.

Requirements for district trials. Before describing the results, we must consider the basic requirements for district trials and see how nearly these requirements were adequately met, or could be met in the future. The first requirement is to be able to test a wide range of selections simultaneously at any one centre. Expansion of the trials in Uganda from the original 5×5 Latin Squares to trials of 4×4 balanced lattice design (with five replications) helped greatly in this respect. Moreover, at some centres it was possible to lay down two such trials and to increase the number of new selections tested by dividing all the trials in the BPA producing area into groups, each group testing a particular group of selections, but all trials having common control varieties. Consideration was given to increasing the size of the testing sites to accommodate trials of 5×5 balanced lattice design with six replications, giving the advantages of testing more selections with somewhat greater precision. With the experience gained by the field staff in managing balanced lattice trials, management of the somewhat larger design should not present very great problems.

A second requirement of any variety trial is that the soil should be uniform or, at any rate, that any gradients in soil fertility or irregularities should be accommodated by the design and layout of the trial. As a generalization, uniform trial sites are difficult to find on tropical soils. The effects on soil fertility of such things as anthills, old house sites, livestock enclosures and erosion gullies persist for many years and when present in a trial serve to increase the experimental error. The greater the experience of the officer laying out the trial, the greater the chances of achieving reasonable uniformity but, even with a sound background of experience, it is not always possible for the trial officer to avoid the irregularities completely. Moreover, under district conditions, there is always the possibility of a trial being damaged by animals, either by grazing or trampling, or by destroying the green bolls as can happen with monkeys and jackals. Greater soil uniformity can be achieved by the liberal use of fertilizers but is achieved only at the expense of testing selections under conditions that are largely atypical of those for which the new varieties are intended. Greater protection can be afforded to trials by expanding the trial centres into larger substations as has already been done at some centres. Plans to implement further improvements of this type were included in the 1971–72/1975–76 national development plan. Nevertheless, when the availability of funds is limited, the urge to increase the precision of individual trials must be balanced against the need for having a large number of sites, well distributed through the producing area. In view of the large genotype–environment interactions that occur, our experience would support the conclusion of Miller, Robinson & Pope (1962) who, in similar circumstances, conclude that 'increasing the number of testing environments would appear to be much more effective in increasing the precision of variety evaluation than would increasing the number of replicates per test'. A map showing the location of district trials in the 1971–72 season is shown in Figure 8.11.

A third consideration is the location of individual trial sites in relation to the geographical importance of the crop. To give an unbiased estimate of the likely contribution of a new variety to increased production, trial sites should theoretically represent a random selection of all possible growing conditions. While this concept is useful to keep in mind, in practice the choice of sites is determined largely by practical considerations. It is however important to keep the proportion of sites in any one area roughly in proportion to the importance of that area in terms of production. In this respect there is sometimes a conflict of priorities between these requirements and those of the

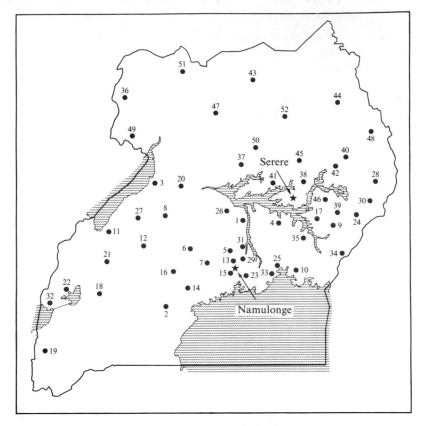

Figure 8.11. District trials in Uganda in the 1971–72 season.

BPA production area		SATU production area
(1) Bale	(19) Kihihi	(36) Abi
(2) Bigasa	(20) Masindi	(37) Aduku
(3) Biso	(21) Matiri	(38) Arapai
(4) Bugaya	(22) Mubuku (Kasese)	(39) Bukedea
(5) Bukalasa	(23) Mukono	(40) Iri-Iri
(6) Bukomero	(24) Nabbongo	(41) Kaberamaido
(7) Busunju	(25) Nakabango	(42) Katakwi
(8) Butemba	(26) Nakasongola	(43) Kitgum
(9) Iki-Iki	(27) Nalweyo	(44) Kotido
(10) Ikulwe	(28) Namalu	(45) Kuju
(11) Kagadi	(29) Namyoya	(46) Kumi
(12) Kakumiro	(30) Ngenge	(47) Labora
(13) Kalagala	(31) Ntenjeru	(48) Moroto
(14) Kanoni	(32) Nyakatonzi	(49) Nebbi
(15) Kawanda	(33) Nyenga	(50) Ngetta
(16) Kawungera	(34) Tororo	(51) Pakelle
(17) Kibale	(35) Vukula	(52) Patongo
(18) Kiburara		

Table 8.6. *Lint yields of BPA and BP52 in district
trials over eight seasons*

Season	BP52[a]	BPA[a]	Increase of BPA over BP52		No. of sites	No. of trials[b]
	(kg per ha)		(kg)	(%)		
1963–64	264	305	+41	+15.5	7	7
1965–66	354	450	+96	+27.1	27	29
1966–67	391	438	+47	+12.0	30	33
1967–68	300	361	+61	+20.3	30	31
1968–69	306	387	+81	+26.5	32	37
1969–70	445	524	+79	+17.8	26	30
1970–71	460	434	−26	−5.7	27	30
1971–72	438	441	+3	+0.7	34	42
Average	370	417	+47	+12.7		(239)

[a] Where BP52 or BPA was represented by more than one seed issue, a mean value was calculated.
[b] Where there was more than one trial at any site, a site mean was used.

extension worker, who often likes to see a variety trial in an area where the crop has only just been introduced and where the contribution to production may, initially, be negligible. With the gradual shift in cotton production that has occurred in Uganda moving broadly from the south to the north, it has been necessary to discontinue certain trial sites and develop new ones. In this respect it is important to know exactly where crop production is concentrated and important work was started at Namulonge in 1970, in conjunction with the Department of Agriculture, to locate more precisely areas of expanding and decreasing cotton production. Such studies could provide a more rational basis for deciding from year to year upon the location of individual trial sites.

Interpretation of results. Table 8.6 summarizes comparisons of BPA and BP52 from 239 trials in eight seasons. The data indicate marked seasonal influences on yield differences, with BPA yielding 27.1 per cent more than BP52 in the 1965–66 season and 5.7 per cent less in the 1970–71 season, even though the average yield did not change markedly from one season to the other.

Part of these differences in relative yields can be attributed both to varietal differences in resistance to bacterial blight and to seasonal differences in its severity. Although all trial seed was dressed with a bactericide, outbreaks of bacterial blight were usually observed towards the end of the season on susceptible varieties. It is impossible to estimate how much influence on yield these outbreaks exerted. As the number of resistant entries in the trials increased, so the noticeable effects of the disease declined and, at many sites, losses from this cause were almost certainly negligible. At others, however, bacterial blight may well have contributed to a reduction in BP52 yields relative to BPA.

Nevertheless, there is little doubt that the explanation of the large differences between these two varieties in relative performance lies mainly in the different patterns of rainfall distribution from place to place and season to season (see Chapter 2). In the 1970–71 season, the early cut-off of the second rains prevented the later-maturing types from realizing their full yield potential. An understanding of differences in cropping patterns in relation to the realization of yield under different environmental conditions is fundamental to the formulation of criteria for selection for improved yields. Consequently, it became routine in the cotton breeding programme, to arrange for trials to be picked at relatively short, regular intervals so that accurate, cumulative picking curves could be constructed. These give a useful picture of the period over which the crop matures and hence a picture of the period over which the bolls were set. On this basis, BPA was shown to produce the main bulk of its crop somewhat later than BP52. Such information can be supplemented by analysing samples of plants to construct plant diagrams (Munro & Farbrother 1969) as, for example, in the elucidation of genotype–environment interactions in the yields of BP52 Local and derived modal bulks (Arnold 1972). Aspects of crop growth in relation to genotype–environment interactions were discussed more fully in Chapter 4.

While it is logical to continue to search for genotypes that show both increased yields and stability of performance relative to contemporaneous commercial varieties, it may well be that it is asking too much to expect a single pattern of crop growth to maximize the yield potential of the environment over the full range of conditions encountered. Breeding specific genotypes for specific localities is feasible only if the environment of one locality, relative to another, is consistent over the majority of seasons. The evidence so far accumulated, with respect to rainfall in particular, suggests that this condition is seldom met to a sufficient degree of precision in Uganda. There are clearly two aspects to the problem: (1) to investigate genotypic differences in relation to the magnitude of the interactions of genotypes with environments, and (2) to study the performance, in similar terms, of heterogeneous varieties made up from mixtures of genotypes that differ widely in their crop physiological characteristics.

Genotype–environment interaction

In routine procedures for selecting lines that show improved performance and good relative stability, it is usual to test a progressively smaller number of genotypes in a progressively greater number of plots spread over a greater number of trial centres. Figure 8.12 illustrates the type of results that are commonly encountered. They show the way in which genotype–environment interactions (together with random experimental error) modify the apparent performance of a new selection, relative to a control variety, when more widely evaluated. Such results lend further emphasis to the dangers of forming opinions of the genotypic worth of new selections on the basis of a single experiment, and point also to the dangers of selecting too intensively in the early stages of a programme.

Final choice of new lines must therefore be based on evaluation in district trials over several seasons. But how long should the breeder be prepared to wait before committing himself? Table 8.6 does not provide the answer to this question. It simply shows that conditions can arise in which the expected yield advantage of a new variety is not realized. It serves to draw attention to the large seasonal influences on relative yields but, in doing so, obscures the equally important fluctuations from place to place. Figure 8.13 shows the frequency distributions of percentage yield increments of BPA over BP52 at all variety trial

Figure 8.12. Performance of two A(66) selections relative to BPA in four seasons. BPA yields in parentheses. * Means of eighteen trials; ** mean of twenty-four trials A(66)22; twenty-three trials A(66)131. RPR, replicated progeny rows; ST strain trials; LT, line trials.

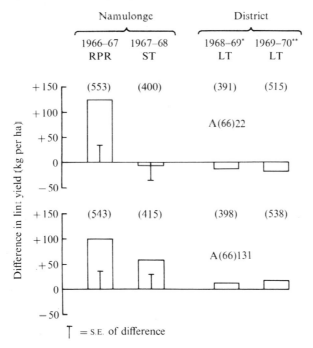

Table 8.7. *Lint yields of BPA and BP52 at seventeen sites for three seasons*

	Site	1966-67				1968-69				1970-71			
		BP52	BPA	Difference		BP52	BPA	Difference		BP52	BPA	Difference	
		(kg per ha)		(kg)	(%)	(kg per ha)		(kg)	(%)	(kg per ha)		(kg)	(%)
Western	Kihihi	451	429	−22±42.8	−4.9	318	313	−5±29.8	−1.6	692	612	−80±46.1	−11.6
	Mubuku	751	835	+84±37.5	+11.2	369	585	+216±40.3	+58.5	585	498	−87±21.6	−14.9
	Matiri	157	174	+17±21.4	+10.8	217	253	+36±40.0	+16.6	166	188	+22±23.9	+13.3
	Average	453	479	+26	+5.7	301	384	+83	+27.6	481	433	−48	−10.0
Central	Bigasa	333	367	+34±34.9	+10.2	78	83	+5±14.7	+6.4	325	287	−38±27.5	−11.7
	Kanoni	455	504	+49±22.6	+10.8	283	303	+40±13.2	+15.2	409	403	−6±36.4	−1.5
	Busunju	397	474	+77±50.5	+19.4	409	547	+138±53.6	+33.7	470	383	−87±38.0	−18.5
	Bukomero	485	584	+99±53.0	+20.4	199	330	+131±55.0	+65.8	250	231	−19±42.2	−7.6
	Namulonge	388	419	+31±24.9	+8.0	405	541	+136±19.9	+33.6	786	697	−82±20.3	−11.3
	Kalagala	323	495	+172±54.2	+53.2	228	293	+65±26.3	+28.5	558	528	−30±41.6	−5.4
	Namyoya	456	495	+39±40.0	+8.6	377	444	+67±28.4	+17.8	432	413	−19±75.4	−4.4
	Ntenjeru	638	642	+4±39.3	+0.6	474	655	+181±37.5	+38.2	469	446	−23±66.4	−4.9
	Bale	603	649	+46±39.5	+7.6	493	532	+39±42.0	+7.9	578	606	+28±68.7	+4.8
	Average	453	514	+61	+13.5	325	414	+89	+27.4	475	444	−31	−6.5
Eastern	Ikulwe	493	673	+180±44.4	+36.5	405	378	−27±22.0	−6.7	464	432	−32±27.7	−6.9
	Nakabango	182	261	+79±28.5	+43.4	249	192	−57±22.3	−22.9	282	254	−28±17.3	−9.9
	Vukula	600	678	+78±61.8	+13.0	383	410	+27±29.1	+7.0	558	623	+65±70.8	+11.6
	Bugaya	257	343	+86±30.1	+33.5	312	414	+102±24.6	+32.7	374	287	−87±32.6	−28.3
	Tororo	251	278	+27±32.5	+10.8	505	575	+70±44.0	+13.9	634	618	−16±32.0	−2.5
	Average	357	447	+90	+25.2	371	394	+23	+6.2	462	443	−19	−4.1
	Overall average	425	488	+63	+14.8	334	403	+69	+20.7	472	442	−30	−6.4

centres for three seasons. It gives a visual representation of variation in the relative difference in yield between the two varieties from place to place in the same season as well as from one season to the next.

Presented in this way, the yield responses of BPA in the 1966–67 season do not appear to be markedly different from those recorded in the 1968–69 season. Nevertheless, closer examination of the data shows that there were marked differences at certain localities. Results from an orthogonal set of seventeen trials from all three seasons are shown in Table 8.7. The trial centres are divided into three main geographical regions (western, central and eastern) but there does not appear to be a very consistent pattern of results associated with this regional grouping. Indeed such a pattern would only be expected to the extent that the pattern of weather is closely consistent within any geographical

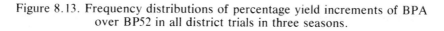

Figure 8.13. Frequency distributions of percentage yield increments of BPA over BP52 in all district trials in three seasons.

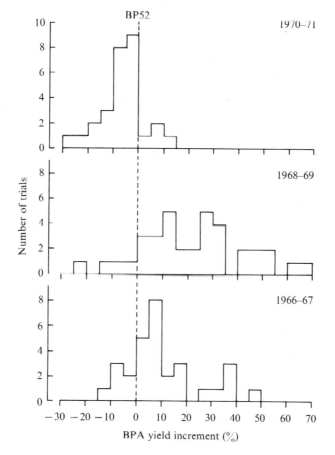

region. From the large effect of relatively small differences in rainfall at critical periods in crop growth (see Chapters 2 and 4) such a situation is unlikely under the prevailing climatic conditions.

It is nevertheless essential for the plant breeder to make decisions on the basis of results of this type about what genotypes to discard and what to carry forward for further breeding. For this purpose, we have found it useful to apply a regression analysis of the type proposed by Yates & Cochran (1938). H. L. Manning used this technique at Namulonge during the 1950s and it later became routine for comparing new breeding material with the established variety. From analyses of this type, it was shown (Walker 1963) that the slope of the regression line for Albar 51 was greater than that for BP52, with the extrapolated cross-over point representing a yield level not far short of the estimated national average for cotton at the time (Figure 8.14). Further selection in the Albar 51 material was therefore devoted to producing material that outyielded BP52 over the whole range of yield levels normally encountered: in terms of the regression analysis, to look for stocks that

Figure 8.14. Genotype–environment interaction. Data from twenty-one trials in the BP cotton area, season 1958–59. For difference in slope, Student's $t = 3.48$ with 38 degrees of freedom ($p < 0.01$). For the Albar equation, co-efficient of determination $r^2 = 0.93$: for BP52 K/51. $r^2 = 0.99$ (from Walker 1963).

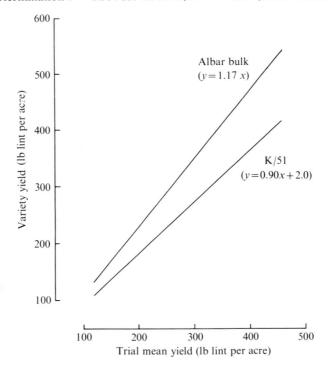

gave regression lines parallel to that of BP52. Figure 8.15 shows a regression analysis of the yield of BPA, relative to BP52 over all trials for four seasons.

Such an analysis illustrates the stability, over a range of yield levels, of the yield of a new variety or line relative to an established one. It does not make use of the concept of actual stability introduced into similar analyses by Finlay & Wilkinson (1963). It would appear that some of the criticisms levelled at their descriptions of stability in terms of the slopes of the regression lines (such as those by Freeman & Perkins 1971) do not apply when only the relative stability of one variety to another is under consideration. As Eberhart & Russell (1966) have pointed out, it is not only the slopes of the regression lines that are important but also the deviations from regression, which can readily be examined as a further aspect of relative stability.

Nevertheless, the presentation in Figure 8.14 is open to criticism in several respects. It can be criticized on the grounds that the entries in the trials were not identical from one season to the next. Although the

Figure 8.15. Regression of variety mean yield of lint ($\log_{10}$ (kg per ha)) on trial mean for the BP cotton area during the period from 1963–64 to 1967–68. BPA, broken line, compared with BP52, unbroken line, from a total of 86 trials (from Innes 1969*b*).

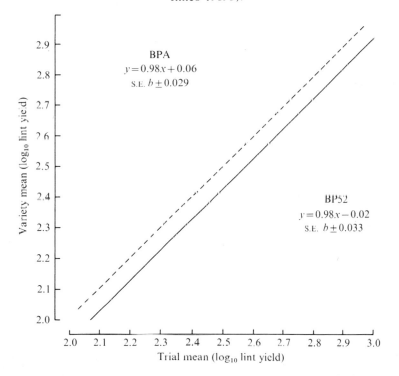

same controls were used from season to season, the lines under test in one season were successively replaced by reselections in ensuing seasons. It is impossible to estimate what effect this might have on the analysis but, in terms of the decisions the breeder has to make, probably not very much. More serious limitations are, first, the extent to which such a set of results can be thought of as adequately sampling all the environmental conditions to which the material is likely to be exposed and, second, the extent to which the mean value of any particular set of genotypes may be taken as an appropriate measure of the environment.

Concerning the first point, reference has already been made in Chapter 2 to cycles or sequences of weather patterns and, in Chapter 3, the consequences of such sequences on the availability of soil nutrients were discussed. The yield results in both Tables 8.1 and 8.6 give indications that there may well be sequences of years that broadly favour one genetically determined pattern of crop growth and others that favour a different pattern. There is, however, no known way of predicting the number of cropping seasons for which any particular cropping type will remain better suited than other available types to exploit the full range of environments encountered.

Concerning the second, clearly the validity of using the mean performance of a group of genotypes as a measure of environment, must be dependent upon the genotypes themselves forming an adequate sample of the range of possible genotypes. But what does 'adequate' mean in this context? How, for example, is it related to the various different ways in which, in crop physiological terms, a population of cotton plants can produce a crop? And how do such things as susceptibility or resistance to diseases and pests affect the choice of genotypes? We cannot yet be sure that we know the answers to these questions.

While it may therefore be dangerous to infer too much from regression analyses of variety and locality means, the extension of the analysis to calculate heritabilities, as has been proposed by Breese (1969), might go a long way towards meeting the criticisms we have made of heritabilities calculated from more restricted data. It could well be that tropical Africa, where large genotype–environment interactions are so often encountered, could become a valuable testing-ground for new ideas of this type. Certainly, experiments with these aims in mind should be laid down and maintained unaltered until a worthwhile run of data can be accumulated. Such experiments should not be limited to the interactions of cropping patterns with weather conditions (as determined by such things as season, locality and sowing date) but should include other factors of crop environment such as population density, competition from weeds and losses caused by pests and diseases.

In a developing country, the task of the plant breeder is made more difficult because even though he may specify when his variety should be sown, at what spacing and what protection against insect attack should be given, he knows that many growers will deviate from the recommended practices (Chapter 10). Consequently he must ensure that any new variety will not do worse than its predecessor even when the recommendations are not followed. This is especially true for cotton, because the grower has no freedom of choice as far as the variety is concerned (see p. 241). The system of variety evaluation in district trials forms a substantial safeguard against this contingency for, whatever the intention, some district trials will be sown late, or will suffer from loss of stand, or will not be correctly sprayed with insecticide, or will be sown on poor soil.

However, further safeguards can be provided by growing additional line trials of the same entries as those in district trials upon which different treatments are superimposed, such as spraying or not spraying, close or wide spacing, early or late sowing. This approach was introduced at Namulonge when it appeared that the management of district trials had reached a general level that was considerably better than the average for the commercial crop. Comparisons among trials subjected to different treatments, show clear evidence of genotype–environment interaction, and serve to draw the breeder's attention to those lines that are suspect for stability of performance.

The approach is thus very different from that of the breeder who is specifically looking for genotypes that respond better than others to a superimposed treatment, as for example in the case of the cereal breeder who is breeding for varieties with short, stiff straw that can successfully use heavier rates of applied nitrogen. Nevertheless, the breeder must always think ahead and keep in mind a range of possibilities so that he has locally-adapted breeding material ready to meet changes as they arise. The development of schemes for agricultural credit, for example, might well mean that it would be possible to introduce a 'package deal' approach for cotton in Uganda, as has been done successfully in other countries and with other crops. The successful extension of the use of hybrid maize in Kenya, for example, was dependent upon the concomitant use of fertilizer and adherence to the recommended sowing period. A cotton variety that would produce heavy yields in dense stands over a shorter period, during which insect pests could be controlled economically, might have a place in a future scheme of this type in Uganda. While limiting extended evaluation to those genotypes which fit the current agricultural situation, therefore, the breeder also retains a range of other genotypes that differ widely in their growth characteristics. It is particularly in this context that he or she must work closely with both the agronomist and the crop physiologist.

Release of useful genetic variation

We have now examined some of the problems of selecting and evaluating material under Uganda conditions. As the picture has unfolded, attention has increasingly become focussed on variability, both in the genotype and in the environment. We have considered criteria for assessing genetic variation and we have seen the limitations of selecting in material in which it is inadequate. We have described some of the difficulties that arise from crossing with introduced varieties and have seen that important characteristics, such as resistance to Alternaria boll rot, may be reduced to an unacceptable level in the process. Attempts have therefore been made to find other ways of producing recombinant types in which all of the characters required in a successful commercial variety are retained in the breeding population.

Recurrent selection

The principle of increasing the frequency of recombinations by allowing open pollination in alternate generations was taken a stage further in BPA stocks by controlled crossing. BPA 66, together with the six most promising lines derived from it and two related lines stemming back to A(61)21 (Figure 8.9), were intercrossed in all combinations. Parents and hybrids were included in replicated trials at Mubuku and Namulonge in the 1970–71 season. Additional plots were grown at Namulonge for selfing. Diallel analysis of the trials confirmed that there were considerable genetic differences among the various lines but genetic variances showed marked genotype×site interaction (H. E. Gridley, personal communication). Selfed seed from the best hybrids, assessed for yield and quality in both trials, was used to raise F_2 progenies in which a further reselection programme was started, while blight-resistant plants from each F_2 were intercrossed to provide a second cycle of selection. F_3 populations from the first cycle were raised at Namulonge in 1972–73 among which striking morphological differences were observed, both from plant to plant in the same population and from one F_3 population to another.

As yet it is too early to assess the usefulness of this approach or that of a similar programme started at Serere in elite SATU stocks. In a recurrent selection programme at Namulonge using twelve AH lines, Innes (1973b) found that, although the average yield of the sixty-six hybrids was 26 per cent greater than that of the mean of the twelve parents, a second cycle of crossing with the best hybrids failed to increase the average yield level above that obtained from the first cycle. In the USA, Miller & Rawlings (1967) have reported a linear response in lint yield after three cycles of recurrent selection in a population derived from a cross between Coker 100 and Acala 1517.

Improvements in lint yield can only be exploited in commercial seed issues if the other important characteristics such as lint quality and resistance to pests and diseases are maintained. The complex inheritance and interrelationships of all these characters, as well as the importance of recessive genes in some of them, impose limitations on the operation of recurrent selection programmes, additional to those arising from the difficulties of large scale crossing in cotton. Nevertheless the wide range of phenotypic variability recorded in populations produced by recurrent selection in BPA, SATU and AH material indicates that this approach is well worth pursuing.

Panmixis

The wide range of recombinant types that was selected from the Albar modal bulks, and material derived from them, influenced the creation of other similar populations. The extent to which the production of recombinants was enhanced by the modal selection procedure cannot be determined. It may well be that they would have arisen in any case, simply from natural outcrossing, but Walker's (1964) suggestion, that modal selection may increase the proportion of heterozygotes in the population, should be kept in mind and examined in greater depth when the opportunity arises.

If such a mechanism operates, however, it would serve only to increase the rate of recombination, which could probably be achieved more efficiently by brush pollination. The value in cotton breeding of creating panmictic populations as material for eventual selection, has been the subject of considerable discussion (Hutchinson 1959; Walker 1969). Examples of failures and successes appear to be related both to the choice of parents and to the intensity of selection. Where the choice of parents is too wide, intercrossing results in a break-down of many desirable characters. In our experience, the greatest success has been achieved when all the parental material has been locally adapted, such as in the SATU populations, or in the selection programme from a panmictic population of Albar and BP52 (Arnold 1970*b*).

Following this reasoning, two new panmictic populations were constituted from the best hybrids in the BPA recurrent selection programme. Both were grown in isolation: one was allowed to pollinate naturally, while outcrossing was increased in the other to an estimated 40 per cent by brush pollination (Arnold *et al.* 1971). Meanwhile other stocks, such as a wide range of Albar and Allen types reintroduced from other countries, were screened with a view to including the best either in future panmictic populations or as additional parents in the BPA recurrent selection programme (Innes 1974*a*).

Wide-range crossing

It has become clear that it is difficult to produce useful segregants from wide-range crossing unless precautions are taken to avoid the break-down of important characteristics that have been built up by prolonged selection in the local environment. At the same time, the chances of producing worthwhile recombinations from a single cross are limited by linkage phenomena. Hanson (1959), for example, has suggested that a breeding programme for self-pollinating species should include at least one and preferably three or four generations of intercrossing among at least four parents in order to bring about the break-up of linkage blocks. Attempts must therefore be made to strike a balance between these two aspects of the problem, although in most cases there is little quantitative evidence to draw upon in planning the programme.

Choice of parents. Parental stocks used in wide-range crossing pro-grammes at Namulonge represented three main categories, although each to some extent overlapped the others. First, were successful com-mercial varieties from Uganda and a range of other countries; second, were stocks possessing a single feature of specific importance; and third, were stocks derived from interspecific crosses. Introduced stocks of all three categories were compared with local varieties in small non-replicated plots for a first assessment of adaptation to the local environment. All except the extremely poor performers were then compared with control varieties in replicated trials for components of yield, quality and resistance to diseases and pests. Only after careful evaluation of their performance under local conditions were decisions made on how to attempt to recombine the various desirable characteristics.

Many commercial varieties from other countries have attributes that are worthy of study under Uganda conditions. In the United States, for example, varieties that have been developed for use under intensive systems of farming differ considerably from African upland types in their cropping characteristics, and have greater ginning percentages. Varietal types such as Acala, Deltapine, Stoneville and Coker come into this category.

In the second category (stocks possessing specific characteristics) were introductions with a wide range of different characters. Some, such as frego bract (Jones & Andries 1969), nectariless (Meyer & Meyer 1961) and okra leaf-shape (Andries, Jones, Sloane & Marshall 1969) are of potential importance in breeding for resistance to insect pests and diseases. Others, such as the various 'storm-proof' charac-ters (Low 1962; Niles & Richmond 1962) are of importance in crop management, while the glandless character is the key to the production of seed products that are free from gossypol (McMichael 1960).

The third category (material derived from interspecific crosses) was represented in the programme almost entirely by introductions from other countries. With the exception of crosses between upland (*G. hirsutum*) and *G. barbadense* types such as Pima S_1, little work has been done in Uganda on interspecific crossing and no attempts have been made to transfer desirable characters from the diploid species. In general, in other countries, crossing programmes involving diploids have met with only limited success. Nevertheless, in the USA increased fibre strength has been successfully transferred to a range of upland stocks from a synthetic amphidiploid, produced from a cross between *G. arboreum* and *G. thurberi* (Kerr 1969), and several of the resultant high-quality varieties such as Atlas, Delcerro, Acala SJ-1 and TH 149 were imported into Uganda and incorporated into the programme.

HAR material, derived from the tri-species cross, *G. hirsutum*× *arboreum*×*raimondii* (Kammacher 1965, 1968; Goebel 1968), gave heavy yields of lint with outstanding quality when tested at Namulonge. A programme of pedigree selection was started in those lines which most closely matched BPA in resistance to pests and diseases (Innes 1975). Other lines were used in further crosses with locally-adapted stocks.

Introgression from diploids not only presents opportunities for transferring favourable genes, but also may lead to increased genetic variation in the tetraploid species. Saunders (1970), for example, obtained a wealth of variability in *G. barbadense* by the introduction of chromatin from *G. anomalum*. He attributed the release of variation not to the presence of *anomalum* genes but to *anomalum* chromatin causing increased recombination in adjacent segments of the chromosomes, where there were only *barbadense* genes. It has been suggested that the increased recombination is associated with a shift in the position of chiasmata resulting from the insertion of a segment from the diploid species (Rhyne 1960, 1962). Saunders also transferred *anomalum* chromatin to an Acala 4-42 background. When some of these upland selections were tested at Namulonge, they were poorly adapted and suffered badly from *Alternaria* attack (Innes 1975).

Crossing systems. Stocks that were poorly adapted to the local environment but had one or two desirable characteristics were backcrossed to a local variety once or twice before selecting types to be included in the main programme. Single cross hybrids (A×B), triple-cross (A×B)×C as well as double-cross hybrids (A×B)(C×D) were produced and the material was then handled following the general scheme shown in Figure 8.16. A few generations of selfing and selection, designed both to allow recessive genes to express themselves and to restore complex polygenic characters to an acceptable level, were followed by inter-crossing and diallel analysis to select crosses for inclusion in cycles of recurrent selection. Meanwhile pedigree selection was continued in outstanding progenies and the best lines evaluated in district trials.

Lint quality

As an export commodity, cotton is essential to many developing countries in earning foreign exchange. In Uganda, cotton and its products accounted for about one-quarter of the total value of annual

Figure 8.16. Breeding system for wide-range crosses. BBI, bacterial blight inoculation; BIT, blight inoculation trial; FB, farm bulk; LB, line bulk; LT, line trial; RPR, replicated progeny rows; SB, strain bulk; ST, strain trial; —, selfed; - - -, open-pollinated.

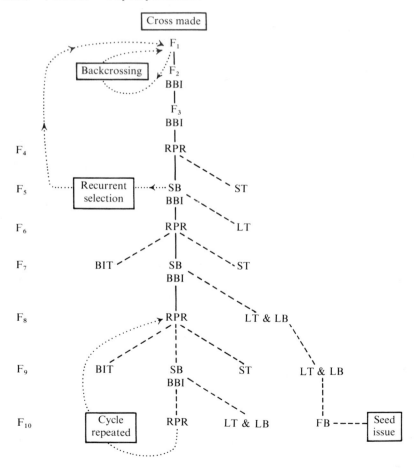

exports for the ten-year period ending in 1971. The maintenance of the competitive ability of cotton lint in world markets, therefore, both in relation to cottons produced in other exporting countries, and in the face of growing competition from man-made fibres, was of vital importance to the economy. The breeder carries a heavy responsibility in this respect. He must keep closely in touch with technological developments in the textile industry and plan his programmes accordingly. Equally, he must have at his disposal the means of evaluating large numbers of different genotypes in terms of the more important components of quality.

For many years, cotton breeders at Namulonge depended for quality evaluation on the services of the Shirley Institute in Manchester. This central testing in an internationally recognized laboratory, was supplemented to an increasing extent by tests carried out locally. A small fibre-testing laboratory was started in 1951 (Farbrother 1961a) and in subsequent years some 3000 samples were tested annually for staple length (using a Shirley photo-electric stapler), fibre bundle strength (on the Stelometer instrument) and micronaire value. An extensive, new, testing laboratory, built by the Uganda Lint Marketing Board in 1972, added the facility for small scale spinning tests to a much increased capacity for fibre testing. It gave promise of filling an important gap in the cotton breeders' immediate resources at Namulonge.

Traditionally, quality in cotton has been defined largely in terms of fibre length, strength and fineness. With the addition of fibre maturity and an assessment of seed coat damage (Evenson 1955) these were the main quality criteria used in breeding BPA, although final selection of components of seed issues was always based on the results of spinning tests. The quality standards set for BPA, however, particularly with regard to staple length, greatly restricted the range of genetic material with which the breeder could work. There is no doubt that if these quality standards could have been relaxed, a variety with a greater yielding potential could readily have been produced, because of the negative relationship between yield and quality that has already been mentioned (p. 217).

The evaluation of factors affecting quality, the development of co-operation between cotton producers and consumers to produce a more acceptable raw material, and the possibilities for further improvement, have been reviewed in detail elsewhere (Arnold 1969b). It is clear that cotton has now entered a period when further changes in quality requirements have to be met. Describing research undertaken by the International Institute for Cotton (IIC), Burkitt (1972) has identified three main aspects of these changes: competition from man-made fibres, processing methods in the textile industry, and consumer demands.

Changes in all three respects have fundamental implications for the maintenance of the export potential of cotton. Individually, the developing countries lack the resources to solve the problems involved and it was to meet this need that a number of producing countries jointly founded the IIC. It is concerned both with research and promotion. On the research side, new methods have already been developed, for example, for chemically treating cotton fibres so that easy-care properties can be imparted, while retaining strength and durability of the fabric. Tests carried out by the IIC on treated and untreated cotton fibres show clearly that there is genetic variation for responsiveness to this type of treatment. Given the development of suitable screening techniques, therefore, the cotton breeder should be able to produce locally-adapted varieties that are better suited than existing ones to this new type of processing. In such ways, genetic change and industrial technology will combine to give a product that will compete more successfully with synthetic fibres.

Burkitt also draws attention to the great expansion in the use of break-spinning, another development that may well have implications for the choice of variety. It is almost certain that some types of cotton will be spun more effectively than others by this process. The breeder must therefore ensure that he carries a wide range of quality types in his locally-adapted breeding stocks. Studies of variability in the components of lint quality and of the correlations among them, such as those reported by Innes (1973a), must be expanded to include the new types of test as soon as suitable procedures have been developed.

These changes, together with changes in harvesting, cleaning and ginning methods (Chapter 10) might well mean that the cotton breeder of the future could break away from the strict requirements of staple length that have so limited his approach in the past, at any rate as far as the BP types were concerned. Nevertheless, we should not lose sight of the important role the cotton breeder has played, and must continue to play, in developing and maintaining confidence on world markets in the quality standards of Uganda cottons.

Seed issues

Production zones

In Uganda, as in many African countries, cotton ginning is entirely under Government control. The producing area is divided into ginning zones, each with its own ginnery or ginneries and associated network of buying posts. Seed cotton is purchased from the grower by a co-operative society acting in association with the Lint and Seed Marketing Board, itself a para-statal body. The co-operative societies are organized into unions which own and operate the ginneries. The

Board markets the lint and seed, and the co-operatives are remunerated for all the operations associated with buying, transporting and ginning the raw cotton.

As a consequence of this centralized system of ginning, the grower is left at the end of the season with no seed of his own, except perhaps for a small quantity of the original seed he acquired for sowing. As the new season approaches, seed is issued to him, free of charge, from the appropriate buying post, now used as a seed distribution centre. The system is well suited to controlling the release and spread of new varieties and has been used for this purpose since the 1920s.

New seed issues were constituted annually by the breeders and nucleus stocks grown on the research stations. These nucleus stocks were multiplied in carefully chosen areas, where the risks from contamination with cotton from neighbouring areas were small and where sowing and marketing could be controlled. Multiplication in the BPA producing area began on government seed farms, prison farms and settlement schemes and proceeded to larger areas until enough was available for allocation to a zone or group of zones (Figure 8.17). In this way, an annual wave of new seed was constantly passing through the whole producing area, each new seed issue taking about seven years from initial release until it was finally withdrawn from the multiplication scheme and marketed for crushing.

Constitution of seed issues

The reasons for constituting seed issues as mixtures of selected lines have been reviewed by Walker (1963). In the early years of cotton breeding in East Africa, mixtures were issued instead of single improved lines largely as a matter of expediency, either to obtain a larger bulk of seed for initial multiplication or because no single line showed the appropriate balance of characteristics. There was also the possibility that components of mixtures might exploit the environment in somewhat different ways so that a mixture might not only outyield a single line, but might also give greater stability of performance over a range of environments. To these advantages, practical and theoretical, Walker added that of exploiting heterosis produced by hybrids among component lines. Walker's conclusion that the BP52 seed issue NC54 did not remain static but showed advances in yielding potential, through heterosis, with successive generations of multiplication, is open to question (Arnold 1970c). Nonetheless the principle has been exploited in various out-pollinating crops and is clearly a possibility with cotton when grown in areas where there is substantial, natural outcrossing, and provided that the components of the multi-line are suitably chosen.

The first release of BPA took the form of a single, though somewhat

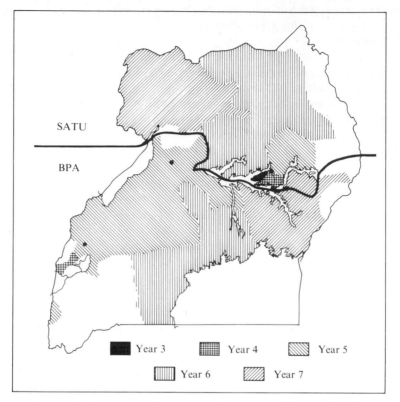

Figure 8.17. Seed multiplication schemes for BPA and SATU.

heterogeneous, line. Subsequently, reselections were issued as multi-lines (Table 8.3). Seed of successive multi-lines, and others derived from similar material, were sent to East Malling Research Station for long-term storage at low temperature. It would eventually be possible, therefore, to test a series of generations of a commercial multi-line simultaneously over a range of environmental conditions. At the same time combining ability studies were initiated using diallel sets of crosses so that further multi-lines could be constituted from those lines showing the best general combining ability.

Evidence on the possible advantages of heterogenous varieties, with respect either to greater yielding potential or to greater stability, is inconclusive. Although some evidence has been adduced to indicate that intrapopulation diversity may improve stability (Simmonds 1962) it is not possible to generalize (Marshall & Brown 1973). Results obtained on cotton by Riggs (1970) at Serere from trials in which four upland varieties were grown in dual mixtures at three ratios led him to conclude that the yield advantage of mixtures, relative to pure stands of their components, was sufficiently promising to warrant further study.

Our own observations on BPA material, based on the results of district trials in southern and eastern Uganda in the 1969–70 and 1970–71 seasons, showed that BPA multi-lines outyielded the weighted means of their components on average, but the advantage was small and not consistent. The multi-lines were not obviously more stable than their component lines and considerably more extensive data would be necessary to draw firm conclusions.

The components of BPA seed issues were all fairly similar in their growth characteristics. In order to test more heterogeneous material, a mixture was constituted from equal weights of seed from an early-maturing AH line and a late-maturing Albar 51 line, and tested in twelve district trials in the 1971–72 season. Overall, the mixture outyielded the mean of its components by 4 per cent, but there were large differences in yield level from site to site and considerable variety×site interaction.

G. B. Jones and N. L. Innes (unpublished) tested a wide range of cotton varieties in 1:1 seed mixtures at three sites over two seasons and the results were similar in that observed and calculated differences of the mixtures were relatively small and inconsistent. Comparison of SATU 71 with the derived mean of its two component lines SA(66)41 and SA(66)51 measured in fifty-seven trials over three seasons gave identical yields of 451 kg of lint per ha. Nevertheless, the yields of the two component lines relative to each other varied considerably from season to season, indicating greater stability for the mixture than for either component line (Jones 1973).

Clearly differences between the yield of a mixture and the weighted mean of its components would be expected only when physiological differences among the components are great enough to result in measurable differences in relative yields under different environmental conditions. More information is needed on the magnitude of the genetic and environmental effects that are likely to give rise to this situation.

Conclusions

In the light of the Namulonge experience what major changes would we suggest for a future breeding programme?

We have drawn attention to the magnitude of the genotype–environment interaction, to the striking differences in relative yields of two genotypes from season to season and place to place. We have shown that in the face of these interactions small differences among genotypes are difficult to evaluate and that the concept of genetic advance is fraught with danger, unless defined in the context of the full range of environmental conditions to which the variety will be exposed. The breeder is concerned with genetic *change*: he must be very careful before claiming that any particular change necessarily represents a genetic *advance*. Indeed, the partitioning of a phenotypic effect into its genotypic and environmental components has limited meaning in this context, where an advance is determined neither by the one nor by the other, but by a delicately balanced combination of both. Perhaps herein lies a principle that has restricted the useful application to plant breeding of biometrical genetics in general – as we have seen for the particular case of the selection index.

Consequently, we would, in the future, more readily sacrifice resources, devoted to the apparent demands of the short-term, for the benefit of more sustained and carefully planned, long-term investigations. In this we would look to a multi-disciplinary approach to give greater understanding of the interplay of genotypes with environments and to define, more precisely in biological and economic terms, where a contribution from plant breeding might best be made.

We would continue our research into breeding methods aimed at a better understanding of how best to strike a balance between the requirement to outbreed for variability and recombination, and the requirement to inbreed, in order to maintain and enhance characters controlled by recessive polygenic complexes. In our routine breeding programmes we would build on the successes already achieved. What would we rate these to be?

First, the maintenance of quality standards that have safeguarded the position of cotton in Uganda's export markets.

Second, the incorporation of resistance to diseases and pests that have meant that former hazards like bacterial blight and the cotton jassid are no longer recognized as important factors in crop loss.

Third, the production of varieties that have performed well over a wide range of conditions, so that the realization of improved genetic potential has not depended upon technological innovation beyond the means of the small farmer.

Fourth, the maintenance of reservoirs of genetic diversity so that the breeder of the future would have useful material to work with, and the breeder of the present could respond rapidly to the needs of change in agricultural practice or industrial use.

Many people have contributed to these achievements, not only the cotton breeders but also those in the Uganda Department of Agriculture, Lint Marketing Board and Department of Co-operative Development who, in numerous ways, have given their full and willing co-operation.

APPENDIX 8.1

Key to the names of cotton varieties and code letters for selections mentioned in the text

A	Selections from Albar material made at Namulonge.
Acala	Large-bolled varieties from California, USA.
AH	Selections from the cross Albar×Barhop.
Albacala	Material derived from crosses between Albar and Acala 4-42.
Albar	Blight-resistant selections from Nigerian Allen.
Allen	An upland variety grown in the USA before the boll weevil became an important factor in crop loss. Seed was imported into Uganda in 1909 and 1910 and into Nigeria from Uganda in 1912. Re-imported into Uganda in 1950, Nigerian Allen gave rise to the blight-resistant Albar selections.
Atlas	An upland variety with strong fibre from the USA.
B181	A selection from material derived from Buganda stocks.
Bar	A general term for selections resistant to bacterial blight, bred in the Sudan. A list giving the derivation of Bar selections up to the end of 1963 is given by Innes (1964).
Barhop	Selections bred in the Sudan for increased fibre bundle strength from a cross between Hopi Acala and Bar 11/7.
Barwilds	Sudan selections from a cross between Wilds Sus 16/1 and Bar 11/7.
BP50	A 'Bukalasa pedigree' selection derived from Buganda Local.
BP52	Commercial variety in southern Uganda before the release of BPA from Namulonge. Selected at Bukalasa from Nyasaland Upland. The initials BP which originally stood for 'Bukalasa Pedigree' were retained in the name BPA so that the quality type could be easily recognized on world markets.
BPA	Namulonge varieties from Albar 51 stocks.
Briga	Selections from the cross Bar 12/16×IL 47/10.
C	Selections from BP52 made at Namulonge.
CA	Selections derived from material obtained by intercrossing BP52(C) lines with Albar.
Coker	Upland varieties from the USA.
DE715/6M	Derived from a cross between B181 ('D') and BP50 ('E').
Delcerro	A large-bolled variety with long, fine, strong lint from the USA.
Deltapine	Upland varieties from the USA.
HAR	Material derived from the tri-species cross *G. hirsutum*× *raimondii*×*arboreum*.
IL 47/10	A former commercial variety from eastern Tanzania, highly susceptible to bacterial blight.
K51	A Kawanda selection from BP52.
LA	Selections from crosses made between Lubaga (Tanzania) material and Albar stocks.

MB	Modal bulk made by the procedure described by Manning (1955) and Walker (1964).
Mwanza Local	The original commercial variety of western Tanzania derived from the first importations of seed of American upland origin.
MZ 561	A selection from Mwanza Local.
NC	Namulonge seed issues of BP52 material.
PB	Selections from the cross Pima $S_1 \times$ Bar 12/16.
Pima S_1	*G. barbadense* variety from the USA with long, fine, strong lint.
Punjab	A short-stapled American upland type from the Punjab.
Reba	Varieties bred by the IRCT in the Central African Republic, highly resistant to bacterial blight but with weak lint.
S	Serere selections or seed issues.
SA	Serere selections from Albar 51.
SATU	Serere Albar type Uganda – seed issues from Serere derived from Albar 51.
Stoneville	Upland varieties from the USA.
UK	Ukiriguru, Tanzania selections and seed issues.
UKA	Selections from crosses between Ukiriguru material and Albar 51.
UKAN	Reselections from UKA material made at Namulonge.
UPA	Selections from the cross (MU8 $\times$ BP52^2) $\times$ Albar 51.
Upland	Annual cottons called 'uplands' comprise the most important section of *G. hirsutum* L. race *latifolium* Hutch. The name arose in the USA to distinguish these cottons, grown in the interior of the country, from the crop of the lowland, coastal areas comprising varieties of *G. barbadense*.

9

The Namulonge farm

M. H. ARNOLD, R. G. PASSMORE, E. JONES
AND F. E. TOLLERVEY

Introduction

The basic purpose of the Namulonge farm was to produce nucleus stocks of pedigree cotton seed large enough for release into the seed multiplication scheme (Chapter 8). At the same time it was planned as an investigation of intensive land use, not simply for producing cotton, but in a much wider context of tropical agriculture. The initial development of the farm and the agricultural problems encountered have been described in detail by Hutchinson *et al.* (1959) but, in order to understand the lessons to be learned from its operation over some twenty years, we must briefly recapitulate some of the information given in their account.

Namulonge is situated about half a degree north of the equator at an altitude of 1143 m measured at the main laboratories. Topographically the estate comprises several broad ridges separated by valleys that characteristically embrace a network of papyrus swamps in the gently rolling countryside. The land had been settled by smallholders until 1946 when, upon its acquisition by the Corporation, the tenants were compensated and moved. There followed a period during which the patches of cultivation reverted to elephant grass (*Pennisetum purpureum* Schumach), and the house sites and plantations became overgrown with shrubs and vines. A start was made on clearing the main farm area in 1948 and was completed by 1957. At first, the intention was to limit cultivation to areas where the slope was not steeper than 6 per cent (or 3½°) but subsequently steeper slopes, down to the edges of the swamps, were brought into use. Prevailing weather conditions were described in Chapter 2 and the soils in Chapter 3.

The cultivation areas were laid out in straight, parallel-sided strips (Figure 9.1). Each block of strips was arranged around a master line that was derived empirically to run along an 'average' of the contour. The strips were delimited on either side by 10 ft (3.05 m) grass paths and each cultivation strip was 103½ ft (31.57 m) wide, designed to accommodate thirty-four rows of cotton. In laying out successive strips above or below the master line, as soon as the alignment of the strip deviated by more than 2.5 per cent from the contour, a new alignment was struck for the next block of strips, leaving a small area of land that was usually put down to permanent pasture (Figure 9.1). In this way blocks of strips, or fields, of about 10 ha were developed. To reduce further the risk of soil erosion, tie-ridging was adopted as the standard method of cultivation for all arable crops.

The farming system was based on recommendations made by the Uganda Department of Agriculture and involved resting the land under grass for three years after intensive arable cropping for three years in most fields, but for six in some. Rhodes grass (*Chloris gayana* Kunth.) in pure stand was adopted for the resting phase and cropping involved two crops per year in the sequence maize–cotton, groundnuts–beans, maize–cotton. Cattle were brought into the system to make use of the large areas of grass.

Owing to the uncertainties of maintaining a regular and adequate labour force, a high degree of mechanization was considered essential and, after clearing the land using crawler tractors, it was shown that medium-powered, wheeled tractors were entirely satisfactory for all routine farm operations. Tie-ridging equipment was developed and used for both land preparation and weeding (Hawkins 1959; Farbrother 1960a). Other tillage implements, as well as seeders and harvesters, were tested and, those that were appropriate, adopted. Much of this work was undertaken in collaboration with other organizations, such as the National Institute of Agricultural Engineering in the UK.

Figure 9.1. Original farm layout, showing blocks of straight, parallel-sided cultivation strips.

Crop processing also received attention. Storage of food crops was recognized as a problem from the outset because of the difficulties of drying crops adequately in the field. Accordingly suitable in-sack and bin driers were designed and installed which remained in service, with little modification until the station was handed over.

Thus a comprehensive farming system was worked out which bore 'promise of high and sustained yields of the kind of produce that the country needs' (Hutchinson *et al.* 1959). In retrospect, however, this initial period of developing the farm can be seen as but the first of three distinct phases in its operation.

The three phases of Namulonge farm

The three phases of Namulonge farm are clearly reflected in the yields of seed cotton per hectare shown in Figure 9.2. The first phase was characterized by progressive expansion in the area under cultivation, during which yield levels appeared to be satisfactory; the second by a greatly increased scale of operation and generally poor yields; the third by strikingly higher levels of yield and a return to the earlier scale of operation.

During the first phase, yields though fluctuating markedly from year to year certainly showed promise of reasonably good levels of productivity. Commenting on the yield potential, Hutchinson (1954) wrote: 'Our range of farming systems is planned to take advantage of the intrinsic fertility of the land and of the favourable climatic regime, and in this work we are more concerned with avoiding or removing limitations on crop production than with devising means of raising an

Figure 9.2. The three phases of Namulonge farm. Yields of seed cotton in kg per ha (solid line) and areas of cotton harvested (broken line) from the main farm bulks from 1951 to 1972.

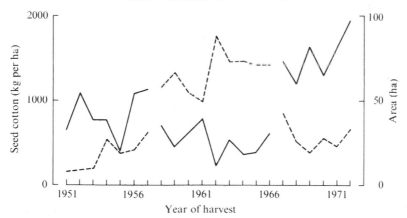

existing yield ceiling'. How different was this view from those expressed during the second phase, when the generally low levels of yield
gave cause for a great deal of debate and no small measure of concern.
At the end of the 1964–65 season, Munro (1966) summarized the main
factors contributing to the poor yields of cotton as uneven emergence,
poor root development, lack of response to fertilizers and insecticides,
and boll rotting. In the face of better yields in earlier years and of good
yields on particular fields even when average yields were poor, it
seemed that some undetected limiting factor, such as a soil condition
causing inhibition of root growth, might be of fundamental importance.

The third phase of the farm began with a re-appraisal of policy. It
had previously been argued that in the absence of worthwhile responses
to fertilizers and insecticides, higher levels of productivity could not
be achieved economically and that yield per unit of input rather than
yield per unit area should be the criterion of success. Attempts to
improve profitability had consequently concentrated upon restricting the
cost of the inputs rather than with improving the performance of crops.
One consequence was that crop establishment frequently suffered
from inadequately prepared seed beds and the resultant poor stands
served only to aggravate an always potentially serious weed problem.
It therefore seemed possible that some of the observed phenomena,
such as poor root growth, might have been the *result* of farming
at low levels of inputs rather than the causes of poor crop yields.

Figure 9.3. Production of cotton for the seed multiplication scheme.

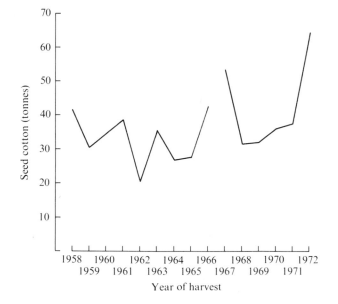

Year of harvest

The new farm policy that was formulated to accompany the reduction in the area of the main farm (referred to in Chapter 1) was aimed at achieving higher levels of productivity without worrying too much in the first instance about the cost of the inputs (Arnold 1967). Nevertheless, the overall cost of running the station remained of overriding importance and net farm running costs could not be allowed to increase. Moreover, in any re-organization of farm activities, it was essential to safeguard the supply of pedigree cotton seed for the multiplication scheme. Consequently, replanning the layout of the farm which was an essential preliminary to many of the changes in crop management that were made during the next few years, had to be effected against a background of these essential requirements. The data summarized in Figures 9.3 and 9.4 show the extent to which this was achieved. Figure 9.3 shows the total production of seed cotton from pedigree stocks during phases two and three, while Figure 9.4 illustrates how gross farm income (from all sources) was stabilized during the main period of re-organization from 1966 to 1969 and progressively increased thereafter. Figure 9.4 also shows net income from the farm, representing gross income after deducting all recurrent expenditure except the salaries of the supervisory staff. These figures, taken from the audited accounts of the station, are impossible to relate to the costs of running a farm as a commercial enterprise, partly because no distinction was made in the accounts between the costs of running the main farm, and the costs of farm operations associated with the research programme. Moreover, no capital items were taken into account so that the figures include

Figure 9.4. Namulonge farm income for the financial years ending on 31 December, from 1957 to 1971.

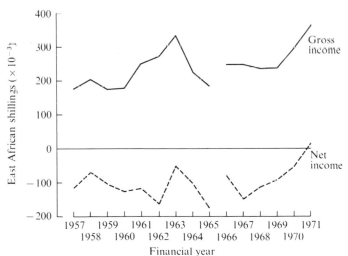

neither depreciation (of tractors, buildings etc.) nor appreciation (of numbers of cattle etc.). Nevertheless, in that they were derived on a similar basis from year to year, the trend of the graphs must give a valid impression of the improvement in profitability that occurred towards the end of the third phase.

The integration of the research programme with the farm

Much of the remainder of this chapter is concerned with analysing problems encountered during the second phase and with describing the inputs needed to reach the higher levels of productivity reached in the third. It was fundamental, however, to the enthusiasm for innovation generated during the third phase, that the research programme and the management of the farm became as fully integrated as possible. From earlier accounts of the work at Namulonge, it is clear that this integration was but an extension of the philosophy propounded by the station's first Director, F. R. Parnell, who laid down the principle that the research should always be done in a farming context, a situation that was described by his successor (J. B. Hutchinson) as 'research against a background of practical farming and farming with a leaven of scientific research'. How different were these attitudes from those all too often found on agricultural research stations, both in the developing and in the developed countries, where the officer-in-charge decides upon the allocation of land and the farm manager is left to get on with growing the crops, with little help or interest from the research staff, except in so far as their own small area is concerned.

During its third phase of operation, the Namulonge farm provided a testing ground for any ideas that could be injected into it. None of the practices developed during the first two phases were held to be sacrosanct and if anyone produced a reasonable case for doing something in a different way, it was tried. Indeed the farm itself came to be regarded by the research staff as one of our most important experimental tools.

The research-station farm as an experimental tool

Reference is made in Chapter 10 to the need for regular feedback of observations from farming practice to research planning. Full recognition is not always given to the fact that a great deal of agricultural research in the developed countries has amounted to little more than accurately defining the limits of applicability of ideas that have already been successfully applied by enterprising farmers. It is only to be regretted, therefore, that a proportion of research workers consider as valid only those observations to which a standard error can be applied, and regard the factorial experiment as the essence of all good field investigation. There are many examples, however, where consideration of management or scale of operation alone are such that small-plot experiments can give misleading results. At Namulonge a great deal was learned from non-replicated comparisons, both large and small in scale, which were subjected to regular observations and discussion by members of staff trained in different disciplines. In this way there was a feedback of questions to the research programme analogous to those stemming from farmers in the developed countries.

As one example we may take the observations made on cultivation strips 9 and 10. These two strips (and certain others) were demarcated in the original farm plan to be maintained under continuous arable cultivation with two crops per year, in order to monitor the run-down in productivity and to discover the types of problem that might arise. Cropping was started in 1950 and, during the first two phases of the farm, yield levels remained well within the range for other farm strips comprising what were referred to as the main farm bulks. (These were the areas used to produce new issues of pedigree cotton seed and other crops in rotation.) After fourteen years of continuous arable cropping, an attempt was made to raise the productivity of one strip by applying farm yard manure, together with additional phosphorus and nitrogen as chemical fertilizers. Strip 9 was chosen for this treatment because the run of the slope was such that it might have permitted some movement of nutrients from strip 10 to strip 9, but not in the reverse direction. Results in the first two years were striking. Although the yields of crops on strip 9 had always tended to be slightly heavier than those on strip 10, these differences were greatly increased during the 1964–65 and 1965–66 seasons (Table 9.1). These observations coincided with the elucidation of the nutrient cycle (described in Chapter 3) and played an important part in the decision to apply fertilizers to all crops in the main farm rotations even though, at that time, no positive results were available from fertilizer experiments. Indeed, the striking increases in yield levels achieved in the farm crops from the 1966–67 season onwards assisted considerably with interpreting results of the new series of fertilizer experiments begun in the same year. A comparison of yields of cotton from strips 9 and 10, together with those from the main farm bulks over a period of ten years, is shown in Figure 9.5. In addition to complementing the results of fertilizer experiments, observations on strips 9 and 10 also had important implications for crop sequences (discussed later) as well as showing the response of crops to better management (even on run-down soils) and the capacity of these soils to continue to produce moderate yields without added nutrients.

Another example of simple, non-replicated comparisons giving valuable information is provided by observation plots laid down, partly to demonstrate to visitors the importance of weeding during the early

Table 9.1. *Continuous arable cropping*

Cropping details for farm strips nos. 9 and 10, each of 0.53 ha. These strips were maintained in continuous arable cropping with two crops per year from the 1950–51 season onwards. Neither strip received any manure or fertilizer until the 1964–65 season when strip 9 only received the first of a series of dressings

Season	Crop	Variety	Crop yields (kg per ha) 9	10	Manure and fertilizers applied to strip 9 (kg per ha) FYM	CAN	SSP	MK	L
1962–63	Maize	K8	1790[a]		—	—	—	—	—
	Cotton	NC62	640	483	—	—	—	—	—
1963–64	Maize	Muratha	2320[a]		—	—	—	—	—
	Cotton	NC63	406	352	—	—	—	—	—
1964–65	Maize	K8	2780	1110	8 960	251	1254	—	—
	Cotton	UPA(57)17	746	400	—	251	—	—	—
1965–66	Maize	K8	1873	808	11 200	251[b]	—	—	1254
	Cotton	UPA(57)17	1388	617	—	251	—	—	—
1966–67	Maize	K8	4050	1030	11 200	251[c]	251	125	1254
	Cotton	BPA 66	2080	967	—	502	502	251	—
1967–68	Beans	Banja 1/2	1750	745	11 200	502	502	251	—
	Cotton	BPA 67	1540	985	—	502	502	251	[e]
1968–69	Beans	Banja 1/2	1710	670	11 200	84	252	84	1254
	Cotton	BPA 68	985	790	—	753[c]	502	251	—
1969–70	Maize	Western Queen	3336	1045	11 200	753[c]	502	251	—
	Cotton	BPA 68	1293	854	—	502	502	—	—
1970–71	Maize	Katumani	4069	870	—	300[d]	200	—	—
	Cotton	BPA 70	1898	694	—	450[d]	250	—	—
1971–72	Maize	Katumani	3850	1085	—	300[d]	200	—	2000
	Cotton	BPA 71	2372	790	—	400[d]	200	—	—

[a] Yields of strips 9 and 10 not recorded separately.
[b] In addition, 125 kg per ha of diammonium phosphate were applied in the seed-bed.
[c] 251 kg per ha top-dressed.
[d] 200 kg per ha top-dressed.
[e] 125 kg per ha of gypsum and a similar amount of a commercial trace element mixture were applied to the seed-bed before drilling the cotton.

FYM, Farm yard manure; CAN, calcium ammonium nitrate (21 per cent nitrogen up to 1969–70; 26 per cent nitrogen thereafter); SSP, single super phosphate; MK, muriate of potash; L, calcitic limestone (85 per cent $CaCO_3$).

period of crop growth, and partly to obtain some quantitative information on the empirical observation that excessive weed growth was one of the main factors contributing to the poor yields of cotton. Results obtained over four seasons (Table 9.2) served to strengthen the arguments both for intensive effort on weed control, and for further investigation into the competitive effects of weeds in relation to the cost of weed control, as well as to the type and timing of farm operations. It was the general recognition of this need that led to a new programme of investigation into some of the biological aspects of weed competition. The work (undertaken by F. E. Tollervey) was only in its infancy when

Table 9.2. *Yields of seed cotton in kg per ha recorded in unreplicated observation plots in which weeding was started at different times. (Harvested area 35 m²)*

| | Season | | | | | % |
	1965–66	1966–67	1967–68	1968–69	Mean	Control
Clean-weeded throughout	1519	2156	1112	1844	1658	100
First weeding delayed until one month after emergence	1022	1719	976	1537	1331	80
First weeding delayed until two months after emergence	335	787	289	398	452	27

Corporation staff left Uganda, but a summary of results is reported later in this chapter.

Examples of inferences where considerations of scale are of overriding importance are the choice of cropping sequences and the layout of cultivation areas (both discussed later in this chapter); and the control of insect pests (discussed in Chapter 5). In all of these respects, the farm itself was used to give meaningful comparisons of things that could either not have been investigated in small plots, or which had already been investigated in small plots and given results that were misleading.

The lesson is of particular relevance to training local graduates for

Figure 9.5. Continuous arable cropping. Yields of seed cotton from strip 9 (left-hand column), strip 10 (middle column) and the main farm bulks (right-hand column), for ten seasons. The number of years of continuous arable cropping for strips 9 and 10 is also shown. The shading represents the more intensive management given to all farm cotton from the 1966–67 season onwards, and fertilizer applications to strip 9 from the 1964–65 season. Strip 10 received no fertilizer or manure from the outset. Further details are given in Table 9.1.

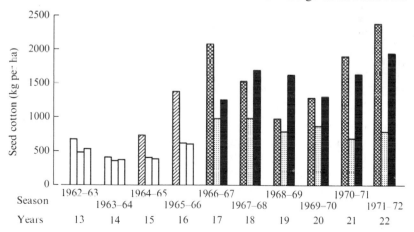

Figure 9.6. Soil erosion. Cotton seedlings after heavy rains had washed away the topsoil, previously loosened during seed-bed preparation. In this instance, the soil wash was caused by overflow from an impeded drainage ditch.

careers in agricultural research in their home countries. At present, many undergo part of their training in the developed countries where their interests become focussed on tiny aspects of agricultural problems rather than upon the broad spectrum of the science and practice of successful farming. What is needed in the developing countries is an approach to agricultural research that does not restrict the interplay of large numbers of variables as is so often necessitated by the requirements of formal experimentation. A bold approach is called for in which critical observation is used against a background of scientific training and farming experience, so that the essentials can be brought into perspective, before investigating limited aspects of them more precisely with finer tools. Perhaps a similar approach is required in the developed countries as well.

Land use and farm layout

The systems of land use and farm layout developed at Namulonge provide good examples of the problems faced when peasant agriculture in the tropics is changed to large-scale arable or mixed farming. The original layout was designed to prevent soil erosion while maintaining areas of arable cropping that were easy to manage and convenient for laying down field experiments. An important feature was that the cultivation strips comprising each main field were to be ploughed in pairs to maintain even cultivation across the width of a strip. All arable areas opcned during the period from 1948 to 1955 were managed in this way as soon as they were cleared. The steeper slopes were put down to permanent pasture with no attempt to erosion control, which was considered unnecessary.

This layout proved generally satisfactory during the 1950s except that some arable land on the steeper slopes had to revert to grass, not because of erosion within the area but because of uncontrolled runoff of water from the upper parts of the slope. By the early 1960s, however, the system was breaking down, with gullies developing across some cultivated strips, and rill erosion occurring along the length of others.

Weakness of the original layout

In retrospect, the breakdown may be attributed to several factors. During the second phase of operation, the farm had attained its maximum area of cultivation; large blocks of sloping land (of up to 16 ha) were cultivated at one time and sown to a single crop. In these large fields, the relatively narrow grass strips were inadequate to prevent accumulated runoff water from moving down the slope. The increased

Figure 9.7. Some effects of ploughing. (*a*) Land profile (exaggerated slope). (*b*) Effect of ploughing to the centre of each cultivation strip. (*c*) Effect of continuous down-hill ploughing.

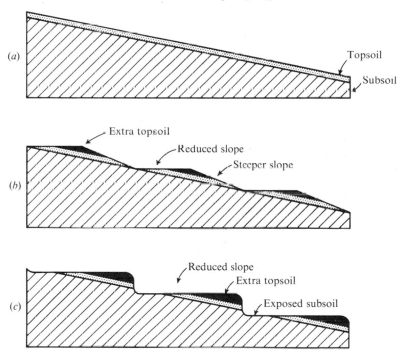

Figure 9.8. Re-forming a vee-ditch with a grader constructed on the farm.

Figure 9.9. Runoff during a heavy rainstorm. A graded contour ditch between two strips of maize can be seen discharging into a broad, grass water-way.

rainfall of the early 1960s aggravated the problem especially because by that time, the soils had lost much of their initial ability to accept water. Deterioration in the physical condition of the soil, as related to water acceptance, was evident on all cultivated land, including that farmed in a ley–arable rotation. Following heavy rainstorms, water remained for long periods in the basins formed by the tie-ridges and deposited fine films of mud which further impeded percolation. Under these conditions it was not surprising that the basins, even if well formed, frequently overflowed, leading to breakage of the ties and a cumulative build-up of runoff water. The use of tie-ridges at Namulonge had been largely inspired by their success on the hill-sand soils of western Tanzania, where percolation rates are presumably faster than those on the run-down soils at Namulonge. The size of the ridge is also important in determining the water-holding capacity of a basin and the tie ridging system in Tanzania was based on five foot (1.525 m) ridges compared with three foot (0.915 m) ridges at Namulonge. In Tanzania cross-ties usually remained intact until the end of the season, whereas during the early 1960s at Namulonge it was not unusual to find little trace of cross-ties in a considerable proportion of the furrows after heavy rain, leaving the fields vulnerable to further erosion.

Other factors also contributed to the erosion problems. Poor crop cover associated with uneven stands and poor growth meant that more soil was exposed to the full impact of the rainfall. Moreover, the initial practice of ploughing the strips in pairs was not maintained, resulting in steeper slopes on the lower side of the strips and increased risk of erosion (Figure 9.7*b*). Observed during heavy rainfall in the mid-1960s, many of the cultivated areas of the farm appeared to be physically unstable and unsuitable for the continued production of arable crops unless drastic remedial measures were taken. This was the background to the modified systems of land use tried during phase 3.

Graded contour layout

At first a very safe system of soil and water conservation was tried on some 40 ha of land. About 6 km of new roads were constructed to serve the area, built to run either along the lines of natural ridges or on a graded contour line with a fall of 1 in 250. Field ditches were constructed with a similar fall at a vertical interval (*VI*) given by the formula $VI = \frac{1}{2}(\% \text{ slope} + 8)\text{ft}$. Thus, for a 6 per cent slope the vertical interval was 7 ft (2.135 m) with a corresponding horizontal interval between ditches of 140 ft (42.7 m) (Riddle 1968). Figure 9.10 shows the cross-section of the ditches that were made to discharge into grassed waterways which increased in width as they progressed down the slope.

This system produced curved fields which varied in width (Figure 9.11), with consequent restrictions on the types of implement that could be used. Nevertheless, ploughing with a reversible plough proved reasonably straightforward. The use of tie-ridging equipment would have been difficult but, in any case, cultivation on the flat was preferred because of the need for flexibility in the choice of row-widths for the various crops. Possible loss of water by runoff was controlled by ripping between the rows up to the time that full crop cover was achieved. This layout was somewhat more wasteful of land than the original layout and up to 15 per cent of the area was taken up with ditches, banks and waterways. However, the fact that it proved possible to cultivate more strongly rounded slopes, to some extent compensated for this wastage, in terms of the total area that could be brought into arable cultivation. In the following year, further areas of the farm were replanned using the graded contour layout.

Figure 9.10. Transverse section of a graded contour ditch (from Riddle 1968).

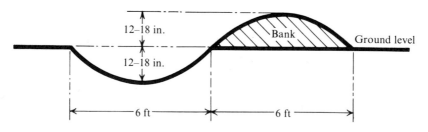

Other layouts

As experience was accumulated modifications of this basic layout were tried on other areas of the farm. For example, the amount of land devoted to ditches was reduced by using simple vee ditches with one very steep side. This shape could be used because an exposed vertical face of Namulonge soil is very resistant to caving-in. On another area of the farm with fairly gentle and regular slopes a layout was tried that attempted to retain the safety of the graded contour system while reducing the proportion of irregularly shaped strips. The field ditches were run in straight lines at twice the standard vertical interval, and the area delimited by successive ditches was divided into two cultivation strips, one with its boundary parallel to one of the ditches and the other a wedge-shaped strip to accommodate the change in slope. The boundary between the two cultivation strips remained either as a narrow grass strip or was formed into a shallow ditch. This layout was less costly to construct but could only be applied to relatively uniform and gentle slopes. A similar principle was applied to an adjacent area where the slope was less regular. Here straight-sided strips could not be incorporated into the layout (Figure 9.12).

A rather different approach was tried on an area of the farm where the original layout fitted the contour more precisely. A problem that had developed with the original cultivation strips was that they were sufficiently wide for runoff water to cause erosion within a single strip. It was therefore decided to try strips of half the original width

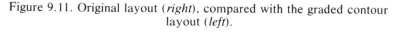

Figure 9.11. Original layout (*right*), compared with the graded contour layout (*left*).

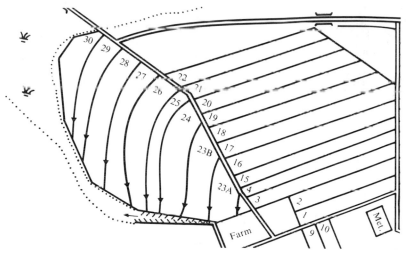

accompanied by a cropping system which ensured that no two adjacent strips were in the same crop or receiving the same cultivation at the same time, so as to avoid the dangers of having a large block of bare soil exposed to rainfall. Within each narrow strip, terrace formation was encouraged by consistently ploughing down-hill (Figure 9.7c). After six years of continuous arable cropping, the development of bench terraces was well advanced, and although some runoff water was occasionally observed to move down the slope during heavy rainstorms, very little soil erosion was caused owing to the slow movement of water across the terraces.

The graded contour layout with field ditches was extended to embrace the permanent pastures, particularly those on the steeper slopes. Even though annual rainfall at Namulonge seldom exceeds the requirement for continuous production of good grass, a great deal of water had previously been lost by runoff. The presence of field ditches enabled the pastures to be ripped across the slope without fear of serious washouts occurring during heavy rainstorms. Regular ripping improved the acceptance of water, resulting in higher responses to applied fertilizers and a rapid improvement in the productivity of the pastures which, in turn, further reduced the problem of runoff.

A feature of all the revised systems of layout for the arable areas was the reduction in slope brought about by the deliberately ploughing downhill (using the reversible plough) resulting in the gradual formation of terraces. While this process aids in reducing the rate at which excess water runs off the land, the movement of topsoil down the slope

Figure 9.12. Modified graded contour layout with straight strips (*right*) and curved strips (*left*).

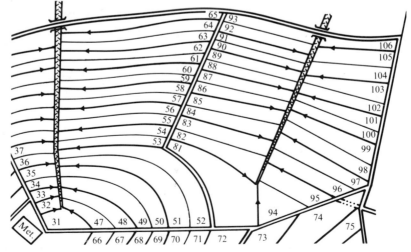

Figure 9.13. A view from the air of the farm layout shown in Figure 9.12.

progressively exposes subsoil along the upper side of each cultivation strip. It was at one time thought that this might have deleterious effects on crop growth and become a major disadvantage of the system. Results of an observation plot, however, indicated that such adverse effects did not occur provided adequate applications of fertilizers and limestone were made. The data in Table 9.3 clearly show how the fertility of the subsoil can be restored even after the removal of 20 cm of topsoil. On some of the terraces formed on the farm by consistently ploughing downhill, a similar amount of topsoil was gradually lost from the upper side of the strips over a five- to six-year period, with no visible effect on the growth or yield of crops.

Looking back on the modified layouts tried during the third phase of the farm, it may well be that the first approach of using the full graded-contour system was an over-cautious reaction to a potentially serious problem which was only partly brought about by limitations of the farm layout itself. With greatly improved crop cover and improved physical condition of the soil the problem of runoff water became very much less. Part of the farm, where serious erosion problems had not developed, was retained under its original layout so that it would be possible to build up long-term comparisons of a range of different layouts, both with respect to convenience of management and to problems of soil and water conservation (Figure 9.14).

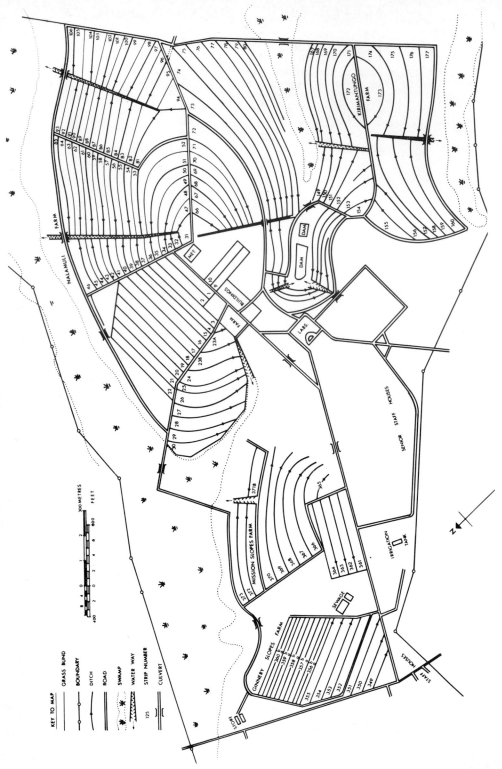

Figure 9.14. Revised layout of the Namulonge farm, showing different types of cultivation strips.

Table 9.3. *Yields in kg per ha of crops grown on subsoil compared with those of crops grown on normal soil*

Each crop received the standard fertilizer mixture (Table 3.10) at the rate of 1504 kg per ha. Cotton yields on the subsoil were lighter than those on the normal soil until lime was applied. Treatments were not replicated; plot size 0.024 ha. (These results may be compared with those in Table 3.4)

| | | | Yield | |
Crop	Year	Season	Subsoil	Normal soil
Beans	1	1	1705	2138
Maize	1	1	5464	5019
Sorghum	1	2	3650	4720
Cotton	1	2	835	2040
Cotton	4	2	1670	1550

Farming systems and crop rotations

In the tropics it is rainfall rather than temperature that determines cropping seasons and it is the distribution and intensity of rainfall that largely governs the approach to farming systems and cropping sequences. As we have seen, the battering effect of rain on bare soil tends to seal the surface and impede water acceptance, leading to problems both of soil erosion and of using the rainfall efficiently. Consequently there are strong arguments for choosing crops and cropping sequences in such a way that the land is covered as densely as possible for as long as possible. Maximum protection can be achieved only from perennial vegetation and, in very wet tropical regions, successful farming systems are usually based on perennial crops.

The ley-farming system and the cropping sequences developed at Namulonge during the first phase of its operation were chosen with these principles in mind, and during the second phase, the system was tested unchanged on an extended scale for nine consecutive years. During this period two major problems were encountered but never entirely solved. One was the problem of controlling perennial weeds; the other was that of timing, particularly with the maize–cotton sequence, which never allowed enough time to harvest all the maize and get all the cotton sown in time for maximum yields.

With the constant scramble to get one crop off the ground and another one established, there was seldom a period when a concerted attack could be mounted on the perennial weeds. It therefore seemed essential to introduce a short fallow break into the rotation even though to do so would be completely contrary to the principle of not exposing bare soil to rainfall. In the event, fallow breaks proved entirely successful, provided that the soil surface was kept in an open condition either

Figure 9.15. Eucalyptus trees were planted in the main valleys to provide
a valuable source of poles and firewood.

by ripping or by ploughing to leave a rough surface. As well as enabling
the weeds to be tackled, these bare fallows allowed much more latitude
in the timing of cultivations for seed-bed preparation for the ensuing
crop. The absence of weeds or crops during the first rainfall season also
allowed water to accumulate in the soil profile giving a reserve to tide
the following crop over a dry spell (see also Chapter 3, p. 56). The
revised 3:3 rotation, which incorporated a bare fallow, is shown in
Table 9.4. in comparison with the original.

Although there were only five crops in the revised rotation, compared
with six in the original, each had the advantage of a wider range of
sowing dates and, particularly with maize, greater flexibility in exploiting the growing season so that a wider range of varieties could be
considered and the crop could be left longer in the field before
harvesting, which helped with the problem of drying. A further advantage was that the area under ley was increased during the driest months
of the year (December and January) when good grazing was often
scarce. This arose from the fact that in the original rotation it was
necessary to break the leys in November and December in order to
prepare the land for drilling maize. In the revised rotation, the ley could
be broken in March or April (when there was usually plenty of good
grazing) and still leave a period to tackle the weeds before drilling
cotton. Thus all the cotton followed either ley or bare fallow which
allowed a good, weed-free, seed bed to be prepared and resulted in
generally improved crop establishment. The ley phase also received

Table 9.4. *Original and revised 3:3 rotations*

	Original 3:3 rotation		Revised 3:3 rotation	
Year	1st rains	2nd rains	1st rains	2nd rains
1	Maize	Cotton	Ley	Cotton
2	Groundnuts	Beans	Groundnuts	Sorghum
3	Maize	Cotton	Fallow	Cotton
			Maize	Ley
4	Ley			
5	Ley		Ley	
6	Ley		Ley	

attention. Instead of sowing only *Chloris gayana*, ley mixtures were introduced which included other, more productive, grass species as well as legumes.

During the next few years, when productivity increased dramatically, it became apparent that the regular triennial turn-round from pasture to arable crops or vice versa was both unnecessary and costly in terms of constructing or dismantling fences and of establishing new pastures. It was clear that it might well be possible to extend arable cropping indefinitely where desirable, and to extend the useful life of pastures beyond the normal three-year period (of which only the two-year period from six months after establishment was fully productive).

The key to continuous arable cropping lies in finding suitable crop sequences which, in effect, means finding pairs of crops that will fit the bimodal pattern of rainfall. The first rains are the more reliable and it is no coincidence that they are the traditional food-crop rains. The problem with the second rains is that it is difficult to find crops other than cotton which can tolerate the dry spells and mature successfully over a range of weather conditions. Neither groundnuts nor beans yield well in the second rains. Sorghum does not yield well when conditions are too humid during ear maturation; soya beans were tried with varying success and showed sufficient promise to support the idea of breeding better adapted varieties. Cowpeas and pigeon peas were also tried, but both need further genetic improvement before they can be incorporated into a highly productive system. Of the range of crops tried at Namulonge in the second rains both sorghum and soya seemed worth growing on an extended scale. Sorghum was chosen partly because of its rapid establishment giving a dense cover which smothers the weeds and protects the soil. Soya was chosen because of its high protein production and importance from a nutritional standpoint. Crops were then paired and put into the following sequence: fallow–cotton, maize–soya, beans–cotton, groundnuts–sorghum, fallow–cotton, etc. This formed the system of continuous arable cropping applied to the half-strip layout already described. Thus in the first rains, the half-strips appeared in the sequence fallow, maize, beans, groundnuts, fallow, maize, etc.; whereas in the second, the sequence was cotton, soya, cotton, sorghum, cotton, etc.

As far as the leys were concerned, it was found that at higher levels of productivity the leys could be maintained in good condition for at least five years. Accordingly, part of the farm was laid down to a 5:5 rotation, using the same basic cropping sequences as follows:

	Year						
	1	2	3	4	5	6	7–10
Crop							
1st rains	Ley	Groundnuts	Fallow	Maize	Beans	Maize	Ley
2nd rains	Cotton	Sorghum	Cotton	Soya	Cotton	Ley	Ley

Yields of crops in these various rotations were determined more by weather conditions during crop establishment and maturation than by any measurable feature of soil fertility, a conclusion that could certainly not be extended to cover soils in a run-down condition. For example, the crop sequence beans–cotton gave rise to relatively poor yields of cotton when tried on strip 9 in the 1967–68 and 1968–69 seasons (Figure 9.4). The explanation almost certainly lay in the difficulty of establishing cotton after beans which, in turn, arose from the poor physical condition of the soil after beans compared with its condition following maize. In this context maize (like sorghum) is a good crop, in that it protects the soil, smothers the weeds, produces a large amount of organic matter and leaves the soil friable; whereas the bean crop is poor, leaving the soil hard and cloddy. With soils in good heart, however, a modicum of maltreatment by growing crops like beans (or leaving a bare fallow) can be tolerated for short periods without adversely affecting the performance of an ensuing crop.

Nonetheless there is a great deal more to be learnt about soil fertility in relation to cropping sequences. The various systems of crop rotation laid down at Namulonge were intended, like the different field layouts, to provide a basis for the long-term investigation of the many problems encountered.

Weeds

In any system of arable farming, weed populations and problems are seldom static. In some circumstances, the system allows the gradual build-up of a weed species from minor to major importance: in others, small changes in management, sometimes with no obvious relevance to weeds, can cause radical changes in the weed flora. While the use of new methods of weed control may allow the farmer more scope in planning his crops, the eradication of one pernicious weed may simply make way for another. Consequently, realistic judgement of the success of weed control measures can only be made over a long period of farming. Such an opportunity is provided by experience at Namulonge where, over the years, an integrated system of mechanical and chemical weed control replaced one based primarily on hand-hoeing. The need for the change was brought about by scarcity of labour, increases in rates of pay and the limitations of hand or mechanical weeding during very wet weather. Encouragement for change was derived from the promising results obtained from the experimental use of herbicides.

Main weed species

The most serious weeds encountered at Namulonge were the perennials *Digitaria scalarum* (Schweinf) Chiov., similar to couch grass and known locally as *lumbugu*; various species of *Cyperus*; and *Oxalis latifolia* H.B. & K. In general, annual weeds presented less serious problems but a few, such as *Galinsoga parviflora* Cav., were capable of increasing very rapidly.

Digitaria scalarum. Lumbugu is a rhizomatous grass weed similar to the noxious weed of temperate climes, *Agropyron repens* L. Beauv., and once introduced, it rapidly colonizes agricultural land. Its competitive nature was well illustrated by Prentice (1957) who reported leaf area indices of up to three for lumbugu, equal to those of the mature maize crop in which it was growing. The observation plots, already referred to (Table 9.2), demonstrated the part that inadequate weeding can play in its establishment. By 1972, the late-weeded plots were heavily infested with lumbugu, whereas the control plot was free of it.

Although not a serious weed of arable crops farmed locally in the traditional way, lumbugu became the dominant weed at Namulonge in the 1950s and early 1960s. It would appear that the traditional practice of resting the land under elephant grass prevents the weed from building up, whereas its spread is less inhibited by a short-grass ley. Similar observations have been made on other large farms near Namulonge where short-grass leys replaced elephant grass in the rotation. Moreover, in experimental plots which carried grass for three years, it was observed that after elephant grass the lumbugu population was sparse, whereas after Rhodes grass it was dense.

It is clear, therefore, that the system of farming initially adopted at Namulonge, and the subsequent attempts to farm with reduced inputs, all contributed to the rapid build-up of lumbugu. Consequently the need to control it economically became an urgent problem.

Early attempts included the use of trichloracetic acid, which was tried either as a blanket application or as a spot treatment on heavily infested land. Neither method was wholly successful and both were extremely expensive. In addition, a residual effect of the herbicide was encountered in subsequent crops. Hand and mechanical grubbing, the accepted methods of controlling the grass, were also expensive and often achieved only limited success. Passmore (1961) reported that on average it required 11 man days per acre per annum to achieve reasonable control of the weed by grubbing. 2, 4-D was shown to suppress lumbugu in sugar cane, but because of the extreme sensitivity of cotton to this herbicide, its use was precluded. In the same year dalapon, used at a rate of 5.6 kg a.i. per ha, in 409 litres of water, was unsuccessful, but when applied both to sugar cane and to a weed fallow in the following season achieved reasonable control (Passmore 1962), probably because the grass was actively growing when the dalapon was applied. Nitrogenous fertilizer applied a few weeks prior to a dalapon spray was shown by Ogborn (1964a) to result in a higher level of control, presumably because of growth stimulation. The addition of aminotriazole to dalapon usually improved control of the lumbugu. It was suggested that a contact spray of 4.5 kg a.i. dalapon plus 1.12 kg a.i. aminotriazole in approximately 300 litres of water per ha was the most suitable method of control (Ogborn 1967).

The original 3:3 rotation allowed insufficient time between one crop and the next for the lumbugu to reach a susceptible state of growth. With the introduction of the revised 3:3 rotation, which included a short fallow period after opening the ley as well as a mid-rotation fallow, it was possible to ensure that the herbicide was sprayed on actively growing lumbugu, and after five years of this treatment the weed was no longer a serious problem on the farm.

Cyperus spp. and Oxalis latifolia H.B. & K. Both *Cyperus* and *Oxalis*, like lumbugu, are perennial weeds spread mainly by vegetative reproduction, *Oxalis* by means of scaly bulbils and *Cyperus* by tubers. The main species of *Cyperus* found on the farm was purple nutgrass (*C. rotundus* L.). Yellow nutgrass (*C. esculentus* L.) was also present, but never achieved dense populations. In fields adjacent to papyrus swamps, large numbers of other *Cyperus* spp. were observed after a ley, but quickly declined with cultivation. In common with lumbugu, nutgrass is not a major weed of small-holdings around Namulonge. Nevertheless, from the mid-1960s, nutgrass replaced lumbugu as the dominant weed at Namulonge. Its rise to prominence coincided with the control of lumbugu and the introduction of more intensive farming methods.

Oxalis was noted as a troublesome weed from the time that Namulonge was first farmed on a large scale, but it remained confined to a few fields. The factors limiting its spread have not been investigated.

Figure 9.16. Cotton seedlings emerging in a dense stand of weeds (almost entirely *Cyperus* spp.).

Purple nutgrass, although presumably present in earlier years, only became a weed of major importance during the mid-1960s, and by 1970 large areas of cultivated land were covered with pure stands of it. Church & Henson (1969) described experiments at Namulonge in which they tried to control nutgrass and *Oxalis* with chemicals, but none of the treatments was successful enough to incorporate into general practice. Repeated applications of paraquat during a fallow break in 1971 reduced a dense infestation of nutgrass to manageable proportions for the subsequent three cropping seasons, but the treatment was expensive.

After ten years of testing herbicides on *Cyperus* spp. Parker, Holly & Hocombe (1969) concluded that all the chemicals available at that time had severe limitations. The most successful method of control so far discovered is by desiccation, as practised in the Sudan (Pothecary & Thomas 1968). Occasionally desiccation worked at Namulonge, but the absence of a dependable dry season precluded its routine use. Tripathis (1969) noted that tubers dried rapidly when exposed to air and, at moisture contents below 16 per cent, viability was lost.

Oxalis populations may be reduced, but seldom eradicated, by frequent weeding. Trifluralin, at rates of application up to 9 kg a.i. per hectare, was found by Ogborn (1964b), and by other workers in East Africa, to be the best available treatment. However, Parker (1967) demonstrated that this herbicide only suppressed the development of *Oxalis* bulbils and did not affect their viability, a result that was

Table 9.5. *List of common weeds of arable crops*

Monocotyledons	Dicotyledons
Brachiaria platynota K. (Schum.) Robyns	*Ageratum conyzoides* L.
Chloris pycnothrix Trin.	*Amaranthus* spp.
Commelina benghalensis L.	*Asystasia schimperi* T. Anders
Cynodon dactylon (L.) Pers.	*Euphorbia hirta* L.
Cyperus rotundus L.	*Euphorbia* spp.
Cyperus spp.	*Galinsoga parviflora* Cav.
Dactyloctenium aegyptium (L.) Beauv.	*Ipomoea tenuirostris* Steud. ex Choisy.
Digitaria adscendens (H. B. & K.) Henr.	*Leucas martinicensis* R. Br.
Digitaria bicornis Roem. & Schult, ex Loud.	*Oxalis latifolia* H. B. & K.
Digitaria scalarum (Schweinf.) Chiov.	*Oxygonum sinuatum* (Meisn.)
Digitaria ternata (Hochst.) Stapf.	*Portulaca oleracea* L.
Digitaria velutina (Forsk.) P. Beauv.	*Senecio discifolius* Oliv.
Eleusine indica (L.) Gaertn.	*Siegesbeckia orientalis* L.
Eragrostis spp.	
Mariscus sieberianus Nees.	
Setaria verticillata (L.) Beauv.	
Sorghum verticilliflorum (Steud.) Stapf.	
Sporobolus spp.	

confirmed at Namulonge by Church & Henson (1969). Thus the success of the treatment depends upon keeping an effective concentration of trifluralin in the soil around the bulbils.

Other weeds. Annual weeds did not present as serious a problem as the perennials, and were usually controlled by the cultivation methods and chemicals used as routine. With a greater reliance on herbicides, however, care had to be taken to ensure that the whole spectrum of the weed flora was controlled by the combination of measures used. The failure of fluorodifen to control *Siegesbeckia orientalis* L. affords an example of a species of minor importance being brought into prominence by virtue of a herbicide giving selective control (Church 1969). Moreover, both fluorodifen and fluometuron failed to control *Commelina benghalensis* L. which has the ability to root from small fragments produced during hoeing, and was clearly becoming more common.

In general, during the arable phase of the rotation, the weed flora gradually became more graminaceous with a greater preponderance of perennial weeds. Nevertheless annual weeds remained important in many fields. A list of the more common weeds is given in Table 9.5. The ubiquitous annual weed *Galinsoga parviflora* was often the primary cause of the need for a further weeding. Although easily controlled, its ability to set seed in a short time under a wide range of conditions gives it a high survival value in sequences of arable crops.

Manual and mechanical weed-control

Neither manual nor mechanical weeding were effective during very wet weather, when hand-hoeing merely resulted in transplanting weeds and the passage of tractors through the crop poached the soil. Delay allowed the weeds to become well-established which made hand-weeding more laborious and tractor weeding less effective. Fortunately, the onset of frequent, heavy rainfall did not usually occur until after full crop cover had been reached, so that only occasionally was weeding during the early part of the cropping season seriously impaired through adverse weather.

Weeding between rows with tractor-drawn implements presented no problems in crops other than cotton. The problem with cotton was that the crop might not cover the ground completely until it had reached a height of about one metre. In such cotton, damage was apparent when inter-row weeders mounted on standard tractors were used, but the problem could almost certainly be overcome by using suitably modified implements together with high-clearance tractors and wheel guards, as is common in countries where cultivation of the crop is fully mechanized. Nevertheless, even with effective inter-row weeding there remains the problem of weeds in the row, removal of which by hand is laborious. Of the time required to weed a cotton crop, 80 per cent may be taken up with in-the-row weeding (Tollervey 1973) and it is here that herbicides have a most important part to play. Band application is all that is required and indeed is to be preferred in a system of management that requires ripping of the inter-row spaces to facilitate acceptance of rainwater.

Because hand labour is likely to be the main means of weeding crops in Uganda for the foreseeable future, however, there is a need for renewed investigation into the most efficient type of hoe to use. The multi-purpose African hoe (*jembe*) is primarily a digging tool which is not very suitable for lightly scraping the soil surface to remove seedling weeds. Consequently weeding tends to be delayed, causing both reduction in yields and greater effort to remove large, deep-rooted weeds (Druyff & Kerkhoven 1970; Tollervey 1973). Light-weight hoes of the type commonly used for row-crops in the UK were tried at Namulonge but proved unpopular with the labourers. This was during the period when crops were grown on ridges, however, and not only is the traditional hoe a more effective implement for rebuilding ridges, but also there is a greater tendency for the ridges to be further destroyed when weeding with a light, swan-necked type of hoe. With cultivation on the flat, the use of lighter hoes might well be more readily adopted.

Experience at Namulonge showed that the method of cultivation and type of implement used in seed-bed preparation markedly affected the weed population encountered in the subsequent crop. In order to obtain some quantitative data on this aspect of weed control, an experiment was started in 1971 and, although not completed by the time CRC staff left Namulonge, preliminary results bore out many of the observations previously made. Details of the experiment and a summary of results are given in Appendix 9.1. The data illustrate how different methods of cultivation not only affect the subsequent effort required to control weeds but also have different effects on individual components of the weed population. For example, rotavation as a primary cultivation, was far more effective than mouldboard ploughing in reducing the nutgrass component of the population but, in the case of lumbugu, the reverse was true. Whereas rotavation chopped up the lumbugu leaving small fragments that could regenerate, mouldboard ploughing, while giving good general weed control by inverting the soil, also served to bring to the surface nutgrass tubers from the deeper levels. Tubers were found to a depth of 0.30 m and those produced in the lower 0.15 m were heavier than those in the upper 0.15 m. It may well be that mouldboard ploughing regularly brought the heavier tubers to the surface and that these, having larger food reserves, were more difficult to eradicate. The data also illustrate the commonly observed consequence of fertilizer application, that vigorous crop growth and vigorous weed growth go together.

Herbicides

Prior to 1967, a number of *ad hoc* experiments had given information on the chemical control of annual weeds in cotton, but no recommendations had been made, and Church (1968) instigated a series of experiments along the lines proposed by Fryer (1960). In 1969, at the end of the consolidation phase, two pre-emergence herbicides, fluorodifen and fluometuron, proved the most promising of those screened. Unlike fluorodifen, fluometuron continued to give a degree of weed control after the soil had been disturbed by hand-weeding. This is important in a relatively slow-growing crop like cotton, but less important in crops that cover the ground more quickly. Consequently, when these herbicides were incorporated into the farming system, fluometuron was applied to cotton and fluorodifen to beans and groundnuts. Diuron showed promise as a lay-by spray in cotton for late weed control, but it was not used on a farm scale.

Little experimental work was done on the use of herbicides in crops other than cotton, but it became farm practice to treat all maize areas with a pre-emergence spray of atrazine, which had proved to be extremely effective under a wide range of conditions in Africa. Sorghum was the only crop in the rotation not to receive herbicides as routine.

Results obtained with pre-emergence herbicides such as fluometuron, fluorodifen and atrazine were not consistently good, however. These herbicides require moist soil to work effectively and, partly because of the short interval between successive crops, it was not always possible to apply the chemicals under suitable conditions. In the 1972 cotton crop, it was observed that fluometuron applied directly after rains gave excellent weed control, whereas that applied only a few hours later gave only partial control.

For successful weed-control using herbicides, therefore, there is a need to devise techniques or to develop chemicals that are less dependent on soil conditions. In this respect, soil-incorporated herbicides such as trifluralin, which was tried with cotton on a small scale, may prove to be more reliable. Moreover, there is a need for more extensive evaluation, under Uganda conditions, of techniques developed in other countries, particularly the use of band-spraying equipment attached to seeders, where the herbicide is applied direct to the moist soil exposed during drilling.

Weed competition in cotton

Cotton, when grown in rows 0.70 m apart, takes approximately three months to produce complete leaf cover. To keep a crop clean for the whole of this period is expensive and it is important to know if weed control can be relaxed without seriously affecting yields. Experimental results obtained in 1971 and 1972 showed that the influence of varying periods of weed control on cotton yields was markedly affected by seasonal conditions. Water is probably the main factor affecting this relationship. In 1971, rain was well distributed and the crop was not stressed until it reached full cover. In 1972, the crop suffered from water stress during the first sixty days after emergence. In both seasons clean-weeded cotton produced the same weight of dry matter per unit area measured at 120 days after emergence, but the effects of varying the period of weeding differed greatly. Effects of some of the weeding treatments are illustrated in Figure 9.17, while data on weeds and labour requirements for hand-weeding are given in Table 9.6.

It is clear that the effects of delayed weeding on the cotton were greater in the drier season (1972) than in the wetter one (1971). Moreover, in 1972, larger quantities of weeds accumulated under the crop after the cessation of weeding, associated with the poorer growth of the cotton during the period of water stress. The data thus illustrate the way in which crop growth and weed growth are negatively correlated.

Data presented in Chapter 3 show that sufficient water is usually available for growth of cotton during the first sixty days after emergence, after which it frequently becomes limiting. Clearly, however, the timing of the onset of water stress can be influenced by the extent to which competition from weeds is allowed to develop. Not only can water stress be induced in cotton during the first sixty days but also water in excess of the crop requirement, which might later have been available to the crop, can be dissipated by excessive weed growth. Weeds can also be important at the time of crop establishment when rainfall is often unreliable. In this respect, perennial weeds such as lumbugu and nutgrass are particularly important. The greater food reserves in the underground storage organs of these weeds give them

Figure 9.17. Effect of clean weeding for different periods on the growth of cotton in two seasons. 1971: ●—●, 0–100; ○—○, 0–30; △---△, 30–100; ▲---▲, 50–75, 50–100. 1972: ●—●, 0–90; ○—○, 0–30; △---△, 30–90; ▲---▲, 60–90.

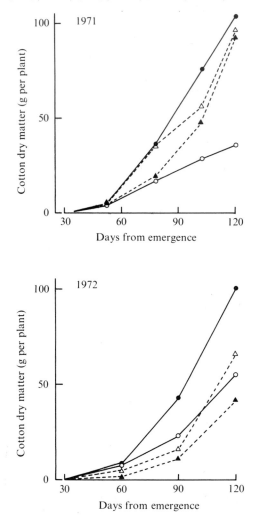

Table 9.6. *Weed development and labour requirement*
under different weeding regimes

Year	Weed-free period (days)	Weeds removed		Labour requirement (man-hours per ha)	Weeds present at day 120	
		Dry matter (g per m²)	Numbers per m²		Dry matter (g per m²)	Numbers per m²
1971	0–100	69	1206	343	2	44
	0–70	62	1263	308	10	110
	0–30	15	638	74	343	87
	30–70	128	1105	294	22	232
	30–100	97	1019	368	6	112
	50–100	336	920	444	7	106
1972	0–90	28	1835	305	12	217
	0–60	22	1090	211	165	290
	0–30	15	975	142	386	237
	30–60	86	920	234	290	272
	30–90	81	1512	350	21	257
	60–90	263	1580	345	462	462

an advantage over the crop seedling. Such weeds can completely cover a crop during the early stages of its establishment, in which case light may become limiting for continued crop growth. Where perennial weeds become established in advance of the crop, water in the top 0.05 m of the soil may be depleted to the extent that emergence of the cotton is patchy.

There is some evidence to show that a late weeding to ensure that the mature cotton crop is free from weeds is beneficial. Leaves and seeds of weeds can greatly increase the amount of hand-sorting required in cotton after picking. Moreover, if the weeds are left to seed, the potential weed population for the following season is increased. The fast-growing annual weed *Galinsoga parviflora* illustrates these points very well. It has a short life cycle and a variable growth-habit. Depending on the circumstances it may, for example, flower a few centimetres above the ground, or it may reach the top of the cotton canopy before flowering. Results of an experiment in which the weeding requirement of cotton grown in rows 0.35 m apart was compared with that in rows 0.70 m apart, showed that although fewer weedings were required in the closer rows during the main period of growth, an extra weeding was required before picking because of the earlier leaf-fall and consequent stimulation of weeds like *G. parviflora* under the crop.

Factors such as these must be kept in mind when making decisions on the management of a cotton crop, and it is clear that both the type and timing of cultivations exert pronounced influences on the effectiveness of weed control. In practice, a balance has to be struck between the desirability of having a weed-free crop and the cost of keeping the crop clean throughout the growing season.

Crop management and performance

A summary of farm practices developed during the third phase is given in Appendix 9.2, together with a table of yields (Table A9.6). Here, we recount some of the experience that led to these practices, amplify certain aspects of them and draw attention to some of the problems that require further investigation.

Flexibility

The basic crop rotations were maintained for comparative purposes as far as practical consideration would allow, but some flexibility was essential in order to adjust to unfavourable weather conditions. For example, in the sequence *cotton–groundnuts–sorghum* a delay in cotton harvesting might well result in delay in land preparation for groundnuts because, by the time all the cotton had been picked, the land might have become too hard to cultivate effectively. If the optimum sowing period for groundnuts had been passed before the land could be prepared, beans (which have a shorter growing season) were substituted for groundnuts in the affected fields. This concept of flexibility in the choice of crops was further extended on other areas of the farm, where cropping was entirely discretionary and decided in the light of the condition of the soil, the requirement for the various types of produce (including forage) and the possibilities for cultivation in relation to the weather at the time. This greater flexibility of cropping undoubtedly contributed to the heavier yields of crops recorded in the third phase, compared with those in the second.

Crop residues

As we have seen, the established philosophy in Uganda was that, because of the overriding importance of the physical condition of the soil, resting periods in perennial vegetation, such as elephant grass or a short-grass ley, were essential in order to maintain productivity. The philosophy developed at Namulonge was that arable crops adequately supplied with nutrients were themselves capable of fulfilling a similar function. To this end it was important that crop residues were returned to the soil. For example, the traditional way of disposing of cotton stalks was by uprooting, piling into heaps and burning. In order to maintain uniform conditions on the cultivation strips at Namulonge, the stalks were carried to the grass borders to be burnt, thus not only removing the organic matter from the cultivated areas but the ash as well. Government sanction was therefore obtained (for burning was required

Figure 9.18. Slashing cotton stalks at the end of the season.

by law) to dispose of cotton stalks by slashing and ploughing-in the residues. No adverse effects of this practice were encountered and the practice must have contributed to the maintenance of soil fertility. Latterly, however, it appeared that the stalks were breaking down more slowly than when the practice was first started, and it was suggested that this might have been associated with the systematic destruction of termite nests, begun when the land was cleared and intensified in subsequent years. (Apart from causing irregular soil conditions for experiments, termite mounds create serious problems for mechanized agriculture.) The break-down of crop residues in the soil is, however, only one of many unexplored aspects of the influence of termites on soil productivity.

Soil structure

Cotton stalks provided food for thought in other ways. During the first phase of the farm, uprooting cotton stalks by hand required considerable physical effort. Stalks of well-grown cotton plants are not easily cut at ground level with the traditional hoe and pulling them by hand was at best extremely hard work and at worst virtually impossible. This led to investigating the feasibility of introducing an implement for levering out the plants, similar to that used in the Sudan. The problem disappeared during the second phase, however, because the cotton roots were so poorly developed that hand pulling was quick and easy. During the third phase, although uprooting by hand was discontinued, root growth was clearly very much better, illustrating another aspect of the changes that were taking place. Nonetheless the extent to which

Figure 9.19. Ripping as a primary cultivation, instead of ploughing, came to be preferred in many circumstances.

Figure 9.20. Discing to prepare a seed bed.

Figure 9.21. Drilling cotton.

Figure 9.22. Seeding a ley.

Figure 9.23. Ripping between rows of cotton to control weeds and improve
water acceptance.

the differences in root growth were caused by changes in the nutrient
status of the soil, or in its physical structure, was never conclusively
resolved. As an aspect of maintaining and improving the physical
condition of the soil much trial and thought were given to methods of
primary cultivation. The advantages of the reversible mouldboard
plough in relation to the gradual formation of terraces have already been
discussed. The mouldboard plough is not the best implement for
maintaining soil structure, however, especially where root penetration
to exploit the deeper reserves of moisture is a primary consideration.
Progressively greater use was therefore made of chisel ploughing,
particularly with the Bomford 'Superflow' implement. The two recognized design features of this implement, the forward rake of the tines
to minimize draft and the positioning of transverse structural members
so as not to impede the flow of clods and trash past the tines, made
it particularly suitable for the Namulonge cropping system, where deep
cultivation was required in the presence of large amounts of crop
residues.

   With any method of primary cultivation there was frequently a
problem of clods, although chisel-ploughing produced fewer clods than
mouldboard ploughing. The problem could be avoided, in so far as the
weather would allow, by careful timing of cultivation in relation to soil

Figure 9.24. Harvesting groundnuts from the windrow, after lifting.

moisture but, once clods had been formed, they were difficult to break down. Exposure to sun and rain had little effect and repeated discing and rolling in these circumstances seldom produced a good tilth.

Fertilizers

The principles of fertilizer use on the farm have been discussed in Chapter 3. When fertilizers were applied to run-down soils there appeared to be an advantage in placing the fertilizer deep in the soil along the line of the plant rows before drilling, particularly with crops grown in rows 0.90 m apart. Where the fertilizer placement deviated from the row, marked depressions in crop growth were clearly visible, suggesting that fertilizer placement was beneficial. Although this inference was not borne out by the results of subsequent experiments (Chapter 3), it is possible that placement of fertilizer was effective in certain circumstances. The question was not of lasting importance, however, because the observed effect disappeared as soil fertility increased and this, together with the closer row-spacings subsequently adopted (Chapter 4), enabled fertilizers to be broadcast, which was easier and quicker to do.

At the beginning of the third phase, heavy dressings of fertilizer were used in order to restore productivity as quickly as possible. With the rapid improvement effected, however, it was possible to be more selective in fertilizer use and to apply fertilizers in proportion to the estimated crop needs. An important aspect of the efficient use of fertilizers is understanding not only the movement of nitrogen and other nutrients in the soil but also the contribution made to soil nitrogen by nitrogen-fixing micro-organisms. Not much information is available on the extent to which, under Uganda conditions, atmospheric nitrogen is fixed either by free-living micro-organisms or by those in the nodules of leguminous plants. The inference from the large responses to nitrogen obtained from beans is that fixation is not very effective, at any rate in some leguminous crops. Phaseolus beans nodulate freely at Namulonge and neither inoculation nor additional molybdenum gave any improvement. Soya beans, a new crop for the farm, did not nodulate freely and the seed was inoculated as routine. It is possible that there is more nitrogen fixation in groundnut nodules because yields of groundnuts did not respond to applied nitrogen. With the increasing cost of nitrogenous fertilizers there is clearly an urgent need for more research into these aspects of the nitrogen supply to the soil in Uganda.

Varieties

An important factor limiting the yield levels of crops other than cotton was the restricted choice of adapted varieties. Maize and soya beans were crops in which there was clearly a need for greater genetic adaptation to the local conditions. For example, hybrid maize varieties bred at Kitale in Kenya had the potential for heavy yields but grew too tall under Uganda conditions and suffered from lodging during convection storms. The early maturing composite variety, bred at Katumani (Kenya) was shorter in the stem and, at dense populations, gave good yields but its potential was clearly less than that of a good hybrid variety. SR52, a hybrid imported from central Africa, was more consistent in yield than the Kenya hybrids and had the potential for very heavy yields when no lodging occurred. The yield of 7666 kg per ha obtained from this variety in 1971 from 1.7 ha was believed to be the heaviest yield recorded for maize on a farm scale in Uganda. Seed of this variety was difficult to obtain in quantity, however, and in any case a variety with shorter, stronger stems was to be preferred. As a step in this direction the new composite variety released by the nearby government station at Kawanda gave promising yields in 1971.

Cattle

Although cattle were introduced into the farming system mainly to make productive use of the leys and permanent pastures, a further consideration was the need for a regular supply of meat and milk for employees. In building up a large herd to achieve these aims many problems were encountered. Attempts to solve them involved frequent exchanges of views with officials responsible for government policy on livestock production and there is little doubt that the pioneering work at Namulonge contributed considerably to the rapid expansion of the dairy industry in Uganda during the late 1960s and early 1970s.

The origins of the Namulonge herd of indigenous Nganda animals have been described by Hutchinson *et al.* (1959). Apart from the difficulties of finding an adequate supply of suitable animals, two main problems were encountered during the first phase of developing the farm. One was the biting fly, *Stomoxys* spp.; the other was the tick-borne disease, East Coast fever (ECF).

In response to extreme irritation from Stomoxys flies cattle stop grazing and bunch for protection. The extent to which they stop grazing and the time of day when they bunch vary with the activity of the flies. It was generally presumed that cattle should be housed for protection during the middle of the day when the flies were thought to be most active. Trials at Namulonge, however, showed that the live-weight gains of housed animals were only slightly greater than those of animals left out in the open (Coaker & Passmore 1958). The flies, predominantly *S. nigra* Macq., were shown to be less active between 11 a.m. and 2 p.m. than at certain other times of the day, and it was found that the accumulation of manure around the day-houses became breeding sites for the flies. The practice of day-housing was therefore abandoned; cattle collecting areas were, as far as possible, kept clean and this, together with carefully planned rotational grazing, kept the Stomoxys problem to a minimum.

As far as ECF was concerned, it was accepted during the 1950s that prevention of infection was impracticable and that the only method of controlling the disease was to expose the young calves to infection so that the survivors would acquire resistance, and differential survival would lead to selection of more resistant stock. This was the policy advocated by the then Uganda Veterinary Department and practised at Namulonge for the first ten years. As the numbers of cattle increased, however, so did the population of ticks. Consequently it became necessary to reduce the tick burden while still allowing sufficient to survive in order to transmit the disease, a condition that was achieved by spraying the animals occasionally, at irregular intervals.

It had been recognized from the outset that the Nganda stock were inefficient producers of meat and milk, and the original plan for developing the herd incorporated a breeding programme for stock improvement. Nevertheless, by the late 1950s it had become clear that progress would be slow and no convincing evidence had come to light, either from work at Namulonge or elsewhere in Uganda, of the existence of worthwhile genetic variability for resistance to ECF. At least a third of the calves born died of ECF and the lack of effective control of ticks meant that tick-borne diseases other than ECF also took their toll. The process of nursing animals suffering from ECF was

Figure 9.25. The provision of a supply of clean water was considered essential for good livestock management. The photograph depicts a typical water-trough with the end-cover removed to show the ball-cock.

time-consuming and costly and even those animals that recovered were found not to have acquired life-long immunity to the disease. Difficulty in controlling ECF was therefore one of the main factors that led to a radical change in policy. Another was the poor milk yields.

Milk 'let-down' was difficult to encourage in the Nganda animals, even in the presence of a calf. A system of milking was developed in which the calf was kept away from the dam during the night, but was present at the morning milking. The calf then remained with its dam until the evening when only the very heaviest of milkers were milked a second time. Typical milk yields ranged from 75–500 litres per lactation with very few cows producing more than the upper figure and many unable to produce milk in excess of the immediate needs of the calf, even though the calves were given supplementary feed of a mixture of ground maize, beans and groundnut cake as soon as they would take it.

After about ten years of only limited success with livestock improve-ment, therefore, the decision was taken to make the farm as free of ticks as possible by twice-weekly spraying of all animals, and to begin a programme of up-grading, by crossing with exotic stock. But a major project of this type lay outside the terms of reference of the Cotton Research Corporation and, accordingly the scheme was started jointly with the Veterinary Department to which 280 ha of the estate were loaned for the purpose (Chapter 1). Females of the Namulonge

Figure 9.26. Mineral licks were provided in all grazing areas.

herd were artificially inseminated with Jersey semen and the first calves
were born in 1961. In all 166 calves were supplied to the Uganda
Government for the initiation of this scheme designed to produce
up-graded animals for milk production on enclosed holdings where ticks
could be controlled.

In order that a start could be made on more intensive milk production
on the research-station farm, fifteen up-graded animals, ten of Guernsey
and five of Fresian stock, were purchased from Kenya to start a new
milking herd. These animals, which had been bred originally from
Kenya Zebu (Boran) animals, readily adapted to conditions at Namulonge and most of them produced more milk than they had done in
Kenya. They formed the nucleus of a dairy herd which had expanded
to over 100 animals at the time the station was handed over. The herd
was expanded by continuing to breed from the Kenya animals by
artificial insemination with imported semen from pure-bred bulls. Later,
a Kenya Red Poll animal was added to the herd and used in a similar
manner, as well as an outstanding animal which arose from crossing
the local Nganda stock with exotic breeds. In the absence of a problem
of 'let-down' the calves could be bucket-fed with whole milk or milk
substitute, because it was no longer necessary for the calves to be
present at milking. Yields of milk, which are summarized in Table 9.7,
bore no comparison with those recorded from the original Nganda herd.

Following the importation of up-graded animals for milk production,
the indigenous animals were used in a breeding programme aimed at
improving beef production. The first cross was made using Red Poll

Table 9.7. *Milk yields of the Namulonge dairy herd*

A summary of records taken from the time the herd was started until the last completed lactation before the station was handed over

Nominal breed	Friesian		Guernsey		Red Poll		Red Poll×Nganda
Where bred	Kenya	Namulonge	Kenya	Namulonge	Kenya	Namulonge	Namulonge
Number of animals recorded	5	16	10	32	1	5	1
Best lactation (kg)	5787	5832	5060	4923	2337	2821	2397
Best lactation at 305 days (kg)	5180	5470	4387	3826	2210	2821	2397
Number of calves born	31	50	79	107	5	30	6
Calendar months between successive calvings	13.8	13.3	13.6	13.2	13.5	12.8	12.2
Number of lactations recorded	29	44	75	88	3	27	6
Average yield per lactation (kg)	4092	3405	2697	2599	2080	2080	2084

Figure 9.27. A group of cross-bred beef cattle.

semen imported from the UK. The use of a dual purpose animal as the male parent was designed to improve the milk yields of the F_1 animals, as well as their beef potential, so that they in turn would be better fitted to support faster-growing offspring. In all sixty heifer and sixty-five bull calves were produced from this cross. The heifers were then crossed with beef breeds, the choice being determined mainly by the availability of semen. Two breeds were used at this stage of the programme: Aberdeen Angus and Brangus (Brahmin × Angus). Female offspring of this backcross to beef breeds were then crossed to Hereford bulls by artificial insemination. All these hybrid animals produced good quality meat, the second backcross animals showing somewhat faster live-weight gains than the first.

Initially this programme was designed to follow a system either of criss-crossing or of rotational crossing (Lush 1945) in which successive generations would involve crossing with bulls of a different breed, including the original Nganda stock. This intention resulted from the belief that some advantage was to be gained from retaining part of the indigenous genotype. However, the success of the Kenya up-grades and later, the similar success of pure stock imported by the Uganda Government from Europe and Canada, suggested that little was to be gained from re-introducing germ-plasm of the indigenous stock into the cross-bred animals. Accordingly, only semen of exotic bulls was used

Table 9.8. *Numbers of cattle, liveweight and milk*
production from 1966 to 1972

	1966	1967	1968	1969	1970	1971	1972
No. of cattle at beginning of year	416	386	387	394[c]	462	514	512
Liveweight at end of year (kg)	92 949	98 045	105 818	122 780	123 646	138 697	
Calves born[a]	78	105	85	140	179	123	
Calf deaths	3	13[b]	8	10	25[d]	11	
Adult deaths	1	2	6	1	6	15[e]	
Sold and slaughtered	104	89	66	61	96	99	
Liveweight sold and slaughtered (kg)	28 203	24 949	23 942	25 018	34 517	34 757	
Milk produced (litres)	69 560	81 066	84 008	129 557	165 266	158 848	

[a] Includes still-born calves.
[b] Deaths caused mainly by foot and mouth disease.
[c] Two animals purchased during 1968.
[d] Some deaths may have been associated with excessive nitrate in the pastures.
[e] Several diagnosed as nitrite poisoning.

and hybrid vigour maintained by rotational crossing within the range of exotic breeds. Some of the later crosses were made using Fresian semen both to Nganda females and to cross-bred females, and the offspring showed great promise both for beef and milk production. In all cases, calves were run with their mothers until eight months old; they were brought into calf-houses at night.

A summary of cattle, beef and milk production on the farm for the years 1966 to the beginning of 1972 is given in Table 9.8.

Pastures

Initially, the only sown component of the leys was Rhodes grass. Seed, collected by hand in the Teso District of eastern Uganda and supplied by the Department of Agriculture, gave very variable germination rates, sometimes as low as 3 per cent. Usually, however, the grass established fairly readily when seed was broadcast under cotton or on freshly prepared land. With careful management, good pastures were produced which could be maintained for two or three years, but production tended to decline owing to the failure of the grass to compete successfully with natural, colonizing species. As part of the intensification of production during the third phase of the farm, efforts were made to improve the leys. Other grasses such as *Panicum maximum* Jacq., *Brachiaria ruzizensis* Germain & Evrard and *Setaria anceps* Stapf ex Massey, better suited to higher levels of production in the elephant grass zone of Uganda (Horrell & Tiley 1970)

Figure 9.28. A well-established ley-mixture.

Figure 9.29. Applying fertilizer to a grass ley.

were included with Rhodes grass in the mixture, when seed was available. Legumes such as *Desmodium intortum* Urb. and *Centrosema pubescens* Benth. were also added and became well established in many of the leys. If an early cut for silage was required, a forage sorghum species was also included and, with good establishment, a ley of this type would reach a height of well over a meter in eight weeks.

Management

Established leys and long-term pastures were managed in a carefully controlled system of rotational grazing. Whenever possible at the end of a grazing period the pasture was slashed to keep down weeds such as *Lantana* spp., *Imperata cylindrica* (L.) Beauv. var *africana* (Anderss.) Hubbard and *Sporobolus pyramidalis* Beauv. During each rainfall season, in late March and late October, nitrogenous fertilizer was applied as 200 kg calcium ammonium nitrate per ha (or about 100 kg nitrogen per ha per year). The long-term pastures were ripped across the slope to a depth of about 175 mm using sprung tines set 0.70 m apart. Ripping was done once a year when the soil was damp and was designed to improve water infiltration and to shatter any compaction caused by cattle grazing in wet conditions. If only a small amount of legume was present in the pasture, 4 kg seed per ha of *Stylosanthes gracilis* H.B. & K., together with 200 kg single super phosphate per ha were applied after ripping. Occasionally, after six to eight years, the poorer pastures were ploughed and a single arable crop such as maize or beans taken, before re-seeding with a pasture mixture. Before cropping the pH of the soil was checked and limestone added to adjust it to 6.0 if necessary; a fertilizer mixture was applied according to the crop requirement. Cropping in this way helped to improve the condition of the soil and to eradicate pasture weeds.

Water availability under grass (Chapter 2) was such that pastures remained productive for much of the year. Dairy animals required very little supplementary feeding except during the driest period of the year (December to February) and beef animals usually continued to put on weight, although at varying rates, throughout the year. Nevertheless, forage conservation was desirable because occasionally (say once in every three to five years) a prolonged dry spell was encountered, when grazing was inadequate and supplementary feeding essential. Attempts to make hay at Namulonge were not very successful, largely because of difficulties of drying in the field, aggravated by the presence of semi-succulent weeds such as *Commelina* spp. Silage-making was very successful, however, and surplus forage or green maize was usually available during the first rains.

Productivity

In order to measure the productivity of pasture with no applied fertilizers, one paddock was maintained as a continuous grazing trial. The paddock, comprising an area of about 5 ha, was grazed continuously by ten animals, which were replaced as they matured so as to maintain from 550 to 675 kg liveweight per ha. The pasture was derived by natural regeneration from abandoned small holdings where there had been considerable planting of *Paspalum notatum* which became the dominant species. The paddock was slashed once during each rainfall season and occasionally the larger weeds such as *Lantana* spp. were grubbed out by hand. No fertilizer was applied to the pasture and no supplementary feeding was given to the animals. Water was continuously available as well as a mineral lick, and the animals were starved before weighing. Summaries of the liveweight records are given in Figure 9.30*a, b*.

These results illustrate the inherent productivity of land in the elephant grass zone of Uganda with its favourable rainfall distribution. They show what any local farmer ought to be able to achieve with only reasonable attention to management. The monthly averages, with peak production rates in April and October, reflect the bimodal rainfall distribution. The annual averages show that the rate of liveweight production progressively increased in the later years. These increases were almost certainly attributable to the up-grading of the livestock.

In contrast, the potential for liveweight production with intensive management was estimated on an area of about 10 ha of established pasture on another area of the farm. The area was divided into two equal parts and grazed rotationally using beef animals from the first backcross generation, ranging in age from 18 to 24 months. The stocking rate was varied according to the amount of grass available and ranged from 2.5 to 3.7 animals per ha. The grazing interval was 42 days (21 days in each half) and the animals were starved before weighing. No supplementary feeding was given but water and mineral licks were continuously available.

The pasture was established in March 1966 with a mixture of *Chloris gayana*, *Panicum maximum* and *Centrosema pubescens*. During the period until the trial started in February 1971, a total of only 880 kg per ha of calcium ammonium nitrate had been applied, or about 40 kg nitrogen per ha per year. The soils were analysed and lime added to about one third of the area, where the pH was less than 6.0. Single super phosphate at the rate of 250 kg per ha was applied to the whole area. A quarter of the area, which was somewhat deficient in potassium, received muriate of potash at the rate of 250 kg per ha as an additional dressing. Initially, and after each grazing, the pasture was topped using a rotary slasher set at a height of 20 to 25 mm and 145 kg calcium ammonium nitrate per ha were added. The results, summarized in Figure 9.31, illustrate the enormous potential for beef production under conditions of intensive management.

At these high levels of productivity, however, problems can arise from the high nitrate concentration in the forage. The flush of growth that occurs after rain, following a dry period, can contain as much as 3 per cent nitrate and if animals are exposed to unrestricted grazing in such conditions, a serious problem of nitrite poisoning can arise. Although not a problem in the grazing trial referred to above, several

Figure 9.30. Average liveweight production of ten beef animals on 4.9 ha of unimproved pasture, over five years with no fertilizer: (*a*) annual averages, (*b*) monthly averages.

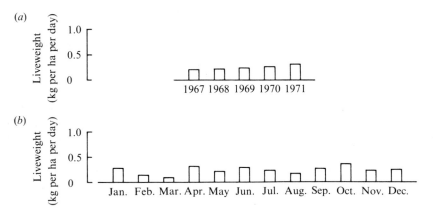

animals were lost from nitrite poisoning on other areas of the farm, before the condition was correctly diagnosed. (Descriptions of nitrate and nitrite poisoning, together with references to original work, are given by Blood & Henderson, 1968.)

Farm records

The accumulated records maintained in the farm office and crop stores at Namulonge represent one of the longest and most comprehensive collections of farm data available in eastern Africa. From the outset, a system of record keeping was instituted and strictly adhered to as a daily routine whereby all operations on each cultivation strip or pasture were recorded. All man-days of labour and all tractor hours were entered as well as all crop produce harvested and disposed of, whether by sale or feeding to livestock. Similarly, detailed records were kept of all the livestock including both the dairy and beef herds.

These data formed the basis of surveys of farm costings undertaken from time to time, and for wider analyses of the economics of large-scale farming in Uganda such as that by Lea & Joy (1963). In a world of changing prices, however, conclusions on the economics of agricultural production are ephemeral, but knowledge of the extent of the inputs required in relation to the magnitude of the yields obtained remains of fundamental importance. It is in this context that the farm records at Namulonge constitute such a valuable source of information for future, comparative studies.

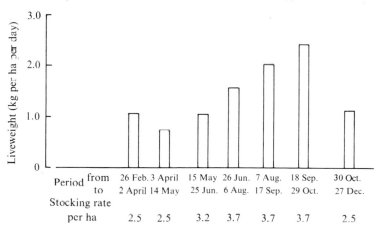

Figure 9.31. Average liveweight production of a beef herd on 9.9 ha of improved pasture with frequent fertilizer applications.

Conclusions

The object of the Namulonge farm was to provide suitable land for field experiments and variety trials; to produce pedigree cotton seed for the multiplication scheme; to help to create an environment for productive agricultural research; and to demonstrate to others the lessons to be learnt. The investigations cn farming methods were not directly related to the immediate problems of the small farmer in Uganda. Rather, the work was complementary to research on the improvement of local small-holdings undertaken by the nearby research farm of Makerere University and the Government research station at Kawanda.

The picture that unfolded at Namulonge was one of contrasts: contrasts within the boundaries of the station, between farming at low levels of productivity and farming at high levels; and, in a broader sense, contrasts between what could be produced in Uganda and what could be produced in the vast areas of the African continent with less favourable rainfall. Its relevance to agricultural development was that it stimulated thought, translated that thought into trial and thus augmented the fund of knowledge on limits to productivity at both ends of the scale.

APPENDIX 9.1

The effects of pre-sowing cultivations on weeds

An experiment was laid down in May 1971 on an area of the farm that was included in the revised 3:3 rotation. The ley was opened by rotavation, and calcitic limestone was applied at the rate of 2000 kg per ha. A mixture of 4.48 kg a.i. dalapon per ha and 1.12 kg a.i. aminotriazale per ha was applied to control lumbugu. Details of the experiment were as follows.

Design	Randomized complete blocks with five replications; plots split for fertilizer application.
Plot-size	14 m × 10 m (main plots); 7 m × 10 m (subplots).
Average depths of cultivation	Mouldboard ploughing, 0.30 m. Chisel ploughing, 0.21 m. Rotavation, 0.16 m.
Fertilizers	Calcium ammonium nitrate (CAN) and single superphosphate (SSP) applied at rates of 200 kg per ha to beans and sorghum in the seed bed. Cotton received muriate of potash in addition at the same rate of application. 200 kg CAN per ha were top-dressed to cotton and sorghum.
Weeding	Plots were weeded when an estimated 25 per cent of the ground was covered by weeds.
Main records	Type and number of weeds removed at each weeding; time required for weeding; crop yields.
Results	Results are summarized in Tables A9.1, A9.2, A9.3.

Table A9.1. *Labour requirement in man-hours per ha*

Crop	Treatment	PD	CD	RD	PR	CR	R	Mean	S.E.
Cotton	No fertilizer	140	182	158	132	120	154	148	±5.2
	Fertilizer	163	180	182	120	133	143	153	
	Mean S.E. +16.7	151	181	170	126	126	148		
Beans	No fertilizer	136	191	180	169	157	171	167	±6.2
	Fertilizer	175	220	161	199	171	229	192	
	Mean S.E. ±12.9	155	205	170	189	164	200		
Sorghum	No fertilizer	128	142	91	107	87	107	110	±4.0
	Fertilizer	108	186	156	135	143	134	144	
	Mean S.E. ±19.7	118	164	123	121	115	120		

Key to main-plot treatments

PD   Mouldboard ploughing (P) followed by disc harrowing (D).
CD   Chisel ploughing, using a Bomford 'Superflow' (C), followed by disc
       harrowing.
RD   Rotavation (R) followed by disc harrowing.
PR   Mouldboard ploughing followed by rotavation.
CR   Chisel ploughing followed by rotavation.
R   Rotavation only.

Table A9.2. *Crop yields in kg per ha* (*sorghum yields not available*)

Crop	Treatment	PD	CD	RD	PR	CR	R	Mean	S.E.
Cotton	No fertilizer	1366	885	1157	1042	966	1376	1132	±23.8
	Fertilizer	1604	1238	1528	1085	1495	1713	1444	
	Mean S.E.±100.0	1485	1061	1342	1063	1230	1544		
Beans	No fertilizer	893	810	714	1089	875	827	868	±22.0
	Fertilizer	1184	1006	1143	1298	1220	1202	1175	
	Mean S.E.±55.0	1038	908	928	1193	1047	1014		

Table A9.3. *Shoot production per m^2 of lumbugu and nutgrass
under sorghum*

Weed	No fertilizer	Fertilizer	First cultivation			Second cultivation	
			R	C	P	D	R
Lumbugu	48	52	62	46	42	44	57
Nutgrass	71	108	51	97	120	91	88

APPENDIX 9.2

Summary of farm practices and crop yields

Land preparation

After leys the grass sward was destroyed by a light rotavation to a depth of 0.025–0.050 m, followed by mouldboard or chisel ploughing to a depth of 0.225–0.325 m. Growth of weeds and regrowth of grass were permitted before applying herbicide. The seed bed was prepared with a heavy disc harrow.

After cotton, sorghum and maize the crop residues were slashed with a gyromower or by light rotavation. This operation was followed by ploughing and discing to give the required tilth or left rough if the land was to enter a fallow period.

After groundnuts and beans, ploughing followed by discing was usually adequate for seed-bed preparation.

Seed-bed operations

All crops except groundnuts were grown on the flat in rows drilled either 0.35 m or 0.70 m apart (Table A9.4). Groundnuts were grown on ridges measuring 0.70 m at the base, the ridges being necessary for successful harvesting with the equipment used. Fertilizer was either applied to the seed bed at a depth of 0.15–0.20 m along the line of sowing or broadcast. Standard unit drills were used for all crops except groundnuts which were drilled with a machine specially constructed to sow two rows, 0.15 m apart, on the top of each ridge.

Inter-row cultivation

Immediately after drilling, the inter-row space was ripped to a depth of 0.20–0.30 m to encourage rainfall acceptance. This operation was repeated, as necessary, until full crop cover was attained. Where herbicides were either not used or proved ineffective, mechanical weeding was done either with a rolling cultivator or with a tined inter-row cultivator.

Nitrogen top-dressing

Where top-dressing with nitrogenous fertilizer was required in young crops, it was applied mechanically at a depth of 0.050–0.075 m in a line running 0.070–0.125 m to the side of the crop row. In older crops the fertilizer was often applied manually by broadcasting it between the rows.

Further details and a summary of crop yields are given in Tables A9.4, A9.5 and A9.6.

Table A9.4. *Cultural practices for arable crops*

	Cotton	Groundnuts	Soya beans	Maize	French beans	Soya beans
Time of sowing	10 June to early July	Mid-February to end March	End August to mid-September	Late January to end February	March	July
Depth of sowing (mm)	25–50	50	12–25	50–75	50	50
Seeding rates per ha (kg)	35–50	90–135	8–11	35–50	65–110	65–110
Seed dressing	Perecot 45	Fernasan D	Dieldrex M	Fernasan D	Fernasan D	Inoculum
Row spacing (m)	0.70	Twin-rows at 0.70	0.70	0.70	0.35	0.35
Spacing in the row (m)	0.30	0.20	0.20	0.30	0.20	0.30
Population per ha	48000	140000	70000	48000	140000	140000
Seed-bed fertilizer (kg per ha)						
CAN	100–200	100	200	200	100	100
SSP	200	200	200	200	200	200
MK	a	a	200	200	a	a
Top dressing (kg/ha)[b]						
CAN	200–400	Nil	200–400	200–400	100	Nil
Pre-emergence herbicide	Fluometuron	Fluorodifen	Nil	Atrazine	Fluorodifen	Fluorodifen
Pesticide applications						
Product	25% DDT+ Rogor 40	Aphox 65	Endosulfan 35	Nil	25% DDT	25% DDT
1st application (days from sowing)	56	10	7	—	40–55	40–55
Interval and number of sprays	14 days×6	7 days×4	—	—	14 days×3	10 days×4
Time of harvesting	Early December to February	Mid-June to mid-July	Late December to early January	Mid-June to late July	May	December

[a] 100 kg per ha applied if soil analysis showed exchangeable potassium content in topsoil less than 0.5 mequiv. per cent.
[b] Always discretionary, depending on crop potential in relation to establishment, water stress etc.
CAN, calcium ammonium nitrate; MK, muriate of potash; SSP, single super phosphate.

Table A9.5. *Harvesting and processing of crops*

Crop	Harvesting	Method of drying	Moisture (%)	Processing	Storage preparation
Cotton	Hand-picking	Sun and air	Ambient	Hand-sorting Roller ginning	Nil
Groundnuts	Two-stage mechanical lifter and stripper	Oil burner and ventilated-floor bins	7	Static decorticator	Nil
Sorghum, maize and beans	Hand-collection	Electrical heater with in-sack grilles in concrete floor	12	Static thresher	2% malathion dusting

Table A9.6. Yields of farm crops in the first rains (upper table) and the second rains (lower table)

The table shows the areas and yields of crops grown in the main farm rotations. It does not include data from experiments, observations plots, variety trials, breeding plots or catch crops taken from pasture areas

Crop and variety	1966–67 (kg per ha)	(ha)	1967–68 (kg per ha)	(ha)	1968–69 (kg per ha)	(ha)	1969–70 (kg per ha)	(ha)	1970–71 (kg per ha)	(ha)	1971–72 (kg per ha)	(ha)
Phaseolus beans												
Banja 1/2	494	13.8	1585	1.6	1423	3.3	780	5.16	1343	3.69	1697	4.11
White Haricot	—	—	1631	7.8	1517	2.5	—	—	—	—	—	—
Groundnuts												
Valencia Bunch (B227)	932	2.1	1372	7.5	1317	8.1	2105	3.16	1984	4.74	1219	1.03
Maize												
Kawanda synthetic	3777	20.4	3582	7.5	3788	0.8	—	—	—	—	—	—
Kawanda composite	—	—	—	—	—	—	4607	1.50	4531	3.13	5210	1.55
Katumani composite	—	—	—	—	—	—	3702	7.84	3384	9.29	5261	1.63
Kitale hybrid	—	—	—	—	5899	11.2	—	—	—	—	4515[a]	3.84
SR52	—	—	4815	1.5	5920	0.6	4685	0.54	5354	0.52	7666	1.66
Soya beans												
Kawanda selections	—	—	—	—	1420	3.0	1464	2.72	831	4.24	n/a[b]	4.48
Sorghum												
Dobbs	5493	2.1	3498	3.3	1702	2.4	1728	1.03	1870	4.36	n/a	3.67
Local beer variety	—	—	3095	1.8	1575	1.0	1317	2.13	742	1.04	—	—
Seed cotton												
BP52	1201	30.5	—	—	—	—	—	—	—	—	—	—
BPA	1429	11.9	1199	26.4	1633	19.6	1295	27.94	1673	22.88	1940	33.37

[a] Damaged by monkeys.

[b] n/a, not available.

10

The application of agricultural science to national development

M. H. ARNOLD

The contribution of science

The striking contribution that agricultural science has made to development needs no emphasis in this day and age when the green revolution, with all its attendant benefits and problems, has captured the imagination of so many. It is easy therefore to jump to the conclusion that it is science that leads the development that follows. Indeed, in some people's minds, a lack of agricultural development is identified with a lack of scientific research and the suggested remedy is to inject more resources into research, sometimes without pausing to analyse the situation in depth or even to take account of research already done. But to adopt such an approach is to invite failure; the contribution that science can make to agricultural development is, in most cases, no greater than the readiness of the stage of development to accept it.

Perhaps this amounts to nothing more than a statement of the obvious. In the case of the Concorde aircraft, we all recognize that it is not only the science that has gone into it that will determine its success, but also the state of readiness of the world at large to accept it. Nevertheless, when we apply ourselves to the problems of the developing countries, we whose views have been formed mainly against a background of a developed situation, can easily make the mistake of taking development for granted and assuming that the main contribution comes from science. So often, however, when we talk of the contribution of science, we do not mean the contribution of science alone but the contribution that results from a complex interacting system involving science, technology, education and organization as well as a wide range of economic and sociological factors.

How then can the results of agricultural research be applied to stimulate development? In the early stages, when farming is generally at the subsistence level, science can best contribute by devising improvements that can be implemented with the minimum amount of investment and that are not dependent for their success upon a large scale of operation. For these reasons some of the most striking initial contributions have come from plant breeding. In contrast, the introduction of mechanization, for example, has often been inhibited both by the relatively large amounts of cash required for the initial purchase of the machines and by the scale of operation required for economic returns. In the sequence of events leading to agricultural development, therefore, the first innovations to be readily accepted have often been things like new seed, either seed of a new crop or improved seed of an established one. For these reasons varietal maintenance and plant breeding have always featured strongly in programmes of agricultural research in the developing countries, the Namulonge programme of cotton research being no exception. Indeed, cotton provides excellent examples of attempts to apply the results of research to agricultural development; some outstandingly successful, others markedly less so.

As an example of the first, we can examine in more detail the increases in production associated with the cotton breeding programme in Tanzania, already mentioned in connection with plant breeding methods (p. 199).

Cotton production in western Tanzania

Cotton was introduced into Tanzania shortly after the turn of the century and rapidly became the most important cash crop in the north-western parts of the country. After the initial successes, however, annual production remained more or less static at levels of between about 30000 and 70000 bales, with fluctuations associated with seasonal conditions and with particular campaigns launched by the Government to grow more cotton.

Two biological factors were identified as limiting production. One was a sucking insect pest, the cotton jassid (p. 194); the other, a disease – bacterial blight (p. 152). The succession of new varieties released from 1946 onwards showed progressively greater resistance to both the pest and the disease and from 1954 further control of the disease was effected by ensuring that all seed issued to the growers was dressed with a bacteriocide. Associated with these changes there were spectacular increases in production that continued in an almost exponential manner from 1953 for the following twelve years. In the 1965–66 season production reached 435 600 bales representing nearly a tenfold increase over average production during the 1930s and 40s (Arnold 1970*b*).

These increases in production were far greater than the measured increases associated with seed dressing and with genetic improvements to the seed stocks. Clearly other factors were at work and one of these was the greater enthusiasm for cotton shown by the grower. He no longer saw his crop crippled with bacterial blight and blasted by jassid but could confidently expect a much bigger return for his labours from a crop that remained relatively healthy throughout the growing season.

There are many examples of this type of reaction by the farmer to an improved variety. The introduction in the USA of the first variety of sorghum suitable for combine-harvesting enabled the crop to be grown more profitably and was immediately followed by a resurgence of production (Doggett 1970). For similar reasons, the introduction of the dwarf wheats into India and West Pakistan, the stiff-strawed rices in the Philippines and Proctor barley in the UK have been associated with far greater increases in production than those implied by the application to existing acreages of the genetic advances alone.

But the crop and the application of science to improve it form only a small part of the whole story. If we return to the example of cotton in Tanzania and ask if similar increases in production would have ensued if the results of the scientific work had been available twenty years earlier, the answer would almost certainly be 'no'. The relatively long period that preceded the impressive upturn in the trend of production was important in many respects. It was a period during which many different problems were encountered and largely solved: problems of seed multiplication and distribution; of primary marketing, collection and transport; of ginning and quality control; of financing the crop and disposing of it profitably on world markets. Expertise in all these diverse aspects was gradually built up so that a structure was created upon which increases in production could be built.

Turning from the organization of crop production to the farmer himself, we have to consider the factors that motivate the subsistence farmer and cause him to want to produce more, to want to introduce a greater degree of professionalism into his practices, and to move away from subsistence farming to farming as a business enterprise. In Tanzania, the period that preceded the striking increase in cotton production saw the introduction to outlying areas of various consumer goods such as the bicycle, corrugated iron as a roofing material and later, the portable radio. In many cases the amount of cash a man could use to purchase such things could be equated to the amount of cotton he grew. Further incentives came from the expanding educational system and the recognition of the importance of education as a means of entering a salaried occupation. This led to an ever increasing need for cash to pay school fees for numerous dependants.

This example of the expansion of cotton production in Tanzania is typical of the progress that has been made through the application of science to many crops, as well as to livestock and their products, in a variety of developing situations. For every example of success, however, there are many where attempts to apply the results of research have been frustrated, and where the desired increases in production have failed to materialize. Cotton production in Uganda forms a good example.

Cotton production in Uganda

In many developing countries in Africa, cotton has become one of the key factors in economic development accounting, as we have seen (p. 238), for a substantial part of the national capacity for earning foreign exchange, and also in providing productive employment for large sections of the community. Uganda cotton, in common with that of most other African countries, has always enjoyed a ready sale on world markets. The basic considerations for expanding the crop have therefore been favourable and, in the light of the world supply and demand situation during the late 1960s and early 1970s, there are good grounds for believing that cotton production in Uganda could have been doubled without seriously affecting world prices.

Recognizing the importance of cotton to national development the Uganda Government made increased production a primary aim of its development plans. The target for the Second Five-Year Plan (1966–1971), for example, was to produce 575 000 bales by the 1970–71 season, representing an increase of more than one-third over levels during the previous three years. It was expected from studies of production trends, particularly in the north and east of the country, that there would be a considerable increase in production even without an intensified promotion programme. The extra effort required to achieve such a large expansion was to be put into the control of pests, the use of fertilizers, the introduction of heavier-yielding varieties and the promotion of more modern cultivation techniques. Later, when it appeared that the target might not be met, vigorous campaigns to encourage farmers to grow more cotton were launched by the Government through the extension services.

Neither attempts to apply the results of research nor the various campaigns to grow more cotton met with signal success, however, and the target was not reached. Indeed average production for the three seasons ending in 1972 was not substantially different from that for the three seasons ending in 1966 (Table 10.1). It is instructive, therefore, to examine the problems more closely and to attempt to identify those factors that limited the application of science to produce greater yields per unit area, as well as those that frustrated efforts to persuade farmers to increase substantially the area planted to cotton.

Factors limiting yields per unit area

On a world basis, the most successful cotton crops are grown under desert or semi-desert conditions in subtropical regions, which have daily sunshine of long duration and where the water supply is ensured by irrigation. In these areas, such as the San Joachim valley of California and the pacific seaboard of Peru, yields of 2000 to 3000 kg seed cotton

Table 10.1. *Cotton production in Ugarda*

Three-year averages (in thousands) of the actual number of bales produced in each ginning zone

Three-year period ending	Ginning zones											Total
	Masaka	Mengo	Mubende	Busoga	Mbale	Teso	Toro	Bunyoro	Lango	Acholi	W. Nile	
1942	14.2	94.2	9.3	69.0	48.0	22.3	2.2	4.4	17.2	7.5	4.3	292.6
1945	5.0	74.3	13.8	43.3	27.8	12.2	1.0	2.5	9.1	2.2	1.9	193.1
1948	8.7	81.1	16.0	37.4	26.9	12.0	1.7	3.3	12.5	6.8	3.5	209.9
1951	18.5	122.4	23.1	68.0	44.1	24.6	4.5	6.8	26.9	13.6	7.9	360.4
1954	8.4	97.1	17.2	83.4	60.3	29.1	4.5	7.5	29.4	17.1	11.5	365.5
1957	7.2	82.7	15.7	88.1	48.5	29.0	5.8	8.1	27.8	20.1	12.2	345.2
1960	5.5	72.4	19.1	97.3	57.5	28.3	8.5	9.2	32.9	24.5	15.6	370.8
1963	3.8	48.3	12.2	81.3	42.6	21.7	6.4	9.0	34.2	25.1	18.7	303.3
1966	4.3	49.2	12.5	87.4	66.2	59.7	5.6	14.8	57.1	35.7	28.6	421.1
1969	3.7	34.7	12.2	73.1	49.6	55.8	8.1	22.0	65.4	35.8	37.3	397.7
1972	1.6	29.7	17.4	88.5	73.0	50.6	12.4	24.1	67.5	31.3	31.9	424.8

per ha can be obtained fairly reliably in commercial practice. In contrast, in Uganda yield levels of only 300 to 500 kg are commonplace. The fact that yield levels of 1000–2000 kg per ha can be readily achieved on research stations shows that the difference can be explained only to a limited extent by the reduced yield potential associated with shorter days, lack of irrigation and uncertain rainfall distribution. By far the greater part of the difference must be explained in terms of the failure on the part of the small farmer to apply improved practices that have been recommended as a result of research.

Perhaps the most striking short-coming is his failure to implement recommended sowing dates when, on the face of it, he could increase his yields very considerably with no extra effort, simply by sowing during the recommended period.

Sowing dates. Recommended sowing dates have been devised so that the growing season of the crop will best fit the expected pattern of rainfall distribution (Chapters 2 and 4). Nevertheless, in any one particular year, sowing outside these periods can, for one reason or another, occasionally appear to be better than sowing within them. This is one of the reasons for the farmers' scepticism about recommended sowing dates. Another and perhaps more cogent reason arises from the fact that, to the layman, a late-sown crop usually looks better than an early-sown one, particularly towards the end of the season, when the plants appear much greener and are often taller and more vigorous in growth. Usually, however, these effects have been induced by a build-up of insect pests which destroy the fruiting bodies and induce excessive vegetative growth (Chapter 5). The earlier-sown crop avoids this build-up of insect pests, and boll counts almost invariably reveal that the later-sown cotton has matured fewer bolls, and that a greater proportion of them have been damaged by pest attack. In consequence, the value of the crop that is picked from the late-sown crop is far less than that from the early-sown.

Even if the farmer accepts the advice given to him on sowing dates, however, there are often telling reasons why he cannot comply with them. In areas where cotton sowing is preceded by a dry season, land cannot be prepared until after the rains break and the soil is softened. This immediately imposes a limit on the amount of land that can be prepared in time, particularly where all cultivations are done by hand. Mechanization, using either tractors or oxen, can help greatly. More land can be prepared in a given time, and land that is too hard to be cultivated by hand can sometimes be prepared by machine although, if discretion is not used, the wear and tear on machinery can be excessive.

At Serere it has been shown that land can be ploughed at the end of one season and left fallow through the ensuing dry season. Only light cultivation is then required when the new rainy season begins, and large areas can be sown in a short time. The system is analogous to the practice of autumn ploughing for spring-sown crops in temperate regions. The wide application of such a system in Uganda, however, must await a greater degree of mechanization because, as the soil dries out towards the end of the cropping season, it is difficult to penetrate with hand implements. Even where a tractor hire service is available, the farmer is unlikely to want to invest money in a future crop, particularly at a time when there are other, more urgent calls on his financial resources.

At cotton sowing time the farmer is also faced with the problem of conflicting demands upon his limited manpower. In many areas food crops are sown before cotton and, faced with the choice of risking the success of his food crops or risking a reduction in his cotton yields through late sowing, there is little doubt as to which he will choose. He has little room for manoeuvre unless he can command the resources either to intensify his methods or to buy his food instead of growing it. In most cases he can do neither.

Spacing. Some of the considerations that apply to sowing dates also apply to spacing. The effects on yield of different plant population densities differ widely, and are related to such factors as soil fertility, pest attack, water availability and variety (Chapter 4). Moreover, the problem of weed control is likely to be more severe in widely spaced than in closely spaced plants, where the leaf canopy closes more rapidly, inhibiting the rapid growth of weeds (Chapter 9). There is considerable latitude in the population density that will maximize yield over a range of conditions and recommendations are made in such a way that farmers can choose a row width and an intra-row spacing that will suit their sowing methods, within limits that give target population densities of about 50 000 plants per ha. Fairly wide deviations from this density can be made without materially affecting yields, under most conditions. Nevertheless, a large proportion of the crop continues to be sown at too wide a spacing, population densities of less than 25 000 plants per ha being commonplace.

The wider the plants are spaced, the greater the crop that is borne on a single plant. It is this that catches the farmer's eye and causes him to argue that the wider-spaced plants produce the greater yields, an assertion that takes no account of the concept of yield per unit area. When sowing is done entirely by hand, closer rows and closer spacing within the row involve more work and this of itself may well be a deterrent. The work can be reduced by using one of a range of simple mechanical seeders, usually based on a wheel that is pushed manually, but relatively few growers possess them.

Figure 10.1. A hand-seeder designed to operate with a bicycle wheel.

Insect pests. Insects can attack the cotton plant at any time during its growth, damaging leaves, stems, flower-buds and mature bolls, If uncontrolled they can set severe limits both to yields per unit area and to the regions where cotton can successfully be produced (Chapter 5). Control measures have been applied with a varying degree of success in Uganda.

Although the pink bollworm (*Pectinophora gossypiella*) can be controlled by a close-season, in practice the recommended control measures fall well short of being entirely effective. The work of uprooting and burning is such that it is not always willingly tackled even though there is a legal requirement to do so. Occasional fields in most areas are left standing too long and provide reservoirs of infestation for the following season. Mechanized destruction of plants, as practised at the research stations and on a few large farms, is beyond the reach of the small farmer.

As far as insecticides are concerned, however, growers are generally enthusiastic about their use. Nevertheless schemes to extend the area of cotton protected by spraying have not met with sustained success. Problems of obtaining an adequate supply of clean water, and carrying it to the site, act as deterrents to the use of sprays: difficulties in obtaining, using and servicing spray pumps act as another and some-times, subsidized insecticide intended for use on cotton, is used on other crops or for other purposes such as killing biting ants in coffee trees. Moreover, the need for repeated spraying has not always been appre-ciated by the grower, and the results of surveys by the Department of Agriculture in Uganda show conclusively that farmers who spray their cotton fewer than the recommended number of times, lose much of the benefit that should be obtained.

Nonetheless, the effective use of control measures against insect pests remains a key factor in the development of cotton production in Uganda. There is a continuing need to devise simple but effective methods of control – ideally, to produce resistant varieties. For the extended use of insecticides there is a need for financial assistance to the farmer to obtain the necessary equipment and chemicals at the beginning of the season and to train him in their correct use. In this connection, schemes for agricultural credit are of great importance and may well prove to be a better basis for helping the farmer than by direct subsidy. Another approach that is rightly receiving attention is the operation of centrally organized spraying teams which can provide a service to the farmer, financed on credit with the cost recovered from the proceeds of his crop.

Fertilizers. Yields of cotton can be increased substantially on all types of soil in East Africa by the judicious use of fertilizers, but simple recommendations involving the use of standard dressings can be given only in certain cases. With the hill-sand soils of north-western Tanzania, for example, a fertilizer application of approximately 20 kg phosphorus and 20 kg nitrogen per ha almost invariably gives a profitable return. In many cases, heavier dressings would give bigger returns, but the risk of not recovering the cost should some other factor, such as drought, intervene would be that much greater.

If the risk factor is important in circumstances where simple recom-mendations can be given, it assumes even greater importance where the situation is more complex. With the red ferrallitic soils of southern Uganda, reliable responses to fertilizer are obtained only when the whole level of inputs can be raised. Liming may be necessary; a carefully balanced fertilizer mixture must be used and applied in the right way; and the crop must be sown at the right time, at the right spacing and protected from insect attack (Chapters 3 and 9). In such circumstances no simple general recommendations can be given and fertilizers can be used profitably only when expert advice is readily available or when the farmer himself has been educated to a far greater level of understanding. It is this interdependence of a number of diverse factors that is one of the major obstacles to rapid improvement in yields in many areas. To implement the recommendations requires a degree of innovation for which the farmer is not yet ready, financial resources that he does not command and risks that he is ill-prepared to take.

Seed. It is through the free issue of improved seed that the greatest benefits to production have been effected. As we have seen, any improvements effected by plant breeding can rapidly and effectively be transferred into production because of the central control of ginning and seed distribution. It is difficult, however, to measure the effects of new seed issues on production in the complicated situation of counteracting influences that exists in Uganda. Furthermore, precise figures for the area of cotton cultivation are not available. We can only assume that the results of district variety trials give reasonable estimates of the contribution made by new varieties to production, even though the trials may partly obscure the benefits of resistance to bacterial blight (Chapter 8).

During the five seasons from 1966–67 to 1970–71 the value of the cotton crop to Uganda averaged £20 million. During the same period the total cost of research on cotton growing in Uganda averaged about £150000 per annum. Although there are marked seasonal fluctuations in the yield increments of the new varieties over the old, an average increase of less than one per cent is all that is required to cover the total cost of all aspects of research into cotton growing. The results of district trials indicate that BPA outyielded its predecessor by an average of 13 per cent over eight seasons and SATU gave a yield increase of 18 per cent averaged over six seasons. Even these increases take no account of the quality factors bred into the new varieties which helped very considerably to maintain the price of Uganda cotton on world markets.

It would therefore appear that the benefits from the new varieties greatly exceeded the cost of the research effort that went into them and entirely vindicated the emphasis placed on plant breeding in the research programme.

Factors limiting the area of production

Because the causes of failure to raise levels of yield per unit area are in many ways related to those associated with failure to expand the area under production, both these aspects of production must be considered if we are to attempt to understand the constraints that limit the application of science. Clearly, important factors in the expansion of the area under cultivation are the availability of suitable land and resources and, particularly in the developing countries, the will of the farmer to expand (Boerma 1973).

Manpower. Pressure on the availability of land, although locally important, is not a major factor limiting cotton production in Uganda. Rather the limit is set by the ability and willingness of the farmer to cultivate a larger area. Increases in production can, in some localities, be accounted for by immigration of people into hitherto uncultivated areas. Moreover there is considerable evidence to show that the area of cotton per head of the population has remained remarkably stable for long periods. A survey of district records suggested that changes in the size of the cotton crop between 1921 and 1964 could be accounted for mainly by changes in the population, after due allowance had been made for the replacement of much of the cotton in Buganda with coffee (MacDonald 1967). In such circumstances, therefore, substantial increases in the area under cotton would be dependent upon a much greater degree of mechanization or much greater incentives to the grower to do more work.

Mechanization. The Uganda Government has given a great deal of attention to the development of mechanization both through the use of oxen and tractors (Brown, Evans-Jones & Innes 1970). It might appear that the use of oxen would be the logical first step in the progression towards a mechanized system, but even where training schemes have been set up and suitable implements devised and marketed at subsidized prices, the use of oxen has sometimes not proved popular among the local people. In other cases, however, such as in the Teso District of eastern Uganda, ox cultivation has become widespread and has contributed a great deal to agricultural production. The failure of ox cultivation to develop in some areas is partly associated with the traditions of the people and partly with such things as the difficulties of providing adequate animal feed and supplying sufficient water, particularly in areas where there is a harsh dry season.

In general, tractors capture the imagination of the African farmer to a greater extent than oxen, and where tractors have become available they have been used enthusiastically. But the problems of the large-scale use of tractors are even greater than those of oxen. The capital investment necessary is beyond the means of the small farmer and hire service schemes, sponsored by the Government, have proved difficult to operate profitably. Serious damage is easily done to tractors and implements through lack of experience in their use and as a result of the difficulties of the terrain over which they have been expected to work. In some areas, tracts of open land of reasonable size are few and far between, so that a tractor belonging to a hire-service scheme may spend far too much of its working life simply travelling from one small piece of land to another. In consequence, hire charges have to be raised to the extent that many small farmers find the use of tractors, even where available, to be worthwhile only for the breaking of new land.

Institutes have been built to train drivers and mechanics, but this is no substitute for the farmer himself gaining experience in the efficient use of tractors and machinery, a process that is bound to be gradual and dependent upon the whole process of economic and technological development.

Primary marketing and ginning. As in many African countries, primary marketing and ginning in Uganda have always been, to a greater or lesser extent, under government control. Prices paid to the grower for seed cotton are fixed each season by the Government and fixed prices are also paid to the ginners for lint and seed. Until 1950 all ginneries in Uganda were owned and operated by commercial companies, which were also responsible for buying from the farmers. In all, 194 ginneries were built, most of them during the initial period of expansion in cotton production, from 1910 to 1930. Spurred on by the apparent profits to be made from buying and ginning, the number of commercial ginners rapidly became greater than the industry could efficiently support. The output from many of these ginneries was less than 2000 bales, whereas later studies showed that an output of 25000 bales should be the target to operate efficiently.

The number of operational ginneries was reduced as a result of voluntary agreement among the companies concerned. Ginning was allocated on a quota system but the work of ginning was concentrated into a smaller number of active ginneries while others (described as 'silent') remained inactive. Even so, more than 100 ginneries were operational in the early 1960s. From 1951 onwards the co-operative movement, backed by legislation, became more and more involved in the ginning industry, progressively bought out the commercial firms and further concentrated the ginning effort. By 1970, the number of operational ginneries had been reduced to fifty-six all owned and operated by co-operative unions.

Thus the co-operative movement largely inherited the legacy of an

Figure 10.2. Transporting cotton to the buying post (courtesy of the East African Railways Corporation).

inefficient ginning industry comprising too many uneconomic units. To this disadvantage were added the changes in the distribution of production over the country as a whole (Table 10.1) leading to a surplus of ginning capacity in some areas and a deficit in others. In such circumstances, campaigns to grow more cotton could only aggravate an already difficult situation, until effective steps could be taken to re-organize and re-equip the ginning industry. The problem was largely one of finance. Faced with repaying the loans with which they originally purchased ginneries, the co-operative unions were unable to find additional funds on a scale large enough to finance the work required. Moreover the inefficiency of ginning meant that the crop could not be disposed of quickly enough. For the co-operatives this resulted in funds being tied up in stocks for uneconomic periods and for the farmer the consequence was difficulty in selling his crop. There could be little more discouraging to a farmer than to transport his cotton by head-load or bicycle to a buying post some miles distant, only to find that buying had temporarily been brought to a halt because of lack of storage space, or because no further funds were immediately available for buying.

In contrast, the take-over of ginning by the co-operative movement in western Tanzania occurred in more favourable circumstances. The main expansion in production in Tanzania took place about twenty-five years later than in Uganda (Figure 10.3). As the crop spread into new areas, new ginneries were built to accommodate the increased production, designed from the outset to take advantage of economies of scale and operate more efficiently. In consequence some of the mistakes made in Uganda were avoided in Tanzania and the grower suffered less from bottlenecks in the ginning industry.

Incentives. Inefficiency in primary marketing and ginning can be seen to be an important factor, therefore, in causing frustration to the producer. But there are other reasons why growers become disen-

Figure 10.3. Cotton production in Uganda (solid line) and Tanzania (dashed line). Five-year averages. (The effect of the second world war is clearly revealed in the marked drop in production during the period from 1941 to 1946.)

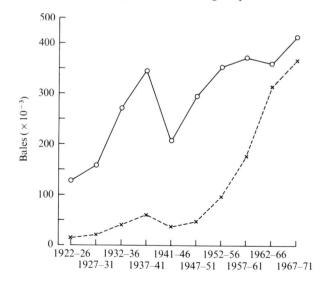

Figure 10.4. In Uganda, cotton is picked entirely by hand (courtesy of the East African Railways Corporation).

chanted with cotton, which can so easily become regarded as the poor man's crop, requiring a lot of effort for small returns. In Uganda, cotton is picked entirely by hand and is purchased from the grower on the basis of two grades: *safi* (clean) and *fifi* (dirty). To reach the standards set for *safi* cotton the grower must first pick carefully and then sort by hand before taking it to the buying post. The work of picking and sorting is slow and tedious and even though it is frequently done by the whole family, it imposes a limit on the amount of cotton that one grower can prepare for marketing. To hire labour for picking and sorting would rob the farmer of a large part of his return.

In consequence of this careful picking and sorting, East African cotton has established a reputation for high grade on world markets, but the maintenance of this high grade, by present methods, is of itself a factor limiting production. It might well be that the standard of cleanliness of seed cotton achieved by the grower could be relaxed somewhat if more extensive cleaning equipment were installed at the ginneries. Moreover, where there is a need for new ginneries, there exists the opportunity to change from roller ginning to saw ginning which for some types of Uganda cotton, particularly that grown in the

Figure 10.5. Seed cotton must be sorted by hand before delivery (courtesy of the East African Railways Corporation).

north of the country, would almost certainly prove to be more efficient. If the combination of greater pre-cleaning and saw ginning were to make it possible for some relaxation in the standards of cleanliness required for primary marketing of *safi* seed cotton, this of itself would act as an incentive to the grower to produce more.

Of the other incentives for growers to produce more cotton, the most important is the price and there is considerable evidence that the small African farmer responds to increases in price in exactly the same way

Figure 10.6. Inside a Uganda ginnery (courtesy of the East African Railways Corporation).

as any other producer. The problem is complicated by the fact that the marketing of lint and seed is controlled by a para-statal marketing board which runs the risk of making a loss if the price fixed for the primary producer proves to be too high in relation to world prices. In consequence the decision on what to pay the grower is left until the likely returns from placing the crop on world markets can be estimated with greater reliability. Consequently, if world prices go up and the marketing board can afford to pay the grower more, knowledge of this comes too late to influence the grower. There is the additional problem in Uganda, unlike in Tanzania, that a tax has traditionally been levied on lint that is exported. One effect of the tax is further to reduce the price that can be paid to the grower, but the problem of finding alternative sources of revenue if the tax were removed has proved impossible to solve in the short term.

Aims of research

Whether development proceeds rapidly or slowly, questions arise that are used to define the aims of research in the various scientific disciplines. As we have seen in the case of cotton production in Uganda, however, some of the most important questions are those that it is quite impossible to define in terms of a single scientific discipline. Questions, for example, that relate to devising farming systems that are best suited to the resources of the farmer, the ecology of the area and the economic needs of the country. To attempt to investigate such problems requires a multi-disciplinary approach that includes not only scientific subjects but sociology and economics as well.

As far as the sociological aspects are concerned, there is little the scientist can contribute. For example, in the closely-knit family unit of African communities, there is an inescapable duty of those with cash to assist with the debts of their less fortunate relations (a phenomenon sometimes referred to as 'social parasitism') and this is only one of many factors that operate to reduce the ability of the cultivator to use his income to improve his methods. In reviewing this type of constraint to agricultural innovation in Uganda, Dunbar & Stephens (1970) describe additional difficulties that the farmer faces. Often he must obtain the approval of members of the family before making a change and, even then, he may fear to put the change into practice because of the enmity that might well arise in his neighbours should be become more successful than they. Furthermore, the incentive to acquire wealth may also be diminished by fear of robbery, particularly in rural areas where armed robbery is common.

There can be little doubt that income is important for innovation and the problem is by no means entirely one of sociology nor peculiar to the developing countries. In the countries of Western Europe, for example, there was little incentive for agricultural innovation during the general depression of the 1930s. The key to development lies in making the primary producer more wealthy while maintaining the balance between supply and demand with respect to the commodities he produces. The problem is related to such things as increasing the size of holding, increasing capital investment and re-deploying surplus labour (Ashton 1973).

Both the sociologist and the economist have a part to play, therefore, in helping to formulate the aims of scientific research. Under their guidance research can be directed to devising new technologies that the farmer is both willing and able to implement. But in attempting to direct research to the immediate needs of development, we must not lose sight of the fact that to make worthwhile, general recommendations the research worker must concern himself with underlying principles. Indeed, research programmes in agriculture cannot usefully be restricted to studying only those innovations that appear to be feasible in the context of the existing stage of development. It is a necessary part of science to ask questions about limits, such as the limits of response in the yield of a crop to a particular factor or group of factors, to ask about the limits to productivity determined by climate, and to ask what might be achieved if the constraints of the existing stage of development could be removed. If the approach required to answer such questions is potentially dangerous in that it could lead to work that is totally unrelated to immediate needs, so also is an approach that is so restricted by sociological considerations that the research is both unsoundly based and loses all sense of adventure. A balanced programme is required in which the prevailing agricultural circumstances are not permitted to cause barriers to creative research and for which the administrative framework is not allowed to inhibit the free interchange of ideas among those involved in widely different disciplines.

How then can realistic programmes be formulated and all concerned imbued with a common purpose? It is here that a carefully structured committee system has an important part to play.

Following studies undertaken by UNESCO, several developing countries in Africa have formed national research councils. In Uganda a National Research Council was inaugurated in 1970. It was established under the Ministry of Planning to be advisory to the Government. It was to operate through six committees (with their associated subcommittees) covering all aspects of research in all disciplines, whether carried out by individuals or institutions and from whatever source financed. It was charged with the responsibilities of determining priorities and drawing up research programmes, of evaluating the results of research and of ensuring that they were published and used, and of advising on the training and terms of service of research staff and on all capital development of research institutions. It was also to be concerned with organizing conferences and seminars and with improving the liaison among individuals and institutions working in related fields. The aims were to make maximum use of available resources and to avoid unnecessary duplication.

It is as yet too early to gauge the contribution to national development that can be made by organizations of this type. Given an effective secretariat and appropriate Government support through changing political circumstances, a National Research Council might well provide essential links in the chain of development that have so often been lacking in the past. It could help to relate research programmes more closely to plans for national development. and make a more positive determination of research priorities. By involving scientists at all levels in thinking about the problems of national development, research committees can help to orientate the attitudes of scientists to be more concerned with national needs. At the level of the council, those most directly concerned with national policy for development can be made more aware of the results of research so that there could well be greater possibilities both of accelerating the application of science to development and of channelling resources for research to the places where they are most likely to be effective.

Research priorities

The determination of priorities for research, when resources are limited and development uncertain, poses very difficult problems. In agriculture, however, it seems useful to distinguish two types of priority:

(i) those related to the long-term development of a sound agriculture;
(ii) those related to the immediate needs of the nation.

Long-term aspects

In a lecture to students in tropical agriculture at Cambridge in 1952, F. B. Wilson stated the aims of agriculture in developing countries as follows:

(i) to ensure the food of the people;
(ii) to preserve the soil;
(iii) to plan the best use of the land;
(iv) to develop crop and animal industries for the production of export commodities;
(v) to broaden the basis of agricultural production;
(vi) to ensure a fair return to the producer and good quality to the consumer.

As a package for long-term agricultural development, it is difficult to improve on these six aims, all of which must clearly be developed simultaneously and all of which might therefore be considered to have a research priority. It is against such a background that immediate national priorities must be viewed, so that in striving to reach short-term objectives, the long-term needs of a sound agriculture are not overlooked. In other words, if it becomes clear that one of these aspects is in risk of neglect it then becomes a priority to fill the gap. To attempt to define priorities in absolute terms is likely to become a sterile exercise. The whole system is dynamic and priority ratings for research must vary with such things as the amount of knowledge already acquired, the changing pattern of economic development and, in practical terms, with such things as the availability of suitable staff and facilities.

Immediate needs

The immediate national need in agriculture is to increase both the efficiency of production and, where feasible, total production in order to provide more wealth for development. The greatest opportunities for rapid increases in wealth lie with those aspects of agricultural production that are already well established. In Uganda these would be cotton, coffee, tea, sugar, staple food crops, livestock and animal products. Research related to increasing the efficiency of production of these commodities is therefore of first priority, even though in some cases, such as with coffee, the possibilities for increasing total production are limited by agreements made in consequence of the world supply and demand situation.

Research effort put into new crops or new livestock products is less likely to have an immediate impact on the national economy. A period of introduction and familiarization is necessary first, in order to identify those products that have the greatest potential for increasing wealth and to define more precisely where research effort is needed. During this introductory period, the work should command a lower priority for research than those aspects of agricultural production that are already known to be capable of greater success.

In this connection the great publicity given internationally to the population explosion and its implications for food supply have sometimes served to obscure important local considerations, particularly in countries like Uganda where population increase relative to the food supply has not yet become a major problem. Visiting experts have tended to stress the need for greater diversity in farming and greater food production in the world context without always taking into account important local factors such as the interdependence of industry and agriculture, an aspect of development that has been so clearly analysed by Hutchinson (1972). Consequently the advice given on priorities for agricultural research by aid organizations has not invariably been what would appear to be best for most rapidly developing the wealth of the country.

Bridging the gap

Wherever the research effort is directed, however, there remains the problem of bridging the gap between research results and agricultural practice, a task that must fall mainly upon the extension services. But it is not always recognized that once the bridge has been established, the traffic over it should be encouraged to flow in both directions. Research done in response to specific questions from farmers is likely to be more profitable than research that leads to farmers being told what to do. Indeed, critical observations by farmers form an essential part of the system for advancing the body of agricultural knowledge. With so many interacting factors, it is impossible for the research worker to cover every possible combination of circumstances in any particular series of experiments. Hence, observations by farmers must continually be fed back to the research worker who can explore and interpret them and, where appropriate, incorporate them into future experiments.

It is for reasons such as these that research needs the stimulus of development just as much as development requires the results of research.

REFERENCES AND AUTHOR INDEX

*The sloping numbers at the end of each entry
refer to the page numbers in this book*

Addicott, F. T. (1970). Plant hormones in the control of abscission. *Biological Reviews, Cambridge*, **45**, 485–524. *104, 105*

Allan, W. (1965). *The African Husbandman*. Edinburgh: Oliver & Boyd. *43, 76*

Andries, J. A., Jones, J. E., Sloane, L. W. & Marshall, J. G. (1969). Effects of okra leaf shape on boll rot, yield and other important characters of upland cotton, *Gossypium hirsutum* L. *Crop Science*, **9**, 705–10. *236*

Anon. (1962). Minutes of the specialist committee on applied meteorological research. Muguga, Nairobi: East African Agriculture and Forestry Organization (mimeo). *30*

Anon. (1965). Minutes of the specialist committee on applied meteorological research. Mugaga, Nairobi: East African Agriculture and Forestry Organization (mimeo). *30*

Arnold, M. H. (1963). The control of bacterial blight in raingrown cotton. I. Breeding for resistance in African upland varieties. *Journal of Agricultural Science, Cambridge*, **60**, 415–27. *156, 179, 182, 184, 211*

(1965). The control of bacterial blight in raingrown cotton. II. Some effects of infection on growth and yield. *Journal of Agricultural Science, Cambridge*, **65**, 29–40. *161*

(1967). *Progress Reports from Experiment Stations, Uganda, 1965–66*. London: Cotton Research Corporation. *10, 128, 251*

(1968). *Progress Reports from Experiment Stations, Uganda, 1967–68*. London: Cotton Research Corporation. *81*

(1969a). *Progress Reports from Experiment Stations, Uganda, 1968–69*. London: Cotton Research Corporation. *38, 83*

(1969b) Developments and prospects in cotton quality. *1er Symposium International de la Recherche Textile Cotonnière*, pp. 23–38. Paris: Institute Textile de France. *239*

(1970a). *Progress Reports from Experiment Stations, Uganda, 1969–70*. London: Cotton Research Corporation. *94*

(1970b). Cotton improvement in East Africa. In *Crop Improvement in East Africa* (ed. C. L. A. Leakey), chap. 7, pp. 178–208. Farnham Royal: Commonwealth Agricultural Bureaux. *102, 163, 179, 182, 194, 235, 304*

(1970c). Cotton: A. The origins and characteristics of Uganda varieties. In *Agriculture in Uganda* (ed. J. D. Jameson), 2nd edn, chap. 12, pp. 154–65. London: Oxford University Press. *82, 102, 241*

(1971). *Progress Reports from Experiment Stations, Uganda, 1970–71*. London: Cotton Research Corporation. *Table 4.1, 111*

(1972). Modal selection in BP 52. *Cotton Growing Review*, **49**, 107–25. *112, 113, 209, 226*

Arnold, M. H. & Arnold, K. M. (1959). *Progress Reports from Experiment Stations, Tanganyika Territory, Lake Province, 1957–58*. London: Empire Cotton Growing Corporation. *192*

(1961). Bacterial blight of cotton: trash-borne infection. *Empire Cotton Growing Review*, **38**, 258–70.   *153*

Arnold, M. H. & Brown, S. J. (1968). Variation in the host–parasite relationship of a crop disease. *Journal of Agricultural Science, Cambridge*, **71**, 19–36.   *40, 156, 180, Figure 7.1, 182, 190, 191, Figure 7.6*

Arnold, M. H. & Church, J. M. F. (1967). *Progress Reports from Experiment Stations, Uganda, 1965–66*. London: Cotton Research Corporation.   *215*

Arnold, M. H., Costelloe, B. E. & Church, J. M. F. (1968). BPA and SATU. Uganda's two new cotton varieties. *Cotton Growing Review*, **45**, 162–74.   *112, 155, 212, 215*

Arnold, M. H., Innes, N. L. & Gridley, H. E. (1971). *Progress Reports from Experiment Stations, Uganda, 1970–71*. London: Cotton Research Corporation.   *197, 235*

Arnold, M. H., Smithson, J. B. & Tollervey, F. E. (1964). *Progress Reports from Experiment Stations, Uganda, 1963–64*. London: Empire Cotton Growing Corporation.   *211*

Arnold, M. H., Walker, J. T. & Tollervey, F. E. (1966). *Progress Reports from Experiment Stations, Uganda, 1964–65*. London: Empire Cotton Growing Corporation.   *163*

Ashley, D. A., Doss, B. D. & Bennett, O. L. (1965). Relation of cotton leaf area index to plant growth and fruiting. *Agronomy Journal*, **57**, 61–4.   *115*

Ashton, J. (1973). Agriculture in developed countries: competition for resources. *Philosophical Transactions of the Royal Society of London, B*, **267**, 13–22.   *320*

Atlas of Uganda (1962, 1967). Uganda: Department of Lands and Surveys.   *22, 26, 31*

Austin, R. B. (1963). A study of the growth and yield of carrots in a long-term manurial experiment. *Journal of Horticultural Science*, **38**, 244–76.   *112*

Bagga, H. S. & Laster, M. L. (1968). A simple technique for evaluating the role of insects in cotton boll rot development. *Phytopathology*, **58**, 1323–4.   *147*

Baker, D. N. & Hesketh, J. D. (1969). Respiration and the carbon balance in cotton (*Gossypium hirsutum* L.). *Beltwide Cotton Production Research Conferences*, 1969, pp. 60–4. Memphis: National Cotton Council, USA.   *96*

Baker, D. N., Hesketh, J. D. & Duncan, W. G. (1972). Simulation of growth and yield in cotton. 1. Gross photosynthesis, respiration and growth. *Crop Science*, **12**, 31–5.   *96*

Balls, W. L. (1912). *The cotton plant in Egypt*. London: Macmillan.   *97*

Basinski, J. J., Beech, D. F., Evenson, J. P. & Wetselaar, R. (1971). Cotton responses to nitrogen: effect of land pre-treatments and fertilizer applications. *Cotton Growing Review*, **48**, 175–93.   *121*

Bierhuizen, J. F. & Slatyer, R. O. (1964). Photosynthesis of cotton leaves under a range of environmental conditions in relation to internal and external diffusive resistances. *Australian Journal of Biological Sciences*, **17**, 348–59.   *95*

Birch, H. F. (1960). Nitrification in soils after different periods of dryness. *Plant and Soil*, **12**, 81–96.   *58*

Bird, L. S. (1950). The bacterial blight disease of cotton. M.S. thesis, A. and M. College of Texas.   *156*

(1974). *Cotton. Breeding Plants for Disease Resistance; Concepts and Applications* (ed. R. R. Nelson). London: Pennsylvania State University Press.   *164*

Bird, L. S. & Hadley, H. H. (1958). A statistical study of the inheritance of Stoneville 20 resistance to the bacterial blight disease of cotton in the presence of *Xanthomonas malvacearum* races 1 and 2. *Genetics*, **43**, 750–67.   *Table 7.1, 178*

Bleasdale, J. K. A. (1966). Plant growths and crop yield. *Annals of Applied Biology*, **57**, 173–82.   *81*

Blood, D. C. & Henderson, J. A. (1968). *Veterinary Medicine*, 3rd edn, p. 791. London: Baillière, Tindall & Cassell.   *295*

Boehring, R. H. & Burnside, C. A. (1956). The effect of light intensity on rate of apparent photosynthesis in leaves of sun and shade plants. *American Journal of Botany*, **43**, 557–61. *95*

Boerma, A. H. (1973). The world food and agricultural situation. *Philosophical Transactions of the Royal Society of London, B*, **267**, 5–12. *331*

Bouyoucos, G. F. (1949). Nylon electrical resistance unit for continuous measurement of soil moisture in the field. *Soil Science*, **67**, 319–30. *91*

Bowden, J. & Ingram, W. R. (1958). A revised interpretation of the causes of loss of crops of cotton in the drier regions of Uganda. *Nature, London*, **182**, 1750. *137*

Bowden, J. & Thomas, D. G. (1970). Cotton: B. Local aspects. In *Agriculture in Uganda* (ed. J. D. Jameson), 2nd edn, chap. 12, pp. 165–87. London: Oxford University Press. *116, 151*

Breese, E. L. (1969). The measurement and significance of genotype–environment interactions in grasses. *Heredity*, **24**, 27–44. *232*

Brinkerhoff, L. A. (1963). Variability of *Xanthomonas malvacearum:* the cotton bacterial blight pathogen. *Oklahoma State University Technical Bulletin*, T-98. *Table 7.1, 178*

 (1970). Variation in *Xanthomonas malvacearum* and its relation to control. *Annual Review of Phytopathology*, **8**, 85–110. *Table 7.1*

Brown, K. J. (1965). Response of three strains of cotton to flower removal. *Cotton Growing Review*, **42**, 279–86. *103*

 (1968). Translocation of carbohydrate in cotton: movement to fruiting bodies. *Annals of Botany*, **32**, 703–13. *100*

Brown, L. H. & Cocheme, J. (1969). *A Study of the Agroclimatology of the Highlands of Eastern Africa*. Rome: FAO. *31*

Brown, S. J. (1964). *Progress Reports from Experiment Stations, Uganda, 1963–64*. London: Empire Cotton Growing Corporation. *180*

 (1966). *Progress Reports from Experiment Stations, Uganda, 1964–65*. London: Empire Cotton Growing Corporation. *163*

 (1968). *Progress Reports from Experiment Stations, Uganda, 1967–68*. London: Cotton Research Corporation. *165*

 (1970). *Progress Reports from Experiment Stations, Uganda, 1969–70*. London: Cotton Research Corporation. *167, 170*

 (1971). *Progress Reports from Experiment Stations, Uganda, 1970–71*. London: Cotton Research Corporation. *156, 169, 170*

Brown, S. J. & Beteise-Hbenye, D. (1973). *Cotton Research Reports, Uganda, 1971–72*. London: Cotton Research Corporation. *169, 170, 193*

Brown, W. T., Evans-Jones, P. & Innes, D. (1970). Implements. In *Agriculture in Uganda* (ed. J. D. Jameson), 2nd edn, chap. 17, pp. 285–96. London: Oxford University Press. *331*

Brunet-Moret, Y. (1969). Etude des quelques lois statiques utilisées en hydrologie *Cahier ORSTOM, série Hydrologique*, **6**, 3. *24*

Bugbee, W. M. & Presley, J. T. (1967). A rapid inoculation technique to evaluate the resistance of cotton to *Verticillium albo-atrum*. *Phytopathology*, **57**, 1264. *165*

Burkitt, F. H. (1972). Cotton in a changing world. In *Cotton*, pp. 32–9. Technical Monograph No. 3. Basle: Ciba-Geigy Ltd. *239*

Carns, H. R., McMeans, J. E. & Addicott, F. T. (1959). An abscission accelerating hormone in cotton and some of its interactions with auxin. *Proceedings of the 9th International Botanical Congress, Montreal*, vol. II, p. 60. Ottawa: Runge Press Ltd. *104*

Channon, J. A. (1968). Memorandum ADU/1413 of 26 September 1968. Kampala: East African Meteorological Department. *29*

Christiansen, M. N. & Justus, N. (1963). Prevention of seed deterioration by an impermeable seed coat. *Crop Science*, **3**, 439–40. *87*

Church, J. M. F. (1968). *Progress Reports from Experiment Stations, Uganda, 1966–67.*
London: Cotton Research Corporation. *274*
  (1969). Evaluation of herbicides for use in cotton. *Cotton Growing Review,*
  46, 245–60. *272*
Church, J. M. F. & Henson, H. M. G. (1969). Chemical control of *Cyperus rotundus* and
Oxalis latifolia in Uganda. *PANS*, **15**, 578–83. *271, 272*
Coaker, T. H. (1956). An experiment on stem borer control on maize. *East African
Agricultural Journal*, **21**, 220–1. *148*
  (1957a). *Progress Reports from Experiment Stations, Uganda, 1956–57.* London:
  Empire Cotton Growing Corporation. *127, 131*
  (1957b). Studies of crop loss following insect attack on cotton in East Africa. II.
  Further experiments in Uganda. *Bulletin of Entomological Research*,
  48, 851–66. *139*
  (1958a). Experiments with a virus disease of the cotton bollworm, *Heliothis armigera*
  (Hbn). *Annals of Applied Biology*, **46**, 536–41. *143*
  (1958b). Palatability of maize under storage in Uganda. *East African Agricultural
  Journal*, **24**, 57–60. *148*
  (1959a). Investigations on *Heliothis armigera* in Uganda. *Bulletin of
  Entomological Research*, **50**, 487–506. *142*
  (1959b). In sack treatment of maize with insecticide for protection against storage
  pests in Uganda. *East African Agricultural Journal*, **24**, 244–57. *148*
Coaker, T. H. & Passmore, R. G. (1958). *Stomoxys* sp. upon cattle in Uganda. *Nature,
London*, **182**, 606–7. *148, 285*
Cognée, M. (1968). Abscission of cotton fruiting organs. *Coton et Fibres Tropicales*,
23, 315–36. *103*
Colonial Insecticides, Fungicides and Herbicides Committee: Fungicides Sub-committee
  (1951). *Survey of the Position Regarding Plant Diseases in the Colonial Empire and
  Other Tropical and Sub-tropical Territories and the Possibilities of their Control.*
  London: Colonial Office. *157*
Costelloe, B. E. & Riggs, T. J. (1967). *Progress Reports from Experiment Stations,
Uganda, 1965–66.* London: Cotton Research Corporation. *220*
Cross, J. E. (1963). Pathogenicity differences in Tanganyika populations of
Xanthomonas malvacearum. Empire Cotton Growing Review, **40**, 125–9. *180*
  (1964). Field differences in pathogenicity between Tanganyika populations of
  Xanthomonas malvacearum. Empire Cotton Growing Review, **41**, 44–8. *180*
Dagg, M., Woodhead, T. & Rijks, D. A. (1970). Evaporation in East Africa. *Bulletin of
the International Association of Scientific Hydrology*, **15**, 1–3. *31*
Dale, J. E. (1957). *Progress Reports from Experiment Stations, Uganda, 1956–57.*
London: Empire Cotton Growing Corporation. *103*
  (1958). *Progress Reports from Experiment Stations, Uganda, 1957–58.* London: Cotton
  Research Corporation. *103*
  (1959). Some effects of the continuous removal of floral buds on the growth of the
  cotton plant. *Annals of Botany*, **23**, 636–49. *97*
  (1960). *Progress Reports from Experiment Stations, Uganda, 1958–59.* London:
  Empire Cotton Growing Corporation. *28, 35*
  (1961). Investigations into the stomatal physiology of upland cotton. The effects of
  hour of day, solar radiation, temperature and leaf water content on stomatal
  behaviour. *Annals of Botany*, **25**, 39–52. *91, 96*
  (1962). Fruit shedding and yield in cotton. *Empire Cotton Growing Review*,
  39, 170–6. *103, 108*
  (1965). Abscission of cotton pedicels following removal of the boll and bracts. *Empire
  Cotton Growing Review*, **42**, 52–5. *104*

Dale, J. E. & Coaker, T. H. (1958). Some effects of feeding by *Lygus vosseleri* on the stem apex of the cotton plant. *Annals of Applied Biology,* **46**, 423–9. *138*

(1961). Growth and yield of cotton sprayed with DDT in east and north Uganda. *Empire Journal of Experimental Agriculture,* **29**, 1–13. *139*

Dale, J. E. & Milford, G. F. (1965). The role of endogenous growth substances in the fruiting of upland cotton. *New Phytologist,* **64**, 28–37. *104*

Davies, J. C. (1970). *Annual Report. Part II.* Kampala: Department of Agriculture. *136*

D'Hoore, J. L. (1964). *Soil Map of Africa* (scale 1–5,000,000). Lagos: Commission for Technical Cooperation in Africa. *75*

Doggett, H. (1970). Sorghum improvement in East Africa. In *Crop Improvement in East Africa* (ed. C. L. A. Leakey), chap. 2, pp. 60–87. Farnham Royal: Commonwealth Agricultural Bureaux. *305*

Donald, C. M. (1968). The breeding of crop ideotypes. *Euphytica,* **17**, 385–403. *113*

Druyff, A. H. & Kerkhoven, G. J. (1970). Effect of efficient weeding on yields of irrigated cotton in eastern Kenya. *PANS,* **16**, 596–605. *273*

Dunbar, A. R. & Stephens, D. (1970). Social background. In *Agriculture in Uganda* (ed. J. D. Jameson), 2nd edn, chap. 8, pp. 98–108. London: Oxford University Press. *320*

East African Meteorological Department (1959). 10% and 20% probability maps of annual rainfall of East Africa. Nairobi: East African Meteorological Department. *31*

(1961). 10% and 20% probability maps of annual rainfall of East Africa. Nairobi: East African Meteorological Department. *22, 31*

Eberhart, S. A. & Russell, W. A. (1966). Stability parameters for comparing varieties. *Crop Science,* **6**, 36–40. *231*

El Sharkway, M. & Hesketh, J. D. (1965). Photosynthesis among species in relation to characteristics of leaf anatomy and CO_2 diffusion resistance. *Crop Science,* **5**, 17–21. *95*

El Sharkaway, M., Hesketh, J. D. & Muramoto, H. (1965). Leaf photosynthetic rates and other growth characteristics among 27 species of *Gossypium. Crop Science,* **5**, 173–5. *95*

Erwin, D. C., Moje, W. & Malca, I. (1965). An assay of the severity of Verticillium wilt on cotton plants inoculated by stem puncture. *Phytopathology,* **55**, 663–5. *165*

Evenson, J. P. (1955). Botanical studies in cotton quality. 1. Morphological factors affecting proneness to nep. *Empire Cotton Growing Review,* **32**, 157–67. *239*

(1960). Intraseasonal variations in boll characteristics in African Upland cotton. *Empire Cotton Growing Review,* **37**, 161–77. *99, 104*

Farbrother, H. G. (1951). A note on the measurement of rainfall intensity. *East African Agricultural Journal,* **17**, 82–4. *25*

(1952). *Progress Reports from Experiment Stations, Uganda, 1951–52.* London: Empire Cotton Growing Corporation. *95, 115*

(1953). *Progress Reports from Experiment Stations, Uganda, 1952–53.* London: Empire Cotton Growing Corporation. *103*

(1954). *Progress Reports from Experiment Stations, Uganda, 1953–54.* London: Empire Cotton Growing Corporation. *30, 81, 87, 97, 103*

(1955). *Progress Reports from Experiment Stations, Uganda, 1954–55.* London: Empire Cotton Growing Corporation. *35*

(1956). *Progress Reports from Experiment Stations, Uganda, 1955–56.* London: Empire Cotton Growing Corporation. *35, 51, 87, 91, 92, 109, Figure 4.7*

(1957). On an electrical resistance technique for the study of soil moisture problems in the field. *Empire Cotton Growing Review,* **34**, 1–19. *33, 91*

(1958). *Progress Reports from Experiment Stations, Uganda, 1957–58.* London: Empire Cotton Growing Corporation. *35, 87, 94*

(1960*a*). Developments in farm mechanization at Namulonge. *Empire Cotton Growing Review*, **37**, 274–9.　*248*

(1960*b*). *Progress Reports from Experiment Stations, Uganda, 1958–59.* London: Empire Cotton Growing Corporation.　*95, 115*

(1960*c*). *Progress Reports from Experiment Stations, Uganda, 1959–60.* London: Empire Cotton Growing Corporation.　*28, 34, 87*

(1961*a*). A note on the air-conditioned laboratory for lint testing instruments at Namulonge. *Empire Cotton Growing Review*, **38**, 42–7.　*239*

(1961*b*). *Progress Reports from Experiment Stations, Uganda, 1960–61.* London: Empire Cotton Growing Corporation.　*35, 87, 97, 98*

(1962). *Progress Reports from Experiment Stations, Uganda, 1961–62.* London: Empire Cotton Growing Corporation.　*34*

(1964). *Progress Reports from Experiment Stations, Uganda, 1963–64.* London: Empire Cotton Growing Corporation.　*91*

Farbrother, H. G. & Munro, J. M. (1970). Water. In *Agriculture in Uganda* (ed. J. D. Jameson), 2nd edn, chap. 4, pp. 30–42. London: Oxford University Press.　*116*

Fergus, E. N. (1936). Shall crops be adapted to soils and soils to crops? *Journal of the American Society of Agronomists*, **28**, 443–6.　*78*

Finlay, K. W. & Wilkinson, G. N. (1963). The analysis of adaption in a plant breeding programme. *Australian Journal of Agricultural Research*, **14**, 742–54.　*181, 231*

Fitzpatrick, E. A. & Nix, H. A. (1969). A model of simulating soil water regimes in alternating fallow–crop systems. *Agricultural Meteorology*, **6**, 303–19.　*90*

Foster, H. L. (1969). *Proceedings of the specialist committee on soil fertility and crop nutrition.* East African Agricultural and Forestry Organization.　*65*

(1972). The identification of potentially potassium deficient soils in Uganda. *East African Agricultural and Forestry Journal*, **37**, 224–33.　*59*

Frazer, Helen L. (1944). Observations on the method of transmission of internal boll disease of cotton by the cotton stainer-bug. *Annals of Applied Biology*, **31**, 271–90.　*170*

Freeman, G. H. & Perkins, Jean M. (1971). Environmental and genotype–environmental components of variability. VIII. Relations between genotypes grown in different environments and measures of these environments. *Heredity*, **27**, 15–23.　*231*

Fryer, J. D. (1960). Field evaluation of selective herbicides. *Span*, **3**, 53–7.　*274*

Gates, C. T. (1968). Water deficits and growth of herbaceous plants. In *Water Deficits and Plant Growth* (ed. T. T. Kozlowski), pp. 135–90. New York: Academic Press.　*92*

Geering, Q. A. (1953). Studies of *Lygus vosseleri*, Popp. (Miridae), a pest of cultivated cotton in East and Central Africa. I. A method for breeding continuous supplies in the laboratory. *Bulletin of Entomological Research*, **44**, 351–62.　*139*

(1956). A method for controlling breeding of cotton stainers. *Dysdercus* spp. (*Pyrhocoridae*). *Bulletin of Entomological Research*, **46**, 743–6.　*141*

Geering, Q. A. & Coaker, T. H. (1960). The effects of different plant foods on the fecundity, fertility and development of the cotton stainer, *Dysdercus superstitiosus*. *Bulletin of Entomological Research*, **51**, 61–75.　*141*

Geering, Q. A., McKinlay, K. S. & Coaker, T. H. (1954). *Progress Reports from Experiment Stations, Uganda, 1953–54.* London: Empire Cotton Growing Corporation.　*143*

Giraud, J. M. (1973). *Recherches des cycles dans les pluies annuelles de Dakar (1901–1972) et du Sénégal (1924–1972).* Publication No. 31, ASECNA, Dakar.　*25*

Glover, J. & Robinson, P. (1953). A simple method for assessing the reliability of rainfall. *Journal of Agricultural Science, Cambridge*, **43**, 275–80.　*24*

Goebel, S. (1968). Travaux de sélection effectués sur les triples hybrides d'origine interspécifique HAR et ATH en Côte d'Ivoire (Station de Bouaké). *Coton et Fibres Tropicales*, **23**, 212–18. *237*

Green, L. (1960). A note on cotton seed dressing in Uganda. *Empire Cotton Growing Review*, **37**, 256–7. *157*

Green, J. M. & Brinkerhoff, L. A. (1956). Inheritance of three genes for bacterial blight resistance in Upland cotton. *Agronomy Journal*, **48**, 481–5. *Table 7.1, 178*

Griffin, D. M. & Nair, N. G. (1968). Growth of *Sclerotium rolfsii* at different concentrations of oxygen and carbon dioxide. *Journal of Experimental Botany*, **19**, 812–16. *169*

Griffiths, J. F. (1968). The climate of East Africa. In *Natural Resources of East Africa* (ed. E. W. Russell), pp. 77–87. Nairobi: Hawkins. *17*

(ed.) (1972). *Climates of Africa*. Amsterdam: Elsevier. *17*

Gubanova, L. G. & Gubanova, G.Ya (1961). The physiology of the dormant period of cotton seed. *Agrobiology*, **130**, 537–43. *87*

Gwynn, A. M. (1936). In *Annual Report of the Department of Agriculture, Uganda*, pp. 12–18. Entebbe: Department of Agriculture. *139*

Hackett, C. (1973). An exploration of the carbon economy of the tobacco plant. I. Inferences from a simulation. *Australian Journal of Biological Sciences*, **26**, 1057–71. *96*

Hancock, G. L. R. (1935). Notes on *Lygus symonyi* Rent. (Capsidae) a cotton pest in Uganda. *Bulletin of Entomological Research*, **26**, 429–38. *137*

Hansford, C. G. (1929). Cotton diseases in Uganda, 1926–28. *Empire Cotton Growing Review*, **6**, 10–26. *166*

(1933). In *Report of the Department of Agriculture, Uganda, 1932*, pp. 55–73. Entebbe: Department of Agriculture. *153, 160, 161*

(1934). In *Report of the Department of Agriculture, Uganda, 1933*, pp. 52–9. Entebbe: Department of Agriculture. *160, 161*

(1938). In *Report of the Department of Agriculture, Uganda, 1936–37*, Entebbe: Department of Agriculture. *164*

Hansford, C. G. & Hosking, H. R. (1938). Recent research in Uganda on blackarm disease. *Empire Cotton Growing Review*, **15**, 7–13. *153*

Hanson, W. D. (1959). The breakup of initial linkage blocks under selected mating systems. *Genetics*, **44**, 867–8. *236*

Harris, P. M. & Farazdaghi, H. (1969). The effect of plant population in the sugar beet crop on the depths of extraction of water from the soil profile. In *Root Growth* (ed. W. T. Whittington), sect. v, pp. 386–7. London: Butterworths. *90*

Harrop, J. F. (1962). Uganda soils. In *Agriculture in Uganda* (ed. J. D. Jameson), 2nd edn, end pocket. London: Oxford University Press. *47*

Hawkins, J. C. (1959). The N.I.A.E. in East Africa. *Empire Cotton Growing Review*, **36**, 35–40. *248*

Hayward, A. C. (1964). Bacteriophage sensitivity and biochemical group in *Xanthomonas malvacearum*. *Journal of General Microbiology*, **35**, 287–98. *180*

Hearn, A. B. (1967). *Progress Reports from Experiment Stations, Uganda, 1965–66*. London: Cotton Research Corporation. *35, 40*

(1968). *Progress Reports from Experiment Stations, Uganda, 1966–67*. London: Cotton Research Corporation. *90, Table 4.1*

(1969). Growth and performance of cotton in a desert environment. I. Morphological development II. Dry matter production III. Crop performance. *Journal of Agricultural Science, Cambridge*, **72**, 65–97. *35, 82, 94, 98, 100, 120*

(1970). *Progress Reports from Experiment Stations, Uganda, 1969–70*. London: Cotton Research Corporation. *Table 3.2, 87, 94*

(1972a). Cotton spacing experiments in Uganda. *Journal of Agricultural Science, Cambridge*, **78**, 13–25. *82, 99, 116*

(1972b). The growth and performance of rain-grown cotton in a tropical upland environment. *Journal of Agricultural Science, Cambridge*, **79**, 121–45.   *51, 82, 87, 91, 92, 94, 98, 99, 100, 109, 116, Figure 4.7*

Hesketh, J. D., Baker, D. N. & Duncan, W. G. (1971). Simulation of growth and yield in cotton: respiration and the carbon balance. *Crop Science*, **11**, 394–8.   *96, Figure 4.7*

Hesketh, J. D. & Low, A. (1968). Effect of temperature on components of yield and fibre quality of cotton varieties of diverse origins. *Cotton Growing Review*, **45**, 243–57.   *99, Figure 4.7*

Holmes, E. (1955). The contribution of commerce to crop protection. *Annals of Applied Biology*, **42**, 325–32.   *157*

Hopkins, J. C. F. (1931). *Alternaria gossypina* (Thum.) Comb. nov. causing a leaf spot and boll rot of cotton. *Transactions of the British Mycological Society*, **16**, 136–44.   *162*

Horrell, C. R. & Tiley, G. E. D. (1970). Grassland. In *Agriculture in Uganda* (ed. J. D. Jameson), 2nd edn, chap. 19, pp. 318–32. London: Oxford University Press.   *290*

Hughes, L. C. (1966). Factors affecting numbers of ovules per loculus in cotton. *Cotton Growing Review*, **43**, 273–85.   *99*

Humphries, E. C. (1967). The dependence of photosynthesis on carbohydrate sinks: current concepts. *Proceedings 1st International Symposium on Tropical Root Crops, Trinidad*, pp. 34–45.   *96*

Hunter, R. E., Brinkerhoff, L. A. & Bird, L. S. (1968). The development of a set of upland cotton lines for differentiating races of *Xanthomonas malvacearum*. *Phytopathology*, **58**, 830–2.   *190*

Hutchinson, J. B. (1950). *Progress Reports from Experiment Stations, Uganda, 1949–50.* London: Empire Cotton Growing Corporation.   *78*

(1954). *Progress Reports from Experiment Stations, Uganda, 1953–54.* London: Empire Cotton Growing Corporation.   *249*

(1959). *The Application of Genetics to Cotton Improvement.* London: Empire Cotton Growing Corporation and Cambridge University Press.   *179, 235*

(1972). *Farming and Food Supply. The Interdependence of Countryside and Town.* London: Cambridge University Press.   *323*

Hutchinson, J. B. & Lawes, D. A. (1953). A note on the estimation of natural crossing in cotton. *Empire Cotton Growing Review*, **30**, 192–3.   *197*

Hutchinson, J. B., Manning, H. L. & Farbrother, H. G. (1958a). On the characterisation of tropical rainstorms in relation to run off and percolation. *Quarterly Journal of the Royal Meteorological Society*, **84**, 250–8.   *25, 38*

(1958b). Crop water requirement of cotton. *Journal of Agricultural Science, Cambridge*, **51**, 177–88.   *33, 34, 78, 81, 83, 87, 90, 92, 101, 113, 114*

Hutchinson, J. B. & Panse, V. G. (1937). Studies in plant breeding techniques. ii. The design of field tests of plant breeding material. *Indian Journal of Agricultural Science*, **7**, 531–564.   *203*

Hutchinson, Sir Joseph, Prentice, A. N., Farbrother, H. G., Lea, J. D. & Stephens, A. L. (1959). The planning of a large farm for the Namulonge Cotton Research Station in Uganda. *Empire Cotton Growing Review*, **36**, 81–134.   *247, 249, 285*

Hutchinson, Sir Joseph & Ruston, D. F. (1965). The Empire Cotton Growing Corporation and the organization of research on raw cotton. In *The Organization of Research Establishments* (ed. Sir John Cockcroft), chap. 8, pp. 129–47. London: Cambridge University Press.   *1*

Huxley, P. A. (1964). Some effects of artificial shading on the growth of upland cotton seedlings. *Empire Cotton Growing Review*, **41**, 100–11.   *95, 96, 115, Figure 4.7*

Ingram, W. R. (1962). Effect of 'Rogor' (Dimethoate) and DDT on cotton mites in Uganda. *Nature, London*, **195**, 1224–5.   *131*

(1965). A survey of the cotton grown in Uganda in 1963. *Empire Cotton Growing Review*, **42**, 1–14. *134*

(1969). Further studies of crop loss following insect attack on cotton in Uganda. *Bulletin of Entomological Research*, **59**, 65–76. *127*

Innes, N. L. (1963). Resistance to bacterial blight of cotton in Albar. *Nature, London*, **200**, 387–8. *178*

(1964). Sudan strains of cotton resistant to bacterial blight. *Empire Cotton Growing Review*, **41**, 285–91. *245*

(1965*a*). Resistance to bacterial blight of cotton: the genes B_9 and B_{10}. *Experimental Agriculture*, **1**, 189–91. *Table 7.1, 178*

(1965*b*). Inheritance of resistance to bacterial blight of cotton. I. Allen (*Gossypium hirsutum*) derivatives. *Journal of Agricultural Science, Cambridge*, **64**, 257–71. *177, Table 7.1, 178*

(1966). Inheritance of resistance to bacterial blight of cotton. III. *Herbaceum* resistance transferred to tetraploid cotton. *Journal of Agricultural Science, Cambridge*, **66**, 433–9. *176, Table 7.1*

(1968). Resistance to bacterial blight in cotton. Ph.D. thesis, University of Aberdeen. *184, Figure 7.3*

(1969*a*). Inheritance of resistance to bacterial blight of cotton. IV. Tanzania selections. *Journal of Agricultural Science, Cambridge*, **72**, 41–57. *Table 7.1, 179*

(1969*b*). Breeding for increased yields in Uganda. In *Seminar on Cotton Production Research* (*Breeding for Improved Yields in Modern Commercial Varieties of Cotton*) *Kampala, Uganda*, pp. 19–37. Washington DC 20250, USA: Secretariat International Cotton Advisory Committee. *Figure 8.15*

(1971). Impressions of cotton production and research in India. *Cotton Growing Review*, **48**, 163–74. *197*

(1973*a*). Selection for fibre characters in upland cotton. *Cotton Growing Review*, **50**, 101–105. *240*

(1973*b*). Promising selections from intervarietal crosses at Namulonge. *Cotton Growing Review*, **50**, 296–306. *222, 234, 235*

(1974*a*). Genetic variability in Albar 51. *Cotton Growing Review*, **51**, 16–25. *220, 235*

(1974*b*). Resistance to bacterial blight of cotton varieties homozygous for combinations of B resistance genes. *Annals of Applied Biology*, **78**, 89–98. *176, 182*

(1975). Upland cotton of triple hybrid origin. *Cotton Growing Review*, **51**, 46–58. *237*

Innes, N. L. & Brown, S. J. (1969). A quantitative study of the inheritance of resistance to bacterial blight in Upland cotton. *Journal of Agricultural Science, Cambridge*, **73**, 15–23. *Table 7.1, 186*

Innes, N. L., Brown, S. J. & Walker, J. T. (1974). Genetical and environmental variation for resistance to bacterial blight of upland cotton. *Heredity*, **32**, 53–71. *186, Figure 7.4*

Innes, N. L. & Busuulwa, I. N. (1973). Plant hairiness in Allen cotton (*Gossypium hirsutum*). *East African Agriculture and Forestry Journal*, **38**, 298–302. *194*

Innes, N. L. & Jones, G. B. (1972). Allen. A source of successful African cotton varieties. *Cotton Growing Review*, **49**, 201–15. *219, Figure 8.10*

Isaac, I. (1967). Speciation in *Verticillium*. *Annual Review of Phytopathology*, **5**, 201–22. *164*

Jameson, J. D. (1952). Report on blackarm disease of cotton. *Records of Investigations by the Department of Agriculture, Uganda*, **2**, 1–3. *153*

Jameson, J. D. & McCallum, D. (1970). Climate. In *Agriculture in Uganda* (ed. J. D. Jameson), 2nd edn, chap. 2, pp. 12–23. London: Oxford University Press. *17*

Jameson, J. D. & Thomas, D. G. (1951). *Progress Reports from Experiment Stations, Uganda, 1950–51*. London: Empire Cotton Growing Corporation. *157*

(1953). The control of blackarm disease in cotton. *Records of Investigations by the Department of Agriculture, Uganda*, **3**, 1–17.   *154, 157*

Johnson, R. E. & Addicott, F. T. (1967). Boll retention in relation to leaf and boll development in cotton. *Crop Science*, **7**, 571–4.   *104*

Jones, E. (1967). *Progress Reports from Experiment Stations, Uganda, 1966–67.* London: Cotton Research Corporation.   *61*

(1968). Nutrient cycle and soil fertility on red ferrallitic soils. *Transactions of the 9th Congress of the International Soil Science Society*, **3**, 419–27.   *46*

(1969). *Progress Reports from Experiment Stations, Uganda, 1968–69.* London: Cotton Research Corporation.   *74*

(1971). *Progress Reports from Experiment Stations, Uganda, 1970–71.* London: Cotton Research Corporation.   *46*

(1972). Principles for using fertilizer to improve red ferralitic soils in Uganda. *Experimental Agriculture*, **8**, 315–332.   *43, 46, 61, 67*

(1973). Improving the SATU variety in Uganda. *Cotton Growing Review*, **50**, 218–24.   *220, 243*

Jones, G. B. & Fielding, J. (1971). *Cotton Research Reports, Uganda, 1970–71.* London: Cotton Research Corporation.   *222*

Jones, J. E. & Andries, J. A. (1969). Effect of frego bract on the incidence of cotton boll rot. *Crop Science*, **9**, 426–8.   *236*

Jones, M. A. H. (1969). *Progress Reports from Experiment Stations, Uganda, 1968–69.* London: Cotton Research Corporation.   *74*

Jong, E. de (1973). *Technical Assistance Report*, No. 813. Geneva: International Atomic Energy Agency.   *Table 3.2*

Joyce, R. J. V. (1956). Insect mobility and design of field experiments. *Nature, London*, **177**, 282.   *126*

Joyce, R. J. V. & Roberts, P. (1959). The determination of the size of plot suitable for cotton spraying experiments in the Sudan Gezira. *Annals of Applied Biology*, **47**, 287–305.   *126*

Juton, M. (1971). Le régime du Fleuve Sénégal de 1965 a 1972. *Etude hydroagricole du Bassin du Fleuve Sénégal*. Dakar: Rapport UNDP.   *25*

Kabaara, A. M. (1965). *Progress Reports from Experiment Stations, Uganda, 1964–65.* London: Empire Cotton Growing Corporation.   *60*

Kammacher, P. (1965). *Etude des relations génétiques et carylogiques entre genomes voisins du genre* Gossypium. Paris. IRCT.   *237*

(1968). Emploi des hybrides d'espèces dans l'amélioration du cotonnier. *Coton et Fibres Tropicales*, **23**, 207–11.   *237*

Kerr, T. (1969). The trispecies hybrid ancestry of high strength cottons. *Beltwide Cotton Production Research Conferences*, 1969, pp. 82–3. Memphis: National Cotton Council, USA.   *237*

Kibukamusoke, D. E. B. (1958). A note on a more precise method of estimation of the Uganda cotton crop. *Cotton Growing Review*, **35**, 91–9.   *83*

(1960). *Progress Reports from Experiment Stations, Uganda, 1958–59.* London: Empire Cotton Growing Corporation.   *30*

(1962). Competitive effects of coffee on cotton production in Buganda. *Cotton Growing Review*, **39**, 106–13.   *151*

King, R. W., Wardlaw, I. F. & Evans, L. T. (1967). Effect of assimilate utilization on photosynthetic rate in wheat. *Planta*, **77**, 261–76.   *96*

Kintukwonka, A. F. (1972). Potassium fixation and release in Namulonge soils, Uganda. M. Agric. Sc. thesis, Reading University.   *49, 59*

Knight, R. L. (1944). The genetics of blackarm resistance. IV. *Gossypium punctatum* Sch. and Thon. crosses. *Journal of Genetics*, **46**, 1–27.   *Table 7.1*

(1946). Breeding cotton resistant to blackarm disease. Part II. Breeding methods. *Empire Journal of Experimental Agriculture*, **14**, 161–74. *178*

(1948a). The role of major genes in the evolution of economic characters. *Journal of Genetics*, **48**, 370–87. *193*

(1948b). The genetics of blackarm resistance. VII. *Gossypium arboreum* L. *Journal of Genetics*, **49**, 109–66. *Table 7.1*

(1950). The genetics of blackarm resistance. VIII. *Gossypium barbadense*. *Journal of Genetics*, **50**, 67–76. *Table 7.1*

(1952). The genetics of jassid resistance in cotton. I. The genes H_1 and H_2. *Journal of Genetics*, **51**, 47–66. *194*

(1953a). The genetics of blackarm resistance. IX. The gene B_{6m} from *Gossypium arboreum*. *Journal of Genetics*, **51**, 270–5. *Table 7.1*

(1953b). The genetics of blackarm resistance. X. The gene B_7 from Stoneville 20. *Journal of Genetics*, **50**, 515–19. *Table 7.1*

(1954). The genetics of blackarm resistance. XI. *Gossypium anomalum*. *Journal of Genetics*, **52**, 466–72. *Table 7.1*

(1957). Blackarm disease of cotton and its control. *Plant Protection Conference, 1956*, p. 53. London: Butterworth. *176, Table 7.1*

(1963). The genetics of blackarm resistance. XII. Transference of resistance from *Gossypium herbaceum* to *G. barbadense*. *Journal of Genetics*, **58**, 328–46. *176, Table 7.1*

Knight, R. L. & Clouston, T. W. (1939). The genetics of blackarm resistance. I. Factors B_1 and B_2. *Journal of Genetics*, **38**, 133–58. *154, 156, 176, Table 7.1, 178*

Kriedeman, P. (1966). The photosynthetic activity of the wheat ear. *Annals of Botany*, **30**, 349–63. *100*

Lagière, R. (1960). *La bactériose du cotonnier dans le monde et en République Centrafricaine* (Oubanquichari). Paris: IRCT. *156, 176, Table 7.1*

Lakhani, D. (1973). *Cotton Research Reports, Uganda, 1971–72*. London: Cotton Research Corporation. *26*

Lamb, H. H. (1966). Climate in the 1960s. *Geographical Journal*, **132**, 183–212. *25*

Lea, J. D. & Joy, J. L. (1963). The development of modern arable farming in Uganda. *Empire Journal of Experimental Agriculture*, **31**, 137–51. *295*

Leakey, C. L. A. & Perry, D. A. (1966). The relation between damage caused by insect pests and boll rot associated with *Glomerella cingulata* (Stonem.) Spauld and von Schrenk (*Colletotrichum gossypii* Southq.) on upland cotton in Uganda. *Annals of Applied Biology*, **37**, 337–44. *141, 170*

Le Mare, P. H. (1953). *Progress Reports from Experiment Stations, Uganda, 1953–54*. London: Empire Cotton Growing Corporation. *44, 60, 66*

(1957). *Progress Reports from Experiment Stations, Uganda, 1956–57*. London: Empire Cotton Growing Corporation. *60, 73*

(1961). *Progress Reports from Experiment Stations, Uganda, 1960–61*. London: Empire Cotton Growing Corporation. *61*

(1963). *Progress Reports from Experiment Stations, Uganda, 1962–63*. London: Empire Cotton Growing Corporation. *61*

(1968). Experiments on the effects of phosphates applied to a Buganda soil. *Journal of Agricultural Science, Cambridge*, **70**, 265–85. *60, 67*

Leonard, E. R. (1962). Inter-relations of vegetative and reproductive growth with special reference to indeterminate plants. *Botanical Review*, **28**, 353–410. *96*

Logan, C. (1958). Bacterial boll rot of cotton (*Xanthomonas malvacearum* E. F. Smith, Dowson). I. A comparison of two inoculation techniques for the assessment of host resistance. *Annals of Applied Biology*, **46**, 230–42. *155*

(1960). An estimate of the effect of seed treatments in reducing cotton crop losses caused by *Xanthomonas malvacearum* E. F. Smith, Dowson, in Uganda. *Empire Cotton Growing Review*, **37**, 241–55. *157, 160*

Logan, C. & Coaker, T. H. (1960). The transmission of bacterial blight of cotton (*Xanthomonas malvacearum* E. F. Smith, Dowson) by the cotton bug, *Lygus vosseleri* Popp. *Empire Cotton Growing Review*, **37**, 26–9.   *138, 152*

Lord, E. & Underwood, C. (1958). The interpretation of spinning test reports. *Empire Cotton Growing Review*, **35**, 26–47.   *Table 8.5*

Low, A. (1962). New cotton varieties of the Sudan. *Empire Cotton Growing Review*, **39**, 95–105.   *236*

   (1964). *Progress Reports from Experiment Stations, Uganda, 1962–63*. London: Empire Cotton Growing Corporation.   *211*

   (1966). *Progress Reports from Experiment Stations, Uganda, 1964–65*. London: Empire Cotton Growing Corporation.   *222*

Low, A., Hesketh, J. D. & Muramoto, M. (1969). Some environmental effects on the varietal node number of the first fruiting branch. *Cotton Growing Review*, **46**, 181–8.   *Figure 4.7*

Ludwig, L. J., Saeki, T. & Evans, L. T. (1965). Photosynthesis in artificial communities of cotton plants in relation to leaf area. i. Experiments with progressive defoliation of mature plants. *Australian Journal of Biological Science*, **18**, 1103–18.   *95*

Lush, J. L. (1945). *Animal Breeding Plans*. 3rd edn. Ames: Iowa State College Press.   *289*

MacDonald, A. S. (1967). Some statistical aspects of cotton production in Uganda. *Cotton Growing Review*, **44**, 105–13.   *331*

Manning, H. L. (1949). Planting data and cotton production in the Buganda Province of Uganda. *Empire Journal of Experimental Agriculture*, **17**, 245–58.   *37, 38*

   (1950). Confidence limits of expected monthly rainfall. *Journal of Agricultural Science, Cambridge*, **40**, 169–76.   *22*

   (1952). Forecasting the Uganda cotton crop. *Empire Cotton Growing Review*, **29**, 241–57.   *83*

   (1956a). The statistical assessment of rainfall probability and its application in Uganda agriculture. *Proceedings of the Royal Society of London*, B, **144**, 460–80.   *23, Figure 2.4, 24*

   (1956b). Yield improvement from a selection index technique with cotton. *Heredity*, **10**, 303–22.   *112, 205, 206, 210*

   (1963). Realized yield improvement from twelve generations of progeny selection in a variety of Upland cotton. In *Statistical Genetics and Plant Breeding* (eds. W. D. Hanson & H. F. Robinson), Publication no. 982, pp. 329–351. Washington DC: NAS–NRC.   *112, 203, 206*

Manning, H. L. & ap Griffith, G. (1949). Fertilizer studies on Uganda soils. *East African Agricultural Journal*, **15**, 87–97.   *44, 60*

Manning, H. L. & Farbrother, H. G. (1953). *Progress Reports from Experiment Stations, Uganda, 1957–58*. London: Empire Cotton Growing Corporation.   *87*

Manning, H. L. & Kibukamusoke, D. E. B. (1958). *Progress Reports from Experiment Stations, Uganda, 1957–58*. London: Empire Cotton Growing Corporation.   *159*

   (1960). *Progress Reports from Experiment Stations, Uganda, 1958–59*. London: Empire Cotton Growing Corporation.   *95*

Marshall, D. R. & Brown, A. H. D. (1973). Stability of performance in mixtures and multilines. *Euphytica*, **22**, 405–12.   *242*

Mason, T. G. (1922). Growth and abscission in Sea Island Cotton. *Annals of Botany*, **36**, 457–84.   *100*

Matthews, G. A. & Tunstall, J. P. (1968). Scouting for pests and the timing of spray applications. *Cotton Growing Review*, **45**, 115–27.   *133, 136*

McKinlay, K. S. (1953). Use of repellants to simplify insecticide field trials. *Nature, London*, **171**, 658.   *126*

(1954). The design of insecticide field trials on cotton in Uganda. *Proceedings of the 6th Symposium of the Colston Research Society, 1953*, 53–9. *126*

McKinlay, K. S. & Geering, Q. A. (1957). Studies of crop loss following insect attack on cotton in East Africa. 1. Experiments in Uganda and Tanganyika. *Bulletin of Entomological Research*, **48**, 833–49. *127, 133, 139*

McKinion, J. M., Baker, D. N., Hesketh, J. D., & Jones, James W. (1975). SIMCOT II: A simulation of cotton growth and yield. In *Computer Simulation of a Cotton Production System*, Part 4, pp. 27–82. Washington, DC: US Department of Agriculture. *120*

McMichael, S. C. (1960). Combined effects of glandless genes gl_2 and gl_3 on pigment glands in the cotton plant. *Agronomy Journal*, **52**, 385–6. *237*

Meyer, J. R. & Meyer, V. G. (1961). Origin and inheritance of nectariless cotton. *Crop Science*, **1**, 167–9. *236*

Miller, P. A. & Rawlings, J. O. (1967). Breakup of initial linkage blocks through intermating in a cotton breeding population. *Crop Science*, **7**, 199–204. *234*

Miller, P. A., Robinson, H. F. & Pope, G. A. (1962). Cotton variety testing: additional information on variety × environment interactions. *Crop Science*, **2**, 349–52. *223*

Mills, W. R. (1953a). Recent fertilizer trials in Uganda. *East African Agricultural Journal*, **19**, 40–2. *44*

(1953b). Nitrate accumulation in Uganda soils. *East African Agricultural Journal*, **19**, 53–4. *71*

Monteith, J. L. (1965). The photosynthesis and transpiration of crops. *Experimental Agriculture*, **2**, 1–14. *94*

(1973). *Principles of Environmental Physics*. Contemporary Biology Series. London: Edward Arnold. *30*

Morris, D. A. (1964a). Varieties in the boll maturation period of cotton. *Cotton Growing Review*, **41**, 114–23. *99*

(1964b). Capsule dehiscence in *Gossypium*. *Cotton Growing Review*, **41**, 167–71. *99, 104*

(1965). Photosynthesis by the capsule wall and bracteoles of the cotton plant. *Cotton Growing Review*, **42**, 49–51. *99*

Munro, J. M. (1966). *progress Reports from Experiment Stations, Uganda, 1964–65*. London: Empire Cotton Growing Corporation. *250*

(1971). An analysis of earliness in cotton. *Cotton Growing Review*, **48**, 28–41. *99, 112*

Munro, J. M. & Farbrother, H. G. (1969). Composite plant diagrams in cotton. *Cotton Growing Review*, **46**, 261–82. *97, 226*

Namken, K. N. (1965). Relative turgidity technique for scheduling cotton (*Gossypium hirsutum* L.) irrigation. *Agronomy Journal*, **57**, 38–41. *92*

Ndegwe, N. A. (1968). *Progress Reports from Experiment Stations, Uganda, 1967–68*. London: Cotton Research Corporation. *86, 87*

Niles, G. A. & Richmond, T. R. (1962). Performance trials with selected strains of storm resistant cotton. Bulletin MP-577. College Station: Texas Agricultural Experiment Station, A. & M. College of Texas. *236*

Nyahoza, F. (1966a). *Progress Reports from Experiment Stations, Uganda, 1964–65*. London: Empire Cotton Growing Corporation. *35*

(1966b). Shedding of fruiting bodies in cotton. M.Sc. thesis, University of East Africa. *103*

Nye, G. W. (1933). *Annual Report, 1932*. Department of Agriculture, Uganda. *163*

(1936). *Annual Report, 1936*. Department of Agriculture, Uganda. *164*

Nye, P. H. & Greenland, D. J. (1960). *The Soil Under Shifting Cultivation*. Technical Communication No. 51. Harpenden: Commonwealth Bureau of Soils. *65, 70*

Nye, G. W. & Hosking, H. R. (1930). *Annual Report, 1929–30*. Department of Agriculture, Uganda. *153*

Ogborn, J. E. A. (1964a). *Progress Reports from Experiment Stations, Uganda, 1962–63.* London: Empire Cotton Growing Corporation.   *270*

(1964b). A preliminary evaluation of herbicides for *Oxalis* control. Proceedings of the 3rd East African Herbicide Conference, pp. 267–8. Muguga, Nairobi: East African Agriculture and Forestry Organization (mimeo).   *271*

(1964c). *Progress Reports from Experiment Stations, Uganda, 1963–64.* London: Empire Cotton Growing Corporation.   *61*

(1965). *Progress Reports from Experiment Stations, Uganda, 1964–65.* London: Empire Cotton Growing Corporation.   *74*

(1967). *Progress Reports from Experiment Stations, 1965–66.* London: Cotton Research Corporation.   *270*

Padwick, G. W. (1956). *Losses Caused by Plant Diseases.* Phytopathological Paper No. 1. Kew: Commonwealth Mycological Institute.   *157*

(1959). Plant diseases in the Colonies. *Outlook on Agriculture*, 2, 122–6.   *157*

Parker, C. (1967). Pot experiments with herbicides on *Oxalis latifolia* Kunth. *Proceedings of the 8th British Weed Control Conference, 1966*, pp. 126–34.   *271*

Parker, C., Holly, K. & Hocombe, S. D. (1969). Herbicides for nutgrass control – conclusions from ten years of testing at Oxford. *PANS*, **15**, 54–63.   *271*

Parnell, F. R., King, H. E. & Ruston, D. F. (1949). Jassid resistance and hairiness of the cotton plant. *Bulletin of Entomological Research*, **39**, 539–75.   *194, 201*

Passmore, R. G. (1961). *Progress Reports from Experiment Stations, Uganda, 1960–61.* London: Empire Cotton Growing Corporation.   *270*

(1962). *Progress Reports from Experiment Stations, Uganda, 1961–62.* London: Empire Cotton Growing Corporation.   *270*

Pearson, E. O. (1958). *The insect Pests of Cotton in Tropical Africa.* London: Commonwealth Institute of Entomology and Empire Cotton Growing Corporation.   *137, 146*

Pearson, E. O., Geering, Q. A. & McKinley, K. S. (1952). *Progress Reports from Experiment Stations, Uganda, 1951–52.* London: Empire Cotton Growing Corporation.   *130*

Peat, J. E. & Brown, K. J. (1961). A record of cotton breeding for the Lake Province of Tanganyika: Seasons 1939–40 to 1957–58. *Empire Journal of Experimental Agriculture*, **29**, 119–35.   *194, 202*

Penman, H. L. (1948). Natural evaporation from open water, bare field and grass. *Proceedings of the Royal Society of London*, A, **193**, 120–45.   *30, 34, 113*

(1956). Evaporation; an introductory survey. *Netherlands Journal of Agricultural Science*, **4**, 9–29.   *30, 34*

Pereira, H. C. (1959). Practical field instrument for estimation of radiation and of evaporation. *Quarterly Journal of the Royal Meteorological Society*, **85**, 253–61.   *27*

Perry, D. A. (1960). *Progress Reports from Experiment Stations, Uganda, 1959–60.* London: Empire Cotton Growing Corporation.   *166*

(1962). Method for determining the reaction of cotton plants to Fusarium wilt. *Empire Cotton Growing Review*, **39**, 22–6.   *167*

(1963). Interaction of root knot and Fusarium wilt of cotton. *Empire Cotton Growing Review*, **40**, 41–7.   *166*

Pothecary, B. P. & Thomas, P. E. L. (1968). Control of *Cyperus rotundus* in the Sudan Gezira. *PANS*, **14**, 236–40.   *271*

Pottie, J. M. (1953). Cotton growing in Uganda. *Plant Protection Overseas Review*, **4**, 26.   *157*

Powell, N. T. & Nusbaum, C. J. (1960). The black shank–root–knot complex in flue-cured tobacco. *Phytopathology*, **50**, 899–906.   *166*

Prentice, A. N. (1957). *Progress Reports from Experiment Stations, Uganda, 1956–57.* London: Empire Cotton Growing Corporation.   *86, 87, 269*

Radwanski, S. A. (1960). *The Soils and Land Use of Buganda.* Memoirs of the Research Division, Department of Agriculture Uganda. Series 1, No. 4. *47*

Reed, W. (1970). *Progress Reports from Experiment Stations, Uganda, 1969–70.* London: Cotton Research Corporation. *145*

(1971). *Progress Reports from Experiment Stations, Uganda, 1970–71.* London: Cotton Research Corporation. *132*

(1972). Uses and abuses of unsprayed controls in spraying trials. *Cotton Growing Review,* **49,** 67–72. *129*

(1973). *Cotton Research Reports, Uganda, 1971–72.* London: Cotton Research Corporation. *146*

(1974). Selection of cotton varieties for resistance to insect pests in Uganda. *Cotton Growing Review,* **51,** 106–23. *141, 194*

Rhyne, C. L. (1960). Linkage studies in *Gossypium.* 2. Altered recombination values in a linkage group of allotetraploid *G. hirsutum* as a result of transferred diploid species genes. *Genetics,* **45,** 673–81. *237*

(1962). Enhancing linkage-block breakup following interspecific hybridization and backcross transference of genes in *Gossypium hirsutum* L. *Genetics,* **47,** 61–9. *237*

Riddle, M. J. (1968). *Progress Reports from Experiment Stations, Uganda, 1966–67.* London: Cotton Research Corporation. *260, Figure 9.10*

Riggs, T. J. (1970). Trials of cotton seed mixtures in Uganda. *Cotton Growing Review,* **47,** 100–11.

Rijks, D. A. (1962). *Progress Reports from Experiment Stations, Uganda, 1961–62.* London: Empire Cotton Growing Corporation. *40, 242*

(1964). *Progress Reports from Experiment Stations, Uganda, 1962–63.* London: Empire Cotton Growing Corporation. *40*

(1965). The use of water by cotton crops in Abyan, South Arabia. *Journal of Applied Ecology,* **2,** 317–43. *34, 35, 39*

(1966). *Progress Reports from Experiment Stations, Uganda, 1964–65.* London: Empire Cotton Growing Corporation. *27*

(1967a). Optimum sowing date for yield: a review of work in the BP52 cotton area of Uganda. *Cotton Growing Review,* **44,** 247–56. *37, 80, 82, 114*

(1967b). Water use by irrigated cotton in Sudan. 1. Reflection of short-wave radiation. *Journal of Applied Ecology,* **4,** 561–8. *34, 39*

(1968a). Agrometeorology in Uganda – a review of methods. *Experimental Agriculture,* **4,** 263–74. *21*

(1968b). Water use by irrigated cotton in Sudan. 2. Net radiation and soil heat flux. *Journal of Applied Ecology,* **5,** 685–706. *31, 34, 39*

(1968c). *Progress Reports from Experiment Stations, Uganda, 1966–67.* London: Cotton Research Corporation. *25, 38*

(1968d). *Progress Reports from Experiment Stations, Uganda, 1967–68.* London: Cotton Research Corporation. *28, 29, 38*

(1969a). Evaporation from a papyrus swamp. *Quarterly Journal of the Royal Meteorological Society,* **95,** 643–9. *34*

(1969b). *Progress Reports from Experiment Stations, Uganda, 1968–69.* London: Cotton Research Corporation. *39*

(1971). Water use by irrigated cotton in Sudan. 3. Bowen ratios and advective energy. *Journal of Applied Ecology,* **8,** 643–63. *34*

Rijks, D. A. & Harrop, J. F. (1969). Irrigation and fertilizer experiments on cotton at Mubuku, Uganda. *Experimental Agriculture,* **5,** 17–24. *38*

Rijks, D. A. & Huxley, P. A. (1964). The empirical relation between solar radiation and hours of bright sunshine near Kampala, Uganda. *Journal of Applied Ecology,* **1,** 339–45. *28*

Rijks, D. A. & Owen, W. G. (1965). *Hydro-meteorological Records from Areas of*

Potential Agricultural Development in Uganda. Entebbe: Ministry of Mineral and Water Resources.   *31, 87*

Rijks, D. A., Owen, W. G. & Hanna, L. W. (1970). *Potential Evaporation in Uganda*. Entebbe: Ministry of Mineral and Water Resources.   *Figure 2.1, 19, 28, 30*

Rijks, D. A. & Walker, J. T. (1968). Evaluation and computation of potential evaporation in the tropics. *Experimental Agriculture*, **4**, 351–7.   *31*

Ritchie, J. T. & Burnett, E. (1971). Dryland evaporative flux in a sub-humid climate. II. Plant influences. *Agronomy Journal*, **63**, 56–62.   *90*

Roane, C. W. (1973). Trends in breeding for disease resistance in crops. *Annual Review of Phytopathology*, **11**, 463–86.   *191*

Robinson, R. A. (1971). Vertical resistance. *Review of Plant Pathology*, **50**, 233–9.   *191*

Russell, K. M. (1955). *Xanthomonas malvacearum*, the cause of bacterial blight of cotton. Ph.D. thesis, University of Cambridge.   *157*

Sansom, H. W. (1954). *The Measurement of Evaporation in East Africa*. Technical Memorandum No. 5. Nairobi: East African Meteorological Department.   *30*

Saunders, J. H. (1961). The mechanism of hairiness in *Gossypium*. I. *Gossypium hirsutum*. *Heredity*, **16**, 331–48.   *194*

(1963). The mechanism of hairiness in *Gossypium*. 2. *Gossypium barbadense* – the inheritance of stem hair. *Empire Cotton Growing Review*, **40**, 104–16.   *194*

(1965*a*). The mechanism of hairiness in *Gossypium*. 3. *Gossypium barbadense* – the inheritance of upper leaf lamina hair. *Empire Cotton Growing Review*, **42**, 15–25.   *194*

(1965*b*). The mechanism of hairiness in *Gossypium*. 4. The inheritance of plant hair length. *Empire Cotton Growing Review*, **42**, 26–32.   *194*

(1965*c*). Genetics of hairiness transferred from *Gossypium raimondii* to *G. hirsutum*. *Euphytica*, **14**, 276–82.   *194*

(1970). Wild diploid species of *Gossypium* in cotton breeding. In *Cotton Growth in the Gezira Environment* (eds. M. A. Siddig & L. C. Hughes), pp. 159–63. Cambridge: Heffers and Sudan Agricultural Research Corporation.   *237*

Saunders, J. H. & Innes, N. L. (1963). The genetics of bacterial blight resistance in cotton. Further evidence on the gene B_{6m}. *Genetical Research, Cambridge*, **4**, 382–8.   *176, Table 7.1*

Simmonds, N. W. (1962). Variability in crop plants, its use and conservation. *Biological Reviews, Cambridge*, **37**, 422–65.   *242*

Simpson, J. R. (1961). The effects of several agricultural treatments on the nitrogen status of red earth in Uganda. *East African Agriculture and Forestry Journal*, **26**, 158–63.   *71*

Slatyer, R. O. (1955). Studies of the water relations of crop plants grown under natural rainfall in Northern Australia. *Australian Journal of Agricultural Research*, **6**, 365–77.   *92*

(1960). *Agricultural Climatology of the Katherine Area*, N.T. technical paper No. 13. Division of Land Research and Regional Survey. Melbourne: CSIRO.   *90*

(1967). *Plant Water Relationships*. New York: Academic Press.   *91, 92*

Smith, G. (1960). *An Introduction to Industrial Mycology*. London: Edward Arnold.   *162*

Smith, H. Fairfield, (1936). A discriminant function for plant selection. *Annals of Eugenics*, **7**, 240–50.   *204*

Snowden, J. D. (1926). *Occurrence of Angular Leaf Spot of Cotton in Uganda*. Circular No. 17. Department of Agriculture, Uganda.   *153*

Spickett, S. G. & Thoday, J. M. (1966). Regular responses to selection. 3. Interaction between located polygenes. *Genetical Research, Cambridge*, **7**, 96–121.   *182*

Stapledon, H. N. (1970). Crop production systems simulation. *Transactions of the American Society of Agricultural Engineers*, **13**, 110–13.   *120*

Stephens, D. (1961). *Annual Report, Part II, 1961–62*. Uganda: Department of Agriculture. *59*

(1966). Two experiments on the effects of heavy application of triple superphosphate on maize and cotton in Buganda clay loam soil. *East African Agriculture and Forestry Journal*, **31**, 283–9. *60*

(1969a). Changes in yields and fertilizer responses with continuous cropping in Uganda. *Experimental Agriculture*, **5**, 263–9. *44, Table 3.1, 59*

(1969b). The effects of fertilizers, manure and trace elements in continuous cropping rotations in southern and western Uganda. *East African Agriculture and Forestry Journal*, **34**, 401–17. *61, 70*

(1970). Soil fertility. In *Agriculture in Uganda* (ed. J. D. Jameson), 2nd edn, chap. 6, pp. 72–89. London: Oxford University Press. *44*

Stern, W. R. (1965). The seasonal growth characteristics of irrigated cotton in a dry monsoonal environment. *Australian Journal of Agricultural Research*, **16**, 347–66. *96*

(1967). Seasonal evapotranspiration of irrigated cotton in a low-latitude environment. *Australian Journal of Agricultural Research*, **18**, 259–69. *90*

Stern, V. M., Mueller, A., Sevacherian, V. & Way, M. (1969). Lygus bug control in cotton through alfalfa interplanting. *California Agriculture*, **23**(2), 8–10. *140*

Stride, G. O. (1964). *Progress Reports from Experiment Stations, Uganda, 1962–63*. London: Empire Cotton Growing Corporation. *139*

(1968). On the biology and ecology of *Lygus vosseleri* (Heteroptera: Miridae) with special reference to its hostplant relationships. *Journal of the Entomological Society of South Africa*, **31**, 17–59. *140*

(1969). Investigations into the use of a trap crop to protect cotton from attack by *Lygus vosseleri* (Heteroptera: Miridae). *Journal of the Entomological Society of South Africa*, **32**, 465–77. *140*

Talboys, P. W. (1960). A culture medium aiding the identification of *Verticillium albo-atrum* and *V. dahliae*. *Plant Pathology*, **9**, 57–8. *164*

Tarr, S. A. J. (1953). Seed treatment against blackarm disease of cotton in the Anglo-Egyptian Sudan. *Empire Cotton Growing Review*, **30**, 117–32. *160*

(1956). Seed treatment against blackarm disease of cotton in the Sudan. 4. Recent research on wet treatments. *Empire Cotton Growing Review*, **33**, 98–104. *160*

Taylor, A. W. & Gurney, E. L. (1965). Precipitation of phosphate from concentrated fertilizer solution by soil clays. *Proceedings of the Soil Science Society of America*, **29**, 94–5. *59*

Taylor, T. H. C. (1945). *Lygus simonyi* as a cotton pest in Uganda. *Bulletin of Entomological Research*, **36**, 121–48. *137, 138, 139, 140*

Thoday, J. M. (1961). Location of polygenes. *Nature, London*, **191**, 368–70. *182*

Thomas, D. G. (1970). In *Agriculture in Uganda* (ed. J. D. Jameson), 2nd edn, chap. 12, pp. 176–8. London: Oxford University Press. *153*

Thorne, G. N. & Evans, A. F. (1964). Influence of tops and roots on net assimilation rate of sugar beet and spinach beet and grafts between them. *Annals of Botany*, **28**, 499–508. *100*

Thorp, T. K. (1973). *Cotton Research Reports, Uganda, 1971–72*. London: Cotton Research Corporation. *24, 37, 40, 51*

Tollervy, F. E. (1973). *Cotton Research Reports, Uganda, 1971–73*. London: Cotton Research Corporation. *273*

Tothill, J. O. (1940). *Agriculture in Uganda*, 1st edn, p. 551. Oxford: Oxford University Press. *143*

Tripathis, R. S. (1969). Ecology of *Cyperus rotundus* L. 3. Population of tubers at different depths of the soil and their sprouting response to air drying. *Proceedings of the National Academy of Science, India, B*, **39**, 140–2. *271*

Troughton, J. H. (1969). Plant water status and carbon dioxide exchange of cotton leaves. *Australian Journal of Biological Science*, **22**, 289–302.   *91, 92*

Van der Plank, J. E. (1968). *Plant Diseases: Epidemics and Control.* New York and London: Academic Press.   *191*

Waddington, C. H. (1961). Genetic assimilation. *Advances in Genetics*, **10**, 257–90.   *182*

Walker, J. T. (1963). Multiline concept and intravarietal heterosis. *Empire Cotton Growing Review*, **40**, 190–215.   *230, 241*

(1964). Modal selection in Upland cotton. *Heredity*, **19**, 559–83.   *206, 235*

(1969). Selection and quantitative characters in field crops. *Biological Reviews, Cambridge*, **44**, 207–43.   *235*

Walker, J. T. & Rijks, D. A. (1967). A computer programme for the calculation of confidence limits of expected rainfall. *Experimental Agriculture*, **3**, 337–41.   *24*

Watson, D. J. (1952). Physiological basis of variation in yield. *Advances in Agronomy*, **4**, 101–45.   *82*

Weatherley, P. E. (1950). Studies in water relations of the cotton plants. 1. The field measurements of water details in the leaves. *New Phytologist*, **49**, 81–97.   *91*

Weaver, J. B. & Ashley, T. (1971). Analysis of a dominant gene for male sterility in upland cotton, *Gossypium hirsutum* L. *Crop Science*, **11**, 596–8.   *197*

Weindling, R. (1948). *Bacterial Blight of Cotton under Conditions of Artificial Inoculation.* Technical Bulletin No. 956. Washington, DC: US Department of Agriculture.   *156*

Wickens, G. M. (1951). *Progress Reports from Experiment Stations, Uganda, 1950–51.* London: Empire Cotton Growing Corporation.   *164*

(1952). *Progress Reports from Experiment Stations, Uganda, 1951–52.* London: Empire Cotton Growing Corporation.   *155*

(1953). Bacterial blight of cotton. A survey of present knowledge, with particular reference to possibilities of control of the disease in African rain-grown cotton. *Empire Cotton Growing Review*, **30**, 81–101.   *155*

(1956). Vascular infection of cotton by *Xanthomonas malvacearum* (E. F. Smith) Dowson. *Annals of Applied Biology*, **44**, 129–37.   *157*

(1957). Treatment of cotton seed against bacterial blight (*Xanthomonas malvacearum* (E. F. Smith) Dowson). *Empire Cotton Growing Review*, **34**, 1–7.   *157, Table 6.1, 160*

(1958). Present practice in the treatment of cotton seed against bacterial blight (*Xanthomonas malvacearum* (E. F. Smith) Dowson). *Empire Cotton Growing Review*, **35**, 1–4.   *159, 160*

(1964). Methods for detection and selection of heritable resistance to Fusarium wilt of cotton. *Empire Cotton Growing Review*, **41**, 172–93.   *167, 192*

Wickens, G. M. & Logan, C. (1956). *Progress Reports from Experiment Stations, Uganda, 1955–56.* London: Empire Cotton Growing Corporation.   *155*

(1957). *Progress Reports from Experiment Stations, Uganda, 1956–57.* London: Empire Cotton Growing Corporation.   *156, 164*

(1958). *Progress Reports from Experiment Stations, Uganda, 1957–58.* London: Empire Cotton Growing Corporation.   *155*

(1960a). *Progress Reports from Experiment Stations, Uganda, 1958–59.* London: Empire Cotton Growing Corporation.   *160*

(1960b). Fusarium wilt and root knot of cotton in Uganda. *Empire Cotton Growing Review*, **37**, 15–25.   *165, 166, 167, 193*

Wit, C. T. de (1965). *Photosynthesis of Leaf Canopies.* Agricultural Research Report No. 663. Wageningen: Centre for Agricultural Publications and Documentation.   *82, 94, Figure 4.7*

Woodhead, T. (1967). Empirical relations between cloud amount, insolation and sunshine duration in East Africa: II. *East African Agriculture and Forestry Journal,* **32**, 474–83. *27*

(1968*a*). *Studies of Potential Evaporation in Kenya.* Nairobi: Government of Kenya. *31*

(1968*b*). *Studies of Potential Evaporation in Tanzania.* Nairobi: East African Agricultural and Forestry Research Organization. *31*

Yates, F. & Cochran, W. G. (1938). The analysis of groups of experiments. *Journal of Agricultural Science, Cambridge,* **28**, 556–80. *181, 230*

SUBJECT INDEX

abscission, hormonal control, 104–5,
108; *see also* boll setting and
shedding
agrometeorological network, 18–22
data analysis, 21
instrumentation, 19–22
agrometeorology, definition, 15
Albar cottons, 4, 155, 178, 219–20,
230–1
Alternaria, 162–3
Alternaria disease, 162–3, 170, 193,
220
assimilates, *see* carbohydrates

B genes, and resistance to bacterial
blight, 176–8
backcrossing, 237–8
and resistance to bacterial blight,
175, 178, 182–3
bacterial blight (*Xanthomonas
malvacearum*)
and compensatory crop growth,
158–9, 161
epidemiology, 152–3, 156–7
history in Uganda, 153–4
host–parasite relationships, 180–4,
186, 188–90
inoculation techniques, 154–6
research programme, 154
resistance: breeding for, 153–6;
genetics, 176–7, 182–6; screening
for, 213–14; *see also* resistance
and resistance breeding
seed dressing: benefits of, 160–1;
large scale, 154, 157–9; and
reduction in seedling blight,
152–3; in Tanzania, 304
surveys, 157–8
bananas, 43
beans, *Phaseolus*
cultural practices and yields,
298–303

in experiments on soil fertility, 45,
67–8
nitrogen fixation, 284
response to fertilizer mixture, 67–8
rotation, place in, 248, 267–8
soil condition following, 268
biological control of pests, 143,
145–6, 148
biometrical analysis and plant
breeding
correlation diagrams, 215
diallel crosses, 186–7
regression analyses, 180–1, 190–1,
230–2
selection index, 199, 203–6, 210
boll rots, 146–7, 157, 170–1
boll setting and shedding
hormonal control, 104–5
non-stress shedding, 101–3
nutritional theory, 82–3, 84, 96–7,
101
stress shedding, 101; in model,
107–8
and water shortage, 101, 103
boll sinks, capacity, 84, 97–100; *see
also* carbohydrates *and*
source–sink relationships
bolls
growth rates, 99–100
morphogenesis, 97–9
photosynthesis in, 99–100
bollworms, *see Cryptophlebia, Earias,
Heliothis and Pectinophora*
BP52 cotton selection
breeding programmes, 202–7,
211–15
commercial evaluation, 207–10
lint yields, 206, 207–9, 225–6, 228
BPA cotton selection
breeding programme, 214–19
district trials, 222–6
lint yields, 215, 217–18, 225–6, 228

345